Thermal Engineering of Nuclear Power Stations

Thermal Engineering of Nuclear Power Stations

Balance-of-Plant Systems

Charles F. Bowman

Seth N. Bowman

CRC Press
Taylor & Francis Group
Boca Raton London New York

CRC Press is an imprint of the
Taylor & Francis Group, an **informa** business

First edition published 2021
by CRC Press
6000 Broken Sound Parkway NW, Suite 300, Boca Raton, FL 33487-2742

and by CRC Press
2 Park Square, Milton Park, Abingdon, Oxon, OX14 4RN

CRC Press is an imprint of Taylor & Francis Group, LLC

ISBN: 978-0-367-82039-8 (hbk)
ISBN: 978-0-367-50236-2 (pbk)
ISBN: 978-1-003-01160-6 (ebk)

Typeset in Times
by Lumina Datamatics Limited

To Nancy

Without whose help this book could not have been possible

Contents

Preface

As the nuclear power industry matures, more and more experienced mechanical design, systems, operations, and maintenance engineers are retiring. With them goes much of the experience and technical know-how gained through their long careers. The principal author, Chuck Bowman, is such a person. Chuck began his career as an instructor at the University of Tennessee where he received his BS and MS degrees in mechanical engineering. He went on to work for 28 years for the Tennessee Valley Authority (TVA) where he participated in the engineering of the fleet of 7 nuclear units, rising from an engineer to a supervisor to a corporate specialist for thermal performance supporting all of the operating nuclear units at TVA. In 1994, Chuck formed Chuck Bowman Associates, Inc., serving the electric power industry and specializing in thermal performance analysis of the power generating cycle and related fields including the analysis and testing of heat exchangers, cooling towers, spray ponds, cooling lakes, etc. The purpose of this book is to impart the knowledge and experience received during more than 50 years of working in the field of thermal performance of balance-of-plant (BOP) systems to the next generation.

The term BOP refers to a large number of systems required to produce electrical energy that are outside of the nuclear steam supply systems (NSSS) that are normally associated with nuclear plants and are furnished by a single vendor as an integral standard package. The BOP systems, which may or may not be safety-related, are normally engineered and procured by an architect\engineering (A\E) firm, and the design of this portion varies widely due to specific site conditions and design philosophies. Whereas the NSSS vendor generally maintains ongoing technical support of the nuclear station, the A\E firm frequently does not.

Although nuclear stations offer system engineering training specific to their plants, that training often does not encompass the daily practical problems encountered by mechanical design, system, and maintenance engineers. Also, in many instances when vacancies occur, the time to turn over responsibilities to the remaining younger engineers is too short to allow for effective training. The scope of this book encompasses the thermal aspects of the entire nuclear station BOP from the source of motive steam to the discharge of the waste heat and beyond. The book is practical in that a qualified engineer may easily follow the logic, and example problems are provided from actual experiences at operating nuclear stations.

Authors

Charles F. (Chuck) Bowman, P.E., is the President of Chuck Bowman Associates, Inc. (CBA), an engineering consulting firm serving the electric power industry since 1994. He received his BS and MS degrees in mechanical engineering from the University of Tennessee and is a registered Professional Engineer in Tennessee. CBA specializes in thermal performance analysis of electric power generating cycles and related fields including the design and analysis of heat exchangers, cooling towers, spray ponds, cooling water systems, etc. Before forming CBA, he was with the Tennessee Valley Authority (TVA) for 28 years. Before his retirement from TVA, he was the Senior Engineering Specialist for thermal performance in TVA's Corporate Engineering Office. He has served as a Consultant to the Electric Power Research Institute and has served on ASME committees that authored PTC 23.1-1983, Code on Spray Cooling Systems and PTC 12.5-2000, Single Phase Heat Exchangers.

Seth N. Bowman received his BS and MS degrees in chemical engineering from the University of Tennessee. He currently serves as the Manager of the Issues Management organization at the Y-12 National Security Complex, managed and operated by Consolidated Nuclear Security, LLC. In prior roles, he has been responsible for the assessment function for the Engineering Division at Y-12 and also served as a Shift Technical Adviser and Shift Manager for Building 9212.

List of Abbreviations

AE	architect engineering
AEC	Atomic Energy Commission
AF	asymptotic fouling
ARM	additive resistance method
ASME	American Society of Mechanical Engineers
AWBT	ambient wet-bulb temperature
AWHX	air-to-water heat exchanger
AWS	ambient wind speed
BFN	Browns Ferry Nuclear
BOP	balance of plant
BWR	boiling water reactor
CAC	containment air cooling units
CTSA	cooling tower simulation algorithm
CCW	condenser circulating water
CGS	Columbia Generating Station
CTI	Cooling Tower Institute
CV	control volume
CWT	cold water temperature
DCA	drain cooler approach
DP	dew-point
dP	pressure drop
ECC	Ecolaire Condenser Company
ELEP	expansion line end point
EPRI	Electric Power Research Institute
FACTS	Fast Analysis Cooling Tower Simulator Algorithm
FBSP	flatbed spray pond
FFW	final feedwater
GLLVHT	Generalized Longitudinal, Lateral, and Vertical Hydrodynamic and Transport Model
HB	heat balance
HDRA	high-density raceway aquaculture
HEI	Heat Exchange Institute
HP	high-pressure (turbine)
HRS	heat rejection system
HWT	hot water temperature
HX	heat exchanger
L/G	liquid to gas
LP	low-pressure (turbine)
LWBT	local wet-bulb temperature
MDCT	mechanical draft cooling tower
MFP	main feedwater pump
MSR	moisture separator reheater

NDCT	natural draft cooling tower
NPSH	net positive suction head
NRC	Nuclear Regulatory Commission
NSSS	nuclear steam supply system
NTU	number of transfer units
OSACT	oriented spray-assisted cooling tower
OSCS	oriented spray cooling system
PEPSE	Performance Evaluation of Power System Efficiencies
PF	power factor
PHX	plate heat exchanger
PSM	power spray module
PWR	pressurized water reactor
RBCU	reactor building cooling units
RH	relative humidity
R–M	Ranz and Marshall
RTD	resistance temperature detector
SG	steam generator
SW	service water
TA	tilt angle
TEMA	Tubular Exchange Manufacturer's Association
T–G	turbine–generator
TTD	terminal temperature difference
TVA	Tennessee Valley Authority
UEEP	utilized energy end point
UHS	ultimate heat sink
USDA	United States Department of Agriculture
VWO	valves wide open
WBNP	Watts Bar Nuclear Plant
WBT	wet-bulb temperature
WHEP	waste heat energy park

1 Thermodynamics of Balance-of-Plant Systems

1.1 TURBINE CYCLE

A turbine cycle consists of the equipment required to convert the thermal energy that is produced in the reactor into electrical energy. In addition to the turbine(s) and generator, it consists of the moisture separator(s) and reheater(s), the main condenser, feedwater heaters, and the associated pumps and piping. The turbine cycle for a nuclear power plant is the Rankine cycle. Figure 1.1 shows the most basic cycle possible without reheat or feedwater heaters.

These steps consist of the following:

- 1–2 Pump the water from the main condenser to the nuclear boiler pressure.
- 2–3 Add heat in the nuclear boiler to bring the water to the saturation temperature.
- 3–4 Convert the water to steam in the nuclear boiler.
- 4–5 Expand the steam through a turbine.
- 5–1 Condense the steam in the condenser.

Figure 1.2 shows the basic thermodynamic efficiency of a Rankine cycle, where entropy is defined as

$$dS = \left(\frac{\delta Q}{T}\right)_{rev} \quad \Rightarrow \quad \delta Q = T\,dS \tag{1.1}$$

where:

δQ = incremental change in heat transfer
T = absolute temperature.

and

$$\eta_{th} = \frac{W}{Q_H} = \frac{W}{W + Q_L} \tag{1.2}$$

where:

η_{th} = thermal efficiency
W = work
Q = heat transfer.

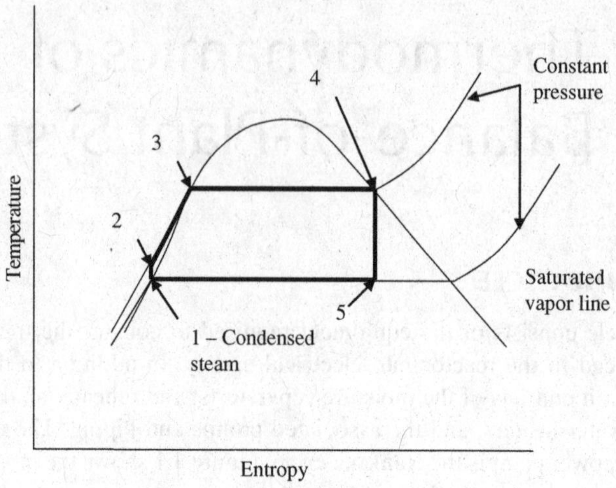

FIGURE 1.1 Basic Rankine cycle for a nuclear plant.

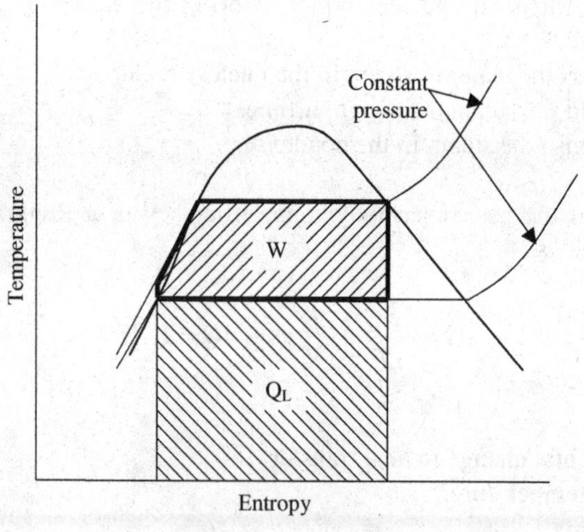

FIGURE 1.2 Thermodynamic efficiency of the basic Rankine cycle.

The efficiency of the basic Rankine cycle may be increased by the addition of feedwater heaters, as shown in Figure 1.3. Feedwater heaters extract a portion of the steam passing through a turbine to heat the condensate and feedwater on its way to the nuclear boiler (i.e., either the reactor in the case of a boiling water reactor [BWR] or the steam generator [SG] in the case of a pressurized water reactor [PWR]). Quite obviously, the heat going to the feedwater heater does not go to the main condenser but is returned to the nuclear boiler, so the amount of work performed by that steam is 100% efficient. Therefore, the more the stages

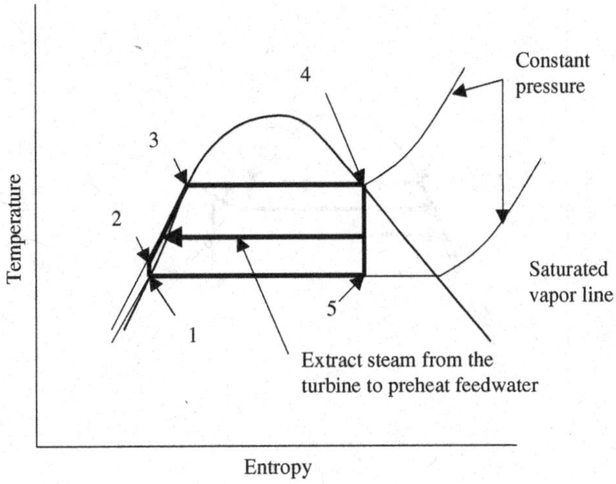

FIGURE 1.3 Rankine cycle with regeneration.

of feedwater heating, the more efficient the Rankine cycle. As a practical matter, most nuclear plants have between five and seven stages of feedwater heating.

Some nuclear plants are equipped with SGs that are capable of providing slightly superheated steam to the turbine. These plants are more efficient owing to the higher temperature, as shown in Figure 1.4 (somewhat exaggerated), because

$$\frac{W'}{Q_L'} > \frac{W}{Q_L} \tag{1.3}$$

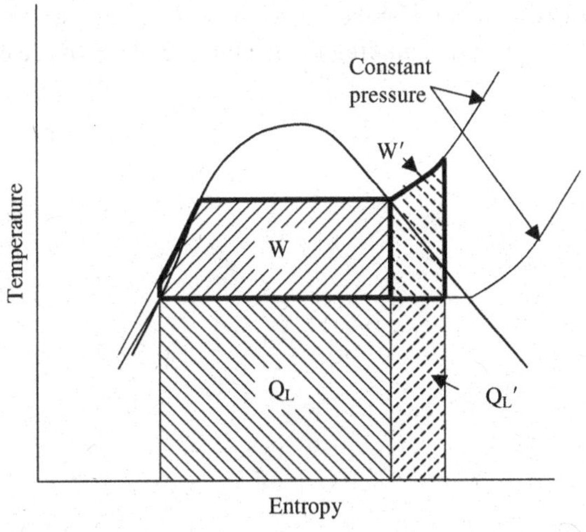

FIGURE 1.4 Higher temperature entering the turbine increases the efficiency of the Rankine cycle.

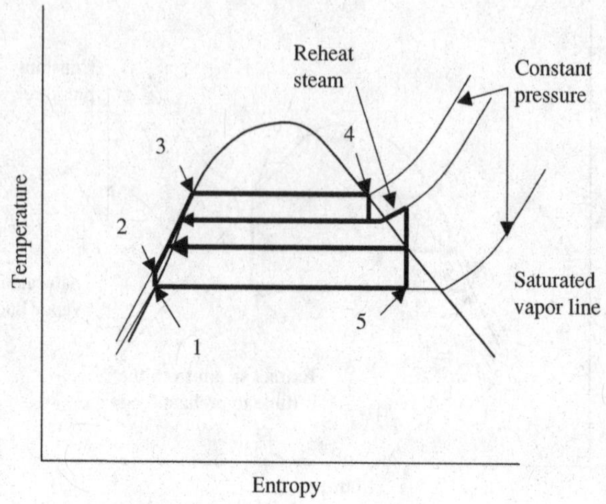

FIGURE 1.5 Rankine cycle with regeneration and reheat.

As shown in Figure 1.3, the turbine cycle in nuclear plants begins with saturated or nearly saturated steam entering the turbine. If this moist steam were permitted to expand all the way through the high-pressure (HP) and low-pressure (LP) turbines, the moisture would be too great, causing damage to the LP turbine blades. Most nuclear plants remove the moisture between the HP and LP turbines by mechanical means with moisture separators. The moisture wrung out of the steam is pumped into the feedwater going to the nuclear boiler. As shown in Figure 1.5, the essentially dry steam leaving the moisture separator may also be reheated by diverting a portion of the steam from the nuclear boiler to one or more reheaters (RH) operating in parallel to slightly superheat the steam going to the LP turbine(s). This process of reheating the steam does not make the Rankine cycle more efficient but does reduce the moisture in the LP turbine, making the LP turbine blades more efficient and less prone to damage.

2 Nuclear Boilers

2.1 BOILING WATER REACTOR

A BWR is a light water reactor in which the feedwater coming from the highest pressure feedwater heater passes directly over the fuel rods in the reactor to capture the heat generated by the fission reaction taking place in the reactor. In so doing, the feedwater is heated to the saturation temperature at the reactor's operating pressure (nominally about 1,000 lbf/in^{2}A (6,895 kPa)) and steam is boiled off through steam separators and dryers at the top of the reactor that wring out a portion of the moisture entrained in the steam. The result is main steam with a quality of approximately 99.75% (0.25% moisture) passing to the HP turbine. Although proven to be highly reliable, one significant disadvantage of a BWR is the fact that the steam, condensate, and feedwater passing through the turbine cycle are everywhere radioactive and thus require special care in performing operation and maintenance activities in the turbine building.

The reactor power is controlled by inserting or withdrawing control rods into or out of the core from below the core and by adjusting the flow through the core. A reactor recirculation system adjusts the flow by varying the speed of the recirculation pumps or adjusting the flow control valves. The reactor pressure is controlled by the HP turbine control valves, and the reactor level is controlled by the feedwater control system.

One of the most important parameters required to evaluate the turbine cycle is how much heat energy is being added to the cycle. Figure 2.1 shows a simplified sketch of the parameters required to calculate the heat addition or calorimetric for a BWR.

The heat added to the turbine cycle is the energy out in the form of main steam flow minus the energy in from the final feedwater (FFW) and the control rod drive flow.

$$Q_{in} = m_{MS}\, h_{MS} - m_{CRD}\, h_{CRD} - m_{FFW}\, h_{FFW} \tag{2.1}$$

$$m_{FFW} = m_{MS} - m_{CRD} \tag{2.2}$$

$$Q_{in} = m_{FFW}\,(h_{MS} - h_{FFW}) + m_{CRD}\,(h_{MS} - h_{CRD}) \tag{2.3}$$

where:

Q = heat transfer rate
m = mass flow rate
h = enthalpy.

The control rod drive system positions the control rods and provides cooling water to the recirculation pump seals. The normal source of the cooling water is the main condenser. Therefore, the enthalpy of the control rod drive can vary widely even during an outage, as the condenser circulating water (CCW) temperature and thus the saturation enthalpy of the water in the condenser hotwell vary accordingly.

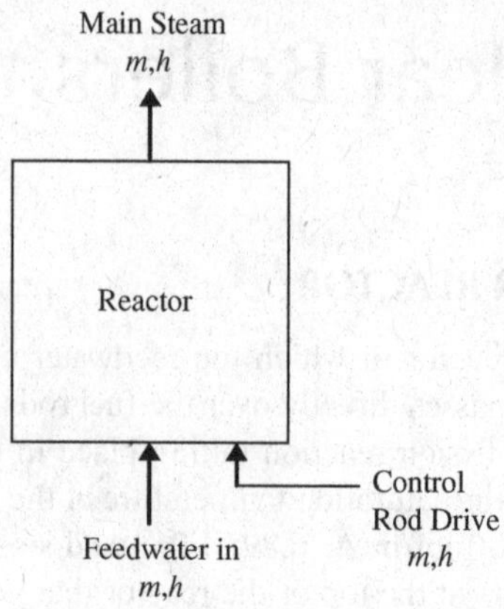

FIGURE 2.1 Boiling water reactor calorimetric.

Therefore, one should avoid using the design value for the control rod drive enthalpy, as it can lead to an inaccurate calculation of reactor power.

2.2 PRESSURIZED WATER REACTOR

A PWR is a light water reactor with the feedwater coming from the highest pressure feedwater heater passing through an SG. An SG is an intermediate heat exchanger (HX) between the reactor and the HP turbine. The great advantage of a PWR is the fact that the steam, condensate, and feedwater passing through the turbine cycle are not normally radioactive, making operation and maintenance activities in the turbine building much less complicated. For this reason, a large percentage of the nuclear plants have NSSS that employs the PWR system.

Primary water is circulated through the reactor under high pressure (normally about 2,300 lbf/in²A (15,858 kPa)) to capture the heat generated by the fission reaction taking place in the reactor and heating the primary water to about 600°F (318.6°C). The primary water remains a compressed liquid as it is circulated to the SG(s), where the heat is transferred to the feedwater, which is heated to the saturation temperature at the SG's operating pressure (nominally about 850 to 1,000 lbf/in²A (5,861 to 6,895 kPa)). An SG is normally a vertical head-down u-tube-type HX containing thousands of normally 0.75 in tubes. A shroud around the tube bundle forces the feedwater to pass over the tubes. The hot compressed liquid primary water flows through the tube side of the HX where the heat from the reactor and reactor recirculating pumps is transferred to the feedwater on the shell side of the HX. As the feedwater is heated, steam bubbles form on the outside of the tubes and rise. A level of boiling water is maintained above the tube bundle, and the resulting steam is boiled off through swirl-vane steam separators and chevrons at the top of the SG, wringing out a portion of the moisture that is entrained in the steam. The result is main steam with a quality of approximately 99.75% (0.25% moisture) that passes to the HP turbine.

SG tubes can degrade over time due to corrosion or denting near the tube sheet and buildup of foreign materials on the outside surface of the tubes such as copper,

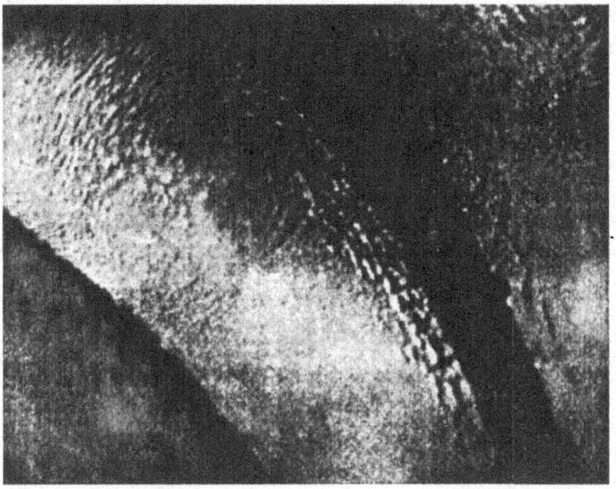

FIGURE 2.2 Fouling on the exterior of SG tubes.

corrosion products such as iron and copper oxides, and soluble materials such as silicates and sulphates. Most nuclear plants have eliminated copper from the secondary side as much as possible. Figure 2.2 shows an example of the fouling that can occur on the exterior of SG tubes. The tubes are frequently eddy-current-tested during outages, and tubes that have the potential for leaking are plugged because a leaking tube could lead to steam that has been contaminated by radioactive material escaping into the turbine cycle. Many SGs have had to be replaced due to the potential for leaking tubes. SG pressure is controlled by the HP turbine control valves, and the SG level is controlled by the feedwater control system.

Figure 2.3 is a simplified sketch of the parameters required to calculate the heat that is added to the turbine cycle from a PWR.

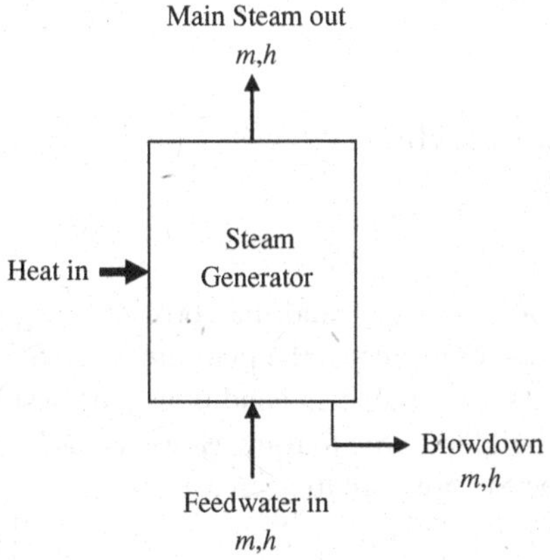

FIGURE 2.3 Pressurized water reactor calorimetric.

For a PWR, the heat added to the turbine cycle is the energy out in the form of main steam flow and blowdown minus the energy in from the FFW.

$$Q_{in} = m_{MS}\ h_{MS} + m_{SGB}\ h_{SGB} - m_{FFW}\ h_{FFW} \tag{2.4}$$

$$m_{FFW} = m_{MS} + m_{SGB} \tag{2.5}$$

$$Q_{in} = m_{MS}\ (h_{MS} - h_{FFW}) + m_{SGB}\ (h_{SGB} - h_{FFW}) \tag{2.6}$$

In calculating the total reactor power, subtract the energy added by the recirculation pumps.

The amount of heat that can be transferred to the turbine cycle is a function of the overall heat transfer coefficient, U, and the log mean temperature difference, *LMTD*, between the primary water and the secondary systems.

$$Q = U\ ALMTD \quad \Rightarrow \quad U = \frac{Q}{ALMTD} \tag{2.7}$$

where:

$$LMTD = \frac{T_h - t_c}{\ln\left(\dfrac{T_h - t_{sat}}{t_c - t_{sat}}\right)} \tag{2.8}$$

and

U = overall heat transfer coefficient
A = reference surface area
t = temperature.

If fouling reduces the value of U, t_{sat} and the corresponding saturation pressure of the SG are reduced. In some instances, the pressure in an SG has been observed to increase following an outage, as transient conditions during startup and/or shutdown may tend to disturb the buildup of the fouling. However, as Figure 2.4 illustrates, the long-term trend is for the SG pressure to decrease.

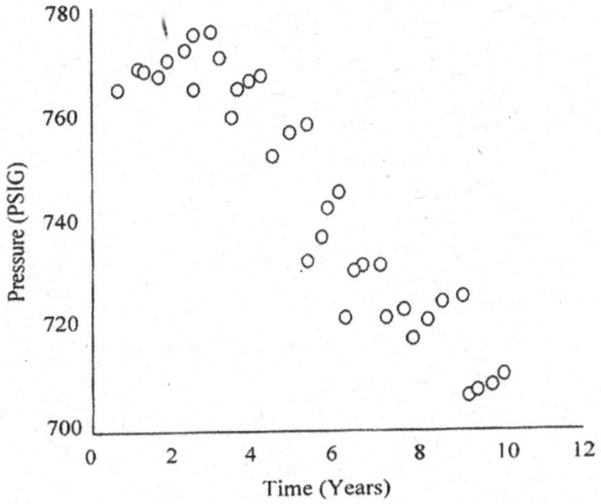

FIGURE 2.4 Steam generator pressure.

3 High-Pressure Turbines

3.1 HIGH-PRESSURE TURBINE CONTROL VALVES

HP turbine control valves limit the amount of main steam flow through the HP turbine and thus control the reactor power by throttling the main steam flow rate. HP turbine control valves are sized to pass the main steam flow that will result in 100% reactor power while not fully open to provide some controllability at full power operation. However, in the case of a PWR, if the SG is fouled, the heat transfer between the primary and secondary sides of the HX is reduced, resulting in a reduced SG pressure. As shown in Figure 3.1, at some point as the fouling increases, the SG pressure is reduced to the point that the HP turbine control valves became wide open (VWO). Subsequent fouling means that 100% reactor power can no longer be maintained. When that happens, the SGs are often replaced.

Historically, the different turbine vendors have offered different HP turbine control valve designs. General Electric, as well as most foreign turbine vendors, offered the full-arc design, whereas Westinghouse offered the partial-arc admission type.

Figure 3.2 illustrates full-arc admission into the HP turbine. With full-arc admission, the steam is admitted from all valves at once to first stage nozzles of the HP turbine, resulting in parallel steam flow through all of the first-stage nozzles at all loads. This design requires more throttling, which results in higher throttling loss if the turbine is operating at less than VWO. Figure 3.3 illustrates partial-arc admission into the HP turbine. With partial-arc, the valves open one at a time as load increases to admit steam from each valve to only a portion of first-stage nozzles of the HP turbine, resulting in full steam flow through some of the first-stage nozzles and less or none through others. This design requires lower throttling loss because some valves are wide open, but it increases cyclic blade loading as the blades alternately pass by nozzles that are admitting steam and those that are not admitting steam.

Although there is a definite pressure drop through the HP turbine control valve, main steam passing through the valve is an adiabatic process (i.e., no heat transfer occurs), so there is no change in the enthalpy. Figure 3.4 shows the expansion of the main steam through the HP turbine in an enthalpy vs. entropy diagram (known as a Mollier diagram). The slightly wet steam is first expanded through the HP turbine control valves(s). If the HP turbine were 100% efficient, the steam would expand in an reversible and adiabatic (or isentropic) manner and the entropy of the steam would be constant. However, only a portion of the energy is converted to work, whereas a portion of the energy excites the steam molecules as entropy is increased.

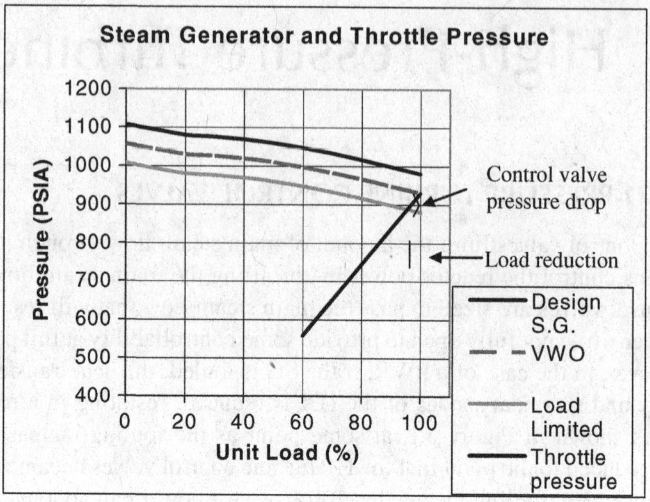

FIGURE 3.1 PWR steam generator and throttle pressures.

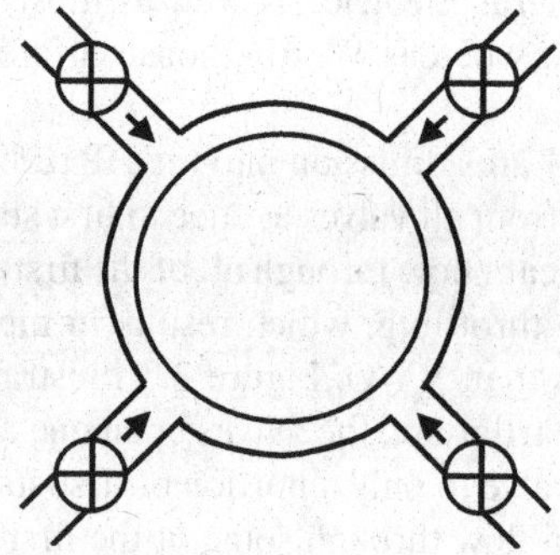

FIGURE 3.2 HP turbine with full-arc admission. (After Steam Turbine Performance Seminar Lecture Notes by Kenneth C. Cotton.)

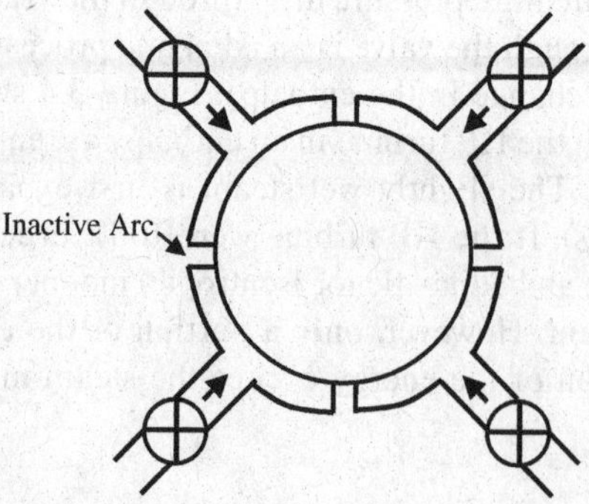

FIGURE 3.3 HP turbine with partial-arc admission. (After Steam Turbine Performance Seminar Lecture Notes by Kenneth C. Cotton.)

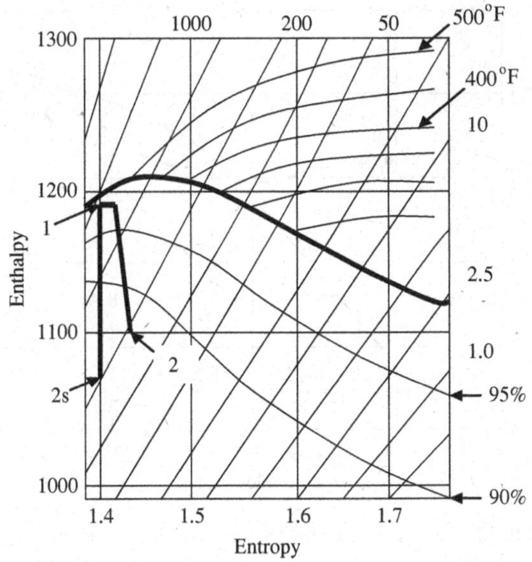

FIGURE 3.4 HP turbine expansion line on the Mollier diagram.

The efficiency of the HP turbine is defined as follows:

$$\eta_{turbine} = \frac{w_{actual}}{w_{isentropic}} = \frac{(h_1 - h_2)}{(h_1 - h_{2S})} \tag{3.1}$$

where:

η = efficiency
w = work
h = enthalpy.

Knowing the steam flow rate and the HP turbine efficiency, one may calculate the work done by the HP turbine.

As Figure 3.5 shows, turbines served by partial-arc control valves are at their point of best efficiency when the valves are either fully open or fully closed, whereas turbines that are served by full-arc control valves are relatively less efficient at partial loads because all of the valves are throttled but may be more efficient when all of the valves are fully open.

Figure 3.6 shows a sectional view of the top half of a typical Westinghouse HP nuclear turbine, which is a two-flow machine. The design philosophy adopted for the Westinghouse HP turbines with their partial-arc control valve design is carried forward into the design of HP turbine's first-stage blades. Steam is first admitted into two donut-shaped bowls (one in each direction of steam flow) that distribute the steam around the turbine casing (the stationary part of the turbine). The steam is then distributed through nozzles to the first stage of blading, which is attached to

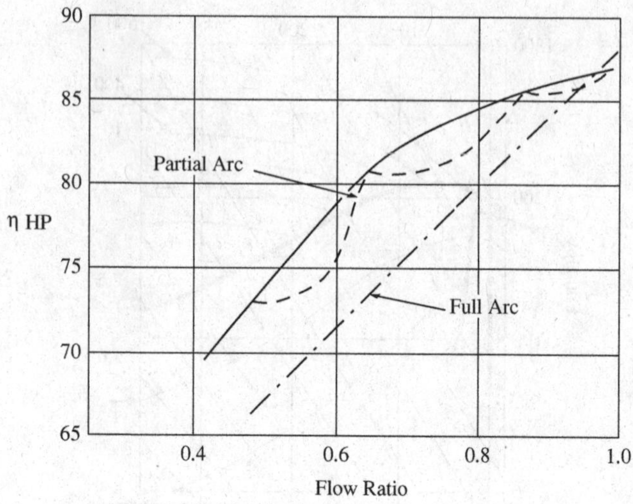

FIGURE 3.5 HP turbine efficiency. (After Steam Turbine Performance Seminar Lecture Notes by Kenneth C. Cotton.)

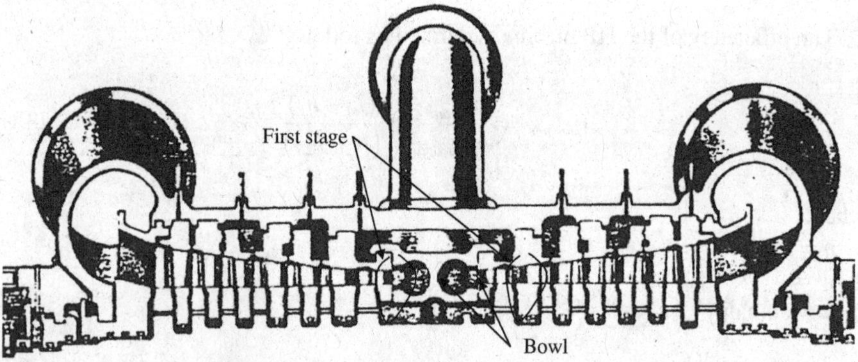

FIGURE 3.6 Westinghouse HP nuclear turbine.

the turbine rotor (the moving part of the turbine). In the Westinghouse design, the first-stage blades are said to be of the impulse design.

Figure 3.7 illustrates the impulse blade design with the blades viewed as if they were rolled out on a table. In this design, the entire pressure drop takes place in the stationary nozzles as the steam pressure is converted into kinetic energy, which is transferred to the moving blades (buckets), depleting the steam velocity as the steam impacts the turbine blades as they pass the nozzles, like a kick in the seat. In this design, the maximum work is achieved when the steam velocity is twice the velocity of the moving blades. The pressure of the steam beyond the first-stage blades

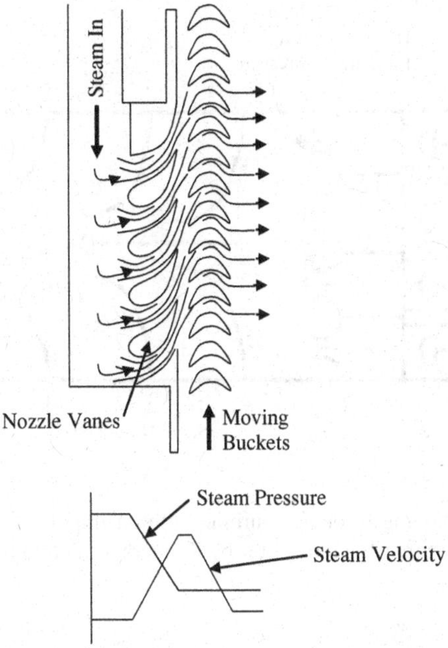

FIGURE 3.7 Impulse blade design. (After Steam Turbine Performance Seminar Lecture Notes by Kenneth C. Cotton.)

is known as the first-stage pressure (a very important parameter in nuclear turbine cycle analysis beyond the first stage blades are designed as reaction blades and the steam flow [and thus the power] through the remainder of the turbine cycle is roughly proportional to the first stage pressure).

Figure 3.8 illustrates the concept of the turbine blading beyond the first stage being thought of as simply a series of orifices. The flow-passing capability of the HP turbine is governed by the following relationship:

$$m = k\sqrt{P_{first\ stage} / v_{first\ stage}} \tag{3.2}$$

where:

k = turbine stage flow coefficient
P = pressure
V = specific volume.

Over a wide range of operating conditions,

$$v_{first\ stage} \approx \frac{1}{P_{first\ stage}} \tag{3.3}$$

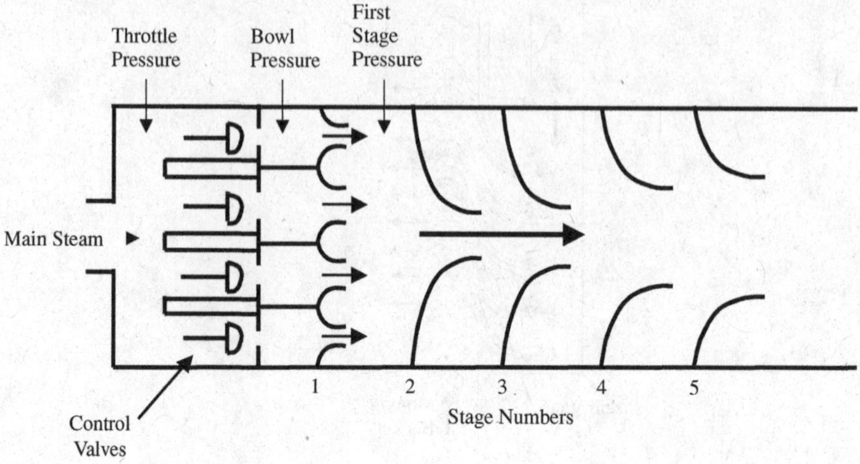

FIGURE 3.8 Steam moving through a turbine with an impulse first stage. (After Steam Turbine Performance Seminar Lecture Notes by Kenneth C. Cotton.)

Therefore,

$$m \approx k\, P_{first\ stage} \tag{3.4}$$

Because unit output is proportional to through flow, the first-stage pressure is an important indicator of unit load and should be trended.

The design philosophy adopted by the rest of the world for HP turbines is to use the full-arc control valve design. Steam is first admitted into the bowls, which distribute the steam around the turbine casing. The steam is then distributed through nozzles to the first stage of blading as with the Westinghouse design except that the first stage is designed as reaction blading similar to the rest of the stages of blading, as shown in Figure 3.9. In this design, the bowl pressure is analogous to the first-stage pressure in the Westinghouse design.

With the reaction blade design, the pressure drop takes place in both the stationary nozzle and the moving blade. Lower fluid velocities are used, resulting in greater efficiency; however, because there is pressure drop across the moving blade, leakage around the tip of the blade is of concern.

In turbine acceptance tests, the measured first-stage pressure is typically higher than the design value shown on the vendor's heat balance (HB) diagram. The reason is that the turbines when delivered typically exhibit tighter clearances between the rotating and stationary blades (i.e., design margin). Other possible reasons why the first-stage pressure might be high would be due to erosion in the first-stage nozzles (impulse blade design only), partially plugged turbine nozzles downstream of the first stage, and speeding (i.e., the reactor power is higher than advertised). Possible reasons for low first-stage pressure include erosion of turbine nozzles and blading in the turbine, degraded turbine seals, poor cycle isolation, high moisture carryover, and lower than advertised reactor power (possibly due to feedwater nozzle fouling).

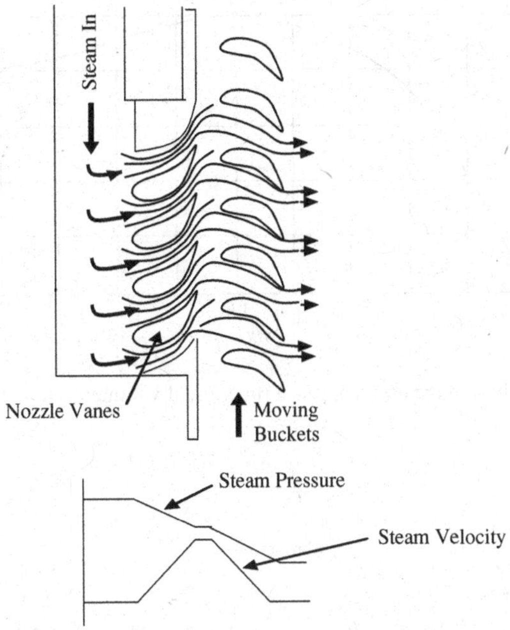

FIGURE 3.9 Reaction blade design. (After Steam Turbine Performance Seminar Lecture Notes by Kenneth C. Cotton.)

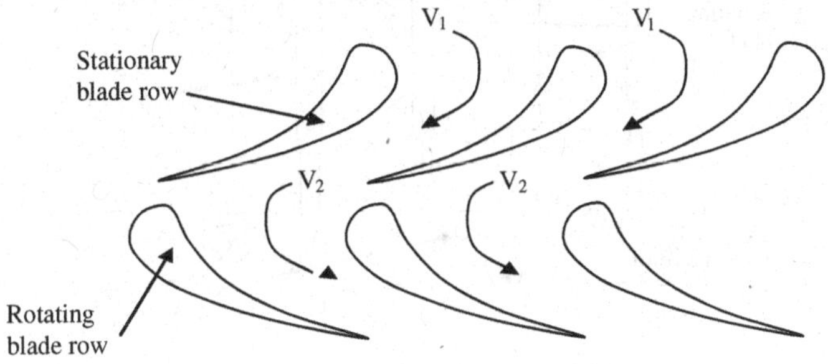

FIGURE 3.10 Each stage of a turbine is designed for a specific volumetric flow rate. (After Steam Turbine Performance Seminar Lecture Notes by Kenneth C. Cotton.)

As shown in Figure 3.10, each stage of a turbine is designed for a specific volumetric flow rate.

As Figure 3.11 shows, when the turbine is operating at some volumetric flow rate other than the design value, the steam flow mismatch results in reduced stage efficiency. Figure 3.12 shows a representative cross section of a turbine with alternating stationary and rotating blades. In an HP turbine, a shroud encircles the rotating blades. The stationary casing has sealing strips at the interfaces between the stationary and rotating parts to reduce the leakage around the tips of the stationary and rotating blades.

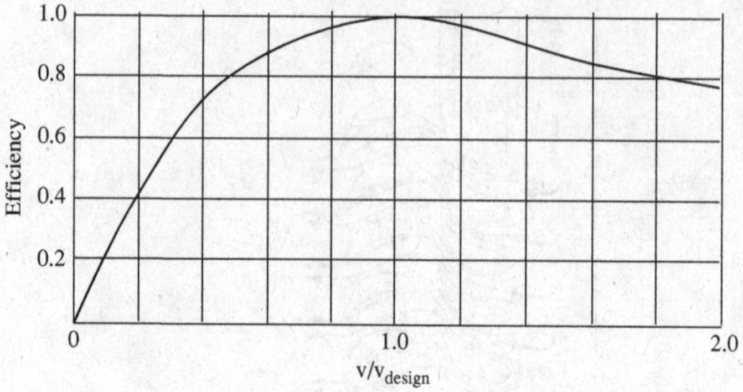

FIGURE 3.11 Turbine stage efficiency as a function of volumetric flow rate.

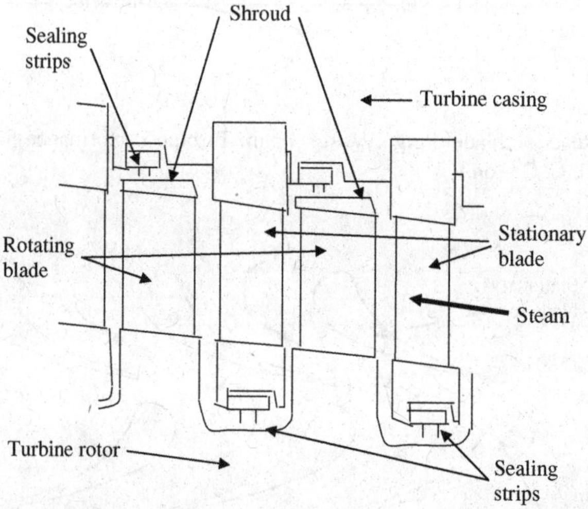

FIGURE 3.12 Cross section of a turbine. (After Schofield, P., Efficient Maintenance of Large Steam Turbines, paper for *Pacific Coast Electric Association 1982 Engineering and Operating Conference*, San Francisco, CA, 1882[1].)

Conventional blade designs are straight with uniform steam flow across the blade. More advanced computer modeling has led designers to develop advanced flow pattern blade designs that concentrate more on the steam flow in the center of the blade and away from the blade tips. The result is less tip end leakage and more efficient blade stages.

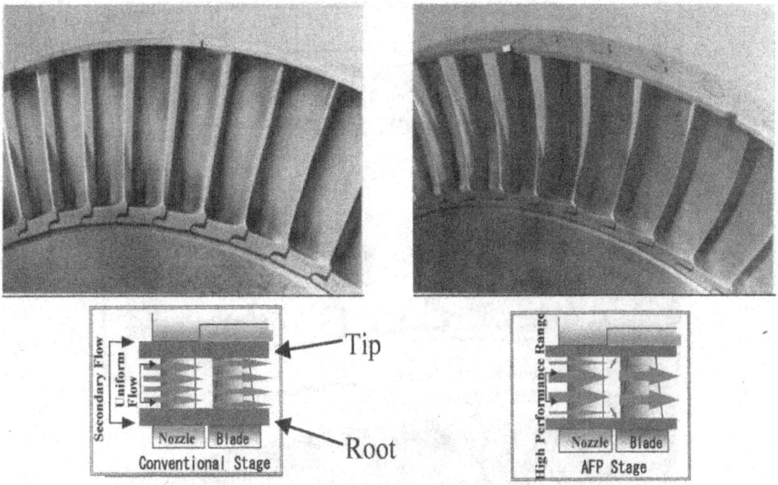

FIGURE 3.13 Conventional and advanced flow pattern turbine blade design.

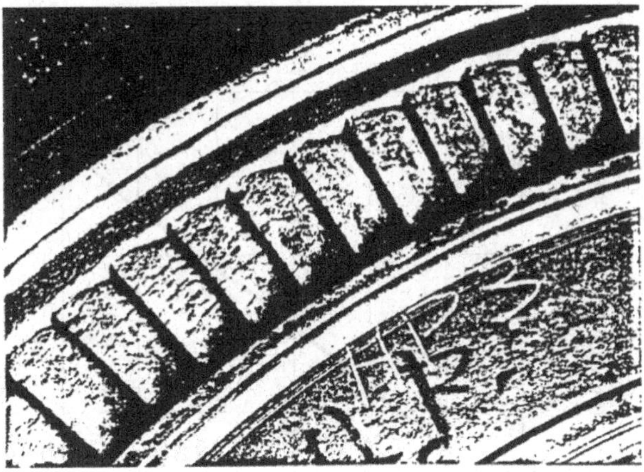

FIGURE 3.14 Deposit buildup on turbine blades. (From Schofield, P., Efficient Maintenance of Large Steam Turbines, paper for *Pacific Coast Electric Association 1982 Engineering and Operating Conference*, San Francisco, CA, 1882.)

Figure 3.13 shows the advanced flow pattern blade design. Turbine blade deposit buildup and blade erosion, as shown in Figures 3.14 and 3.15, are potential sources of reduced turbine blade stage efficiency. These problems are not nearly as prevalent in nuclear turbines as in coal-fired units. However, copper buildup was an issue before most of the nuclear plants made a concerted effort to eliminate copper in their condensate and feedwater systems, and erosion is an issue only where high moisture carryover is a problem.

FIGURE 3.15 Turbine blade erosion. (From Sumner, W.J. et al., Reducing Solid Particle Erosion Damage in Large Steam Turbines, paper for *American Power Conference*, IEEE, 1985[2].)

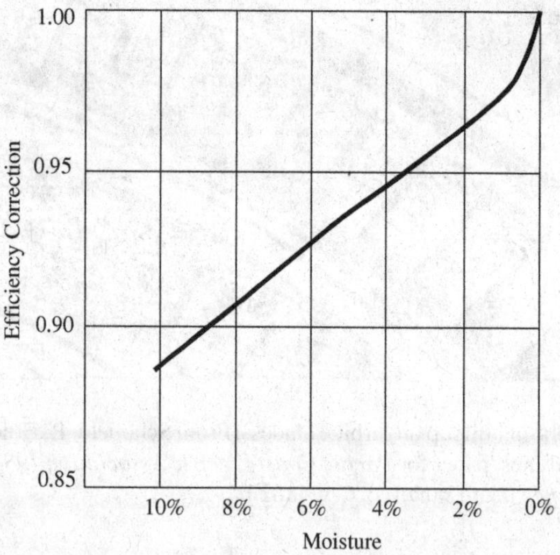

FIGURE 3.16 HP turbine efficiency correction for main steam moisture.

High moisture carryover from the reactor (BWR) or SG (PWR) can reduce the HP Turbine efficiency. As Figure 3.16 shows, each percent of moisture in the main steam reduces the HP turbine efficiency by about 1%.

REFERENCES

1. Schofield, P., Efficient Maintenance of Large Steam Turbines, paper for *Pacific Coast Electric Association 1982 Engineering and Operating Conference*, San Francisco, CA, 1882.
2. Sumner, W. J., et al., Reducing Solid Particle Erosion Damage in Large Steam Turbines, paper for *American Power Conference*, IEEE, 1985.

4 Moisture Separator Reheaters

4.1 MOISTURE SEPARATORS

As shown in Figure 4.1, at State Point 2, the steam that is exhausted from an HP turbine is typically quite wet, requiring that the moisture be separated from the steam prior to entering the LP turbine. This mechanical separation is achieved in a moisture separator (MS). The goal of an MS is to achieve complete removal of moisture from the steam, resulting in a saturated vapor with zero moisture as it exits the MS (State Point 3) while incurring some inevitable pressure drop in the MS. Moisture separation reduces the steam erosion in the LP turbine and increases the mechanical efficiency in the LP turbine due to the lower moisture content in the steam, but it does not increase the thermodynamic efficiency of the turbine cycle as a whole.

As shown in Figure 4.2, the moisture is removed by passing the steam through chevrons. The design shown in Figure 4.2 was commonly used in the construction of an MS in nuclear plants. However, experience demonstrated that the design was insufficient to achieve the desired result, so the improved design shown in Figure 4.3 has been widely adopted as a replacement of the MS. The result has been an increase in the electrical output of the nuclear plant, often justifying the replacement of the MS.

As one may see from Figure 4.3, the only modification to the original design was to add another tab to capture the swirling steam as it passes through the MS. However, this seemingly minor change significantly improves the rate at which moisture is removed from the steam.

There are two basic MS designs: vertical and horizontal. Figure 4.4 shows the vertical design as commonly used in BWR nuclear plants (most notably plants with nuclear steam supplies furnished by General Electric Corporation). As one may surmise, the moisture flows down the chevrons and exits through a drain at the bottom of the MS, whereas any air that may be entrapped is vented through the top of the MS.

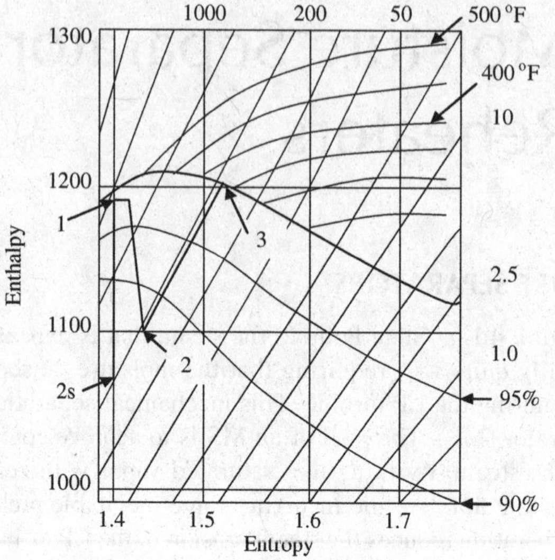

FIGURE 4.1 Moisture separation on the Mollier diagram.

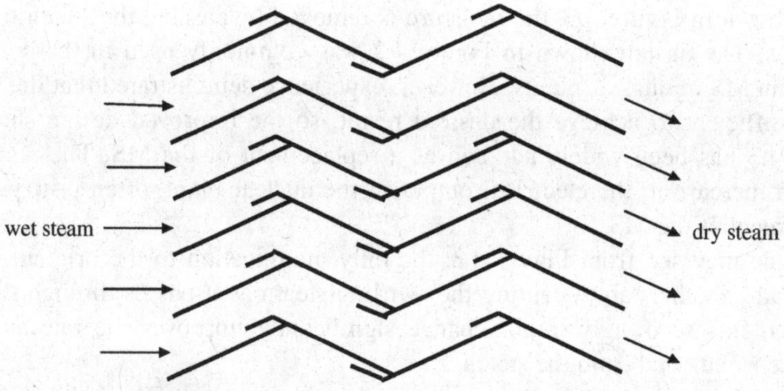

FIGURE 4.2 Original MS chevron design.

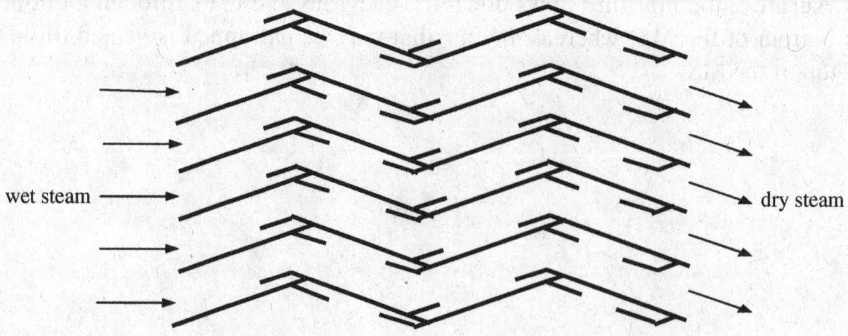

FIGURE 4.3 Current MS chevron design.

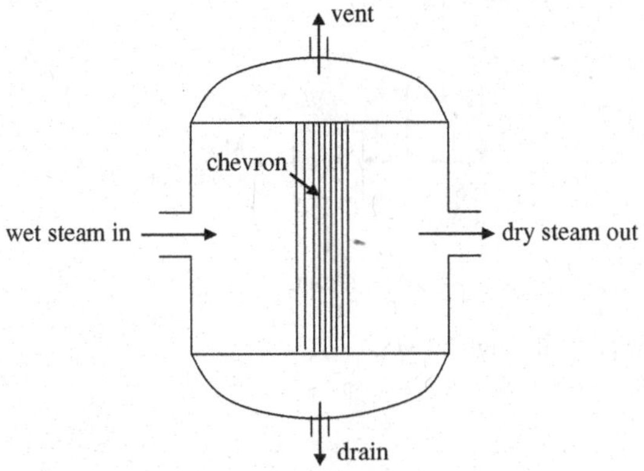

FIGURE 4.4 Vertical moisture separator design.

Figure 4.5 shows the horizontal MS design. As one might surmise, whereas the vertical MS is inside a vertical cylindrical vessel, the vessel containing horizontal MS is oriented horizontally. The vertical MS is normally located just below the HP turbine exhaust, whereas multiple horizontal MSs are normally located on the turbine deck adjacent to the LP turbine(s). The vertical MS design is common when no additional reheat of the steam is employed (perhaps because the steam in a BWR is radioactive).

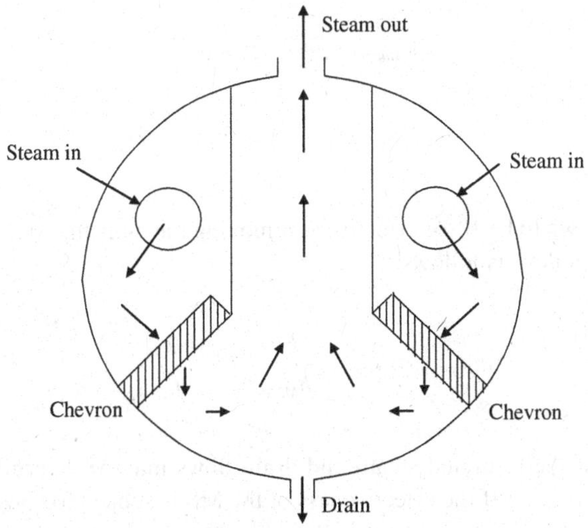

FIGURE 4.5 Horizontal moisture separator design.

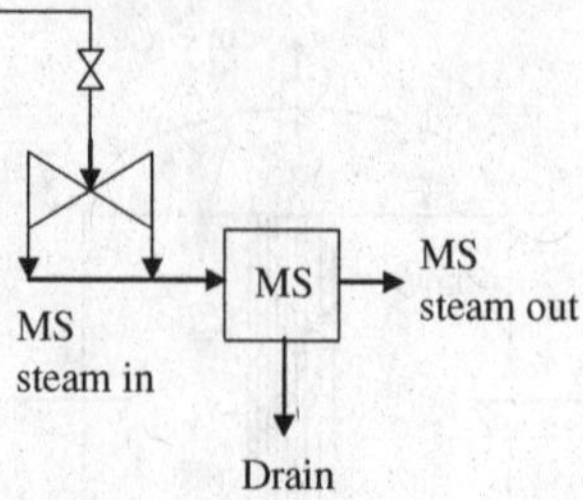

FIGURE 4.6 Flow paths through an MS.

Figure 4.6 illustrates the flow paths through an MS. Assuming that the efficiency of the HP turbine is known, one may calculate the HP turbine exhaust enthalpy as shown in Chapter 3. The HP turbine exhaust flow may be estimated based on the FFW flow rate, which must be accurately measured to determine the reactor power level, as shown in Chapter 2. From the FFW flow rate, the main steam flow rate may be calculated by adding the control rod drive flow rate in the case of a BWR or subtracting the blowdown flow rate in the case of a PWR. From the main steam flow, the HP turbine exhaust flow rate may be determined by subtracting flows to reheater (RH) in the case of a PWR, control valve leak-offs, and extraction flows from the HP turbine and at the HP turbine exhaust to feedwater heaters—to be discussed in later chapters.

Accordingly and assuming that the HP turbine exhaust pressure and pressure drop through the MS are known, one may perform an energy balance around the MS as follows:

$$m_{MS\,in}\,h_{MS\,in} = m_{MS\,stm\,out}\,h_{MS\,stm\,out} + m_{MS\,drain}\,h_{MS\,drain} \tag{4.1}$$

$$m_{MS\,stm\,out} = m_{MS\,in} - m_{MS\,drain} \tag{4.2}$$

$$m_{MS\,in}\,h_{MS\,in} = \left(m_{MS\,in} - m_{MS\,drain}\right) h_{MS\,stm\,out} + m_{MS\,drain}\,h_{MS\,drain} \tag{4.3}$$

If the MS is known to be 100% effective at removing the moisture, one may calculate the "belly" drain flow as follows:

$$m_{MS\,drain} = m_{MS\,in}\,\frac{h_{MS\,stm\,out} - h_{MS\,in}}{h_{MS\,stm\,out} - h_{MS\,drain}} \tag{4.4}$$

The enthalpy of the saturated steam and drain flows may be determined from the *ASME Steam Tables*.[1,2] If the effectiveness of the MS is subject to question, one may determine its effectiveness by measuring the drain flow and calculating the effectiveness by calculating the MS outlet enthalpy as follows:

$$h_{MS\ stm\ out} = \frac{m_{MS\ stm\ in}\, h_{MS\ stm\ in} - m_{drain\ out}\, h_{drain\ out}}{m_{MS\ stm\ out}} \tag{4.5}$$

Knowing the outlet pressure, the effectiveness may be determined by consulting the *ASME Steam Tables.*[1,2]

4.2 REHEATERS

As shown in Figure 4.7, at State Point 3, the steam that exits from the MS is typically dry steam. In some nuclear units (principally those with BWR nuclear steam supplies), the steam might be passed directly to the LP turbine(s) through the "crossover" piping. However, in many nuclear units, the steam is reheated before entering the LP turbine. This reheating is performed integrally to the same vessel containing the MS, thus the term moisture separator reheater (MSR). As shown in Figure 4.7, as State Point 5, this reheating can achieve a significant amount of superheat by extracting a portion of the main steam prior to entering the HP turbine.

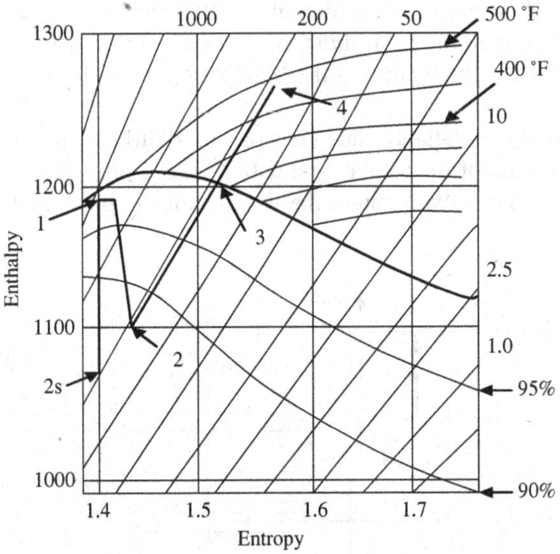

FIGURE 4.7 Moisture separation with reheat on the Mollier diagram.

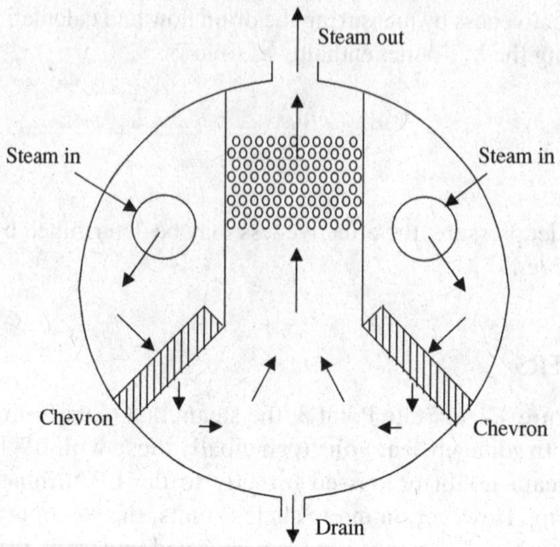

FIGURE 4.8 Moisture separator reheater sectional view.

Reheat further reduces the moisture in the LP turbine, thus reducing the steam erosion of the blading and increasing the mechanical efficiency of the turbine.

Figure 4.8 shows the typical arrangement of the RH in the MSR vessel with the steam entering the vessel from the LP turbine on the end of the vessel and first passing through the MS chevrons, then through the RH, and exiting out from the top of the vessel on the way to the LP turbine, whereas the drains that were wrung out of the MS pass out from the bottom of the MSR vessel to the feedwater system to be pumped back to the SG.

Figure 4.9 shows a schematic side view of an HP turbine and MSR. As one can see, the RH tubes are of the u-tube type with the head protruding out of the MSR vessel where the main steam enters the top half of the tubes and the condensed

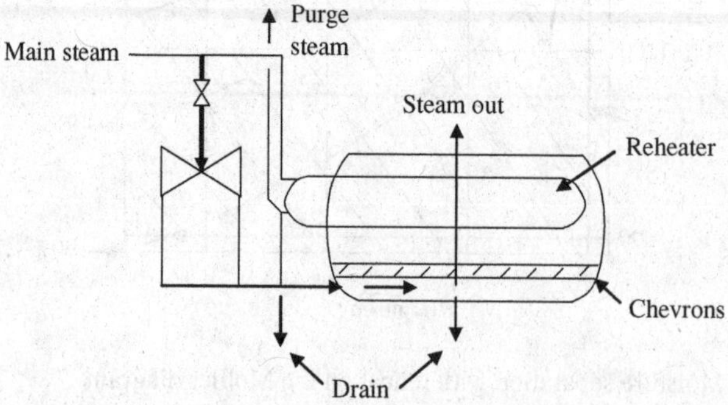

FIGURE 4.9 Schematic side view of an LP turbine and moisture separator reheater.

steam drains from the bottom half of the tubes. A certain amount (approximately 2%) of the main steam entering the tubes becomes "purge steam" that must pass through the tubes to keep the passage through the tubes open and to minimize subcooling of the extracted main steam, thus keeping the temperature of the tube bundle relatively uniform as the steam passes through and is condensed by the steam on the shell side coming from the LP turbine exhaust. Otherwise, the thermal expansion of the tubes in the bundle would not be uniform and the bundle would deform, resulting in tube failures. Because the resistance to heat transfer on the shell side where dry steam exists is greater than that on the tube side where condensing is taking place, the RH tubes are normally of the low-fin design to increase the area available for heat transfer on the shell side to compensate for the higher resistance to heat transfer.

Figure 4.10 shows how one may calculate the amount of heat transferred to the reheat steam and/or to solve for the tube-side drain flow if the outlet pressure and temperature are known or the outlet enthalpy if the tube-side drain flow is known. By drawing a system boundary around the RH and assuming that the purge steam is 2% of the main steam coming into the MSR, one may calculate the heat transfer as follows:

$$m_{tube\text{-}in} = \frac{m_{tube\text{-}drain}}{1-0.02} \tag{4.6}$$

$$m_{purge\ stm} = m_{tube\text{-}in} - m_{tube\text{-}drain} \tag{4.7}$$

$$Q = m_{tube\text{-}in}\ h_{tube\text{-}in} - m_{tube\text{-}drain}\ h_{tube\text{-}drain} - m_{purge\ stm}\ h_{tube\text{-}in} \tag{4.8}$$

or

$$Q = m_{shell\text{-}in}\ h_{shell\text{-}in} - m_{shell\text{-}out}\ h_{shell\text{-}out} \tag{4.9}$$

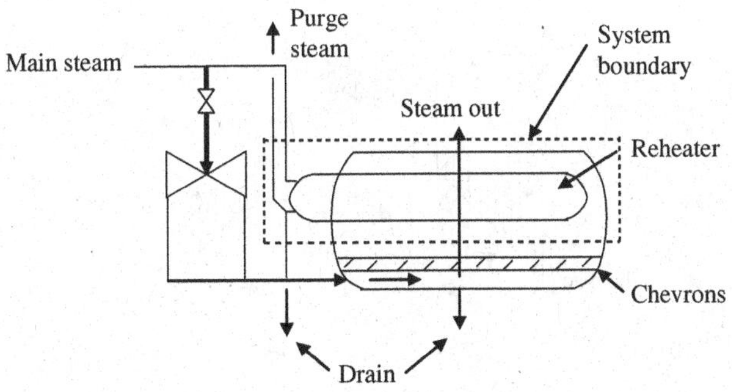

FIGURE 4.10 Schematic side view of an LP turbine and moisture separator reheater with system boundary.

By equating the two such that

$$m_{shell\text{-}in}\, h_{shell\text{-}in} + m_{tube\text{-}in}\, h_{tube\text{-}in} = m_{shell\text{-}out}\, h_{shell\text{-}out} + m_{tube\text{-}drain}\, h_{tube\text{-}drain} + m_{purge\ stm}\, h_{tubeside\ in} \tag{4.10}$$

One may calculate the enthalpy of the hot reheat steam leaving the RH as

$$h_{shell\text{-}out} = \frac{m_{shell\text{-}in}\, h_{shell\text{-}in} + m_{tube\text{-}in}\, h_{tube\text{-}in} - m_{tube\text{-}drain}\, h_{tube\text{-}drain} - m_{purge\ stm}\, h_{tube\text{-}in}}{m_{shell\text{-}out}} \tag{4.11}$$

and the temperature may be determined from the *ASME Steam Tables.*[1,2]

If the hot reheat temperature is known, one may calculate the tube-side drain flow as

$$m_{tube\text{-}drain} = \frac{m_{shell}\,(h_{shell\text{-}out} - h_{shell\text{-}in})}{(h_{tube\text{-}in} - h_{tube\text{-}drain})} \tag{4.12}$$

A common practice is to have two stages of reheat with two tube bundles in series, as shown in Figure 4.11.

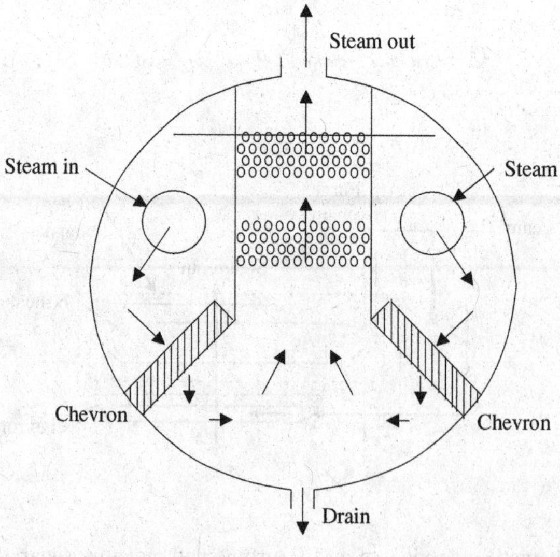

FIGURE 4.11 Moisture separator reheater sectional view with two stages of reheat.

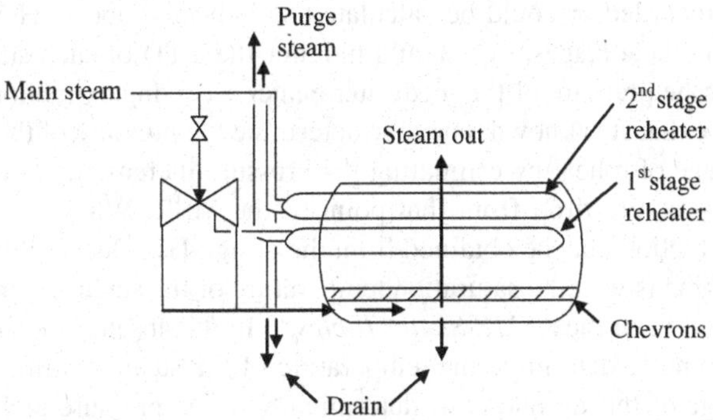

FIGURE 4.12 Schematic side view of an LP turbine and moisture separator reheater with two stages of reheat.

In the case of two stages of reheat, the steam supply for the first stage of reheat is taken from an extraction from the HP turbine with the steam supply to the second stage taken from main steam, as shown in Figure 4.12.

With two stages of reheat, the first stage of extraction steam is able to produce work in the HP turbine before being siphoned off to reheat the steam coming from the MS. This arrangement also enjoys the added benefit of reducing the temperature gradients in the reheat bundles, making tube failure due to thermal expansion less likely.

Because the HP turbine extraction pressures can normally be estimated from the high-pressure feedwater heater shell pressure, the extraction enthalpy can be determined from the design HB diagram or Mollier diagram (see Figure 3.4). The drain flows are saturated liquids, so the enthalpy may be determined based on their temperatures from the *ASME Steam Tables*.[1,2] Therefore, if the tube-side drain flows and temperatures can be measured, the heat transfer occurring in each section of reheat can be measured as follows:

$$Q = m_{tube\text{-}in}\, h_{tube\text{-}in} - m_{tube\text{-}drain}\, h_{tube\text{-}drain} - m_{purge\ steam}\, h_{tube\text{-}in} \tag{4.13}$$

where

$$m_{tube\text{-}in} = \frac{m_{tube\text{-}drain}}{1-0.02} \text{ and } m_{tube\text{-}drain} = 0.02\, m_{tubeside\ in} \tag{4.14}$$

The terminal temperature difference (TTD) of an MSR is a key thermal performance parameter. The TTD is the difference between the tube-side saturation temperature (the maximum temperature that could be achieved) and the temperature of the steam exiting the shell side (referred to as the hot reheat steam). For an MSR with single-stage RH, this parameter is readily determined, as these temperature values may often be recorded or could be calculated as indicated above. However, for an MSR with two reheat stages, one should monitor the TTD of each stage of reheat. Assuming that the pressure of the feedwater heater receiving extraction steam from the LP turbine exhaust is known, one may determine the pressure of the steam entering the first stage of reheat by estimating the pressure increase up to the extraction point and the pressure drop from that point to the MSR. When operating at full power, this correction may be obtained from the design HB. Because the steam exiting the LP turbine is wet, the saturation temperature of the steam entering the tubes may be obtained from the *ASME Steam Tables*.[1,2] In the absence of direct temperature measurement, to determine the temperature of the steam exiting the shell side of the first stage of reheat, one must determine both the pressure and the enthalpy of the steam because it is superheated. If the intermediate pressure between the two stages of reheat is provided on the design HB diagram, the ratio of the pressure drop from the cold reheat inlet to the first reheat stage outlet over the pressure drop from the cold reheat inlet to the second reheat stage outlet may be assumed. The first-stage reheat outlet enthalpy may be calculated as shown above. Having both the first-stage reheat outlet pressure and the enthalpy, the temperature may be determined from the *ASME Steam Tables*.[1,2]

Another RH parameter that should be monitored is the apparent resistance to heat transfer. One of the most basic relationships when evaluating heat transfer in any HX is as follows:

$$Q = UA\ (LMTD) \Rightarrow UA = \frac{Q}{LMTD} \tag{4.15}$$

where

$$LMTD = \frac{t_{shell\text{-}out} - t_{shell\text{-}in}}{t_{tube} - t_{shell\text{-}out}} \tag{4.16}$$

The resistance to heat transfer is simply $1/UA$. This parameter should be compared with the design value and trended to detect apparent fouling in the RH.

For nuclear plants with two stages of reheat, the percentage of total heat transfer in each stage of reheat should be monitored and compared with the design value for each stage. If the overall performance of the RH is deficient, this comparison will be an indication as to which stage of reheat may be deficient.

REFERENCES

1. Meyer, C. A. et al., *ASME Steam Tables*, 6th ed., American Society of Mechanical Engineers, New York, 1993.
2. *ASME Steam Tables Compact Edition*, American Society of Mechanical Engineers, New York, 1993, 2006.

5 Low-Pressure Turbines

5.1 LOW-PRESSURE TURBINE INTERCEPT VALVES

As shown in Figure 5.1, motor-operated intercept valves are positioned between the MS or MSR and the LP turbine.

Intercept valves are designed to automatically close on the turbine trip to prevent over-speeding the LP turbines. These valves do not throttle flow, so the pressure drop through them is negligible.

5.2 LOW-PRESSURE TURBINE MOISTURE REMOVAL

Figure 5.2 shows the expansion of the crossover steam through the LP turbine. The saw-toothed shape of the expansion line indicates moisture removal stages that move the line to the right toward less moisture in the steam. As one might surmise from Figure 5.2, if moisture were not removed, the moisture in the steam would reach unacceptable levels as the expanding steam approaches the very low pressure in the main condenser. Moisture removal is accomplished by designing the last few LP turbine blade stages with grooves in the leading edge to capture the particles of moisture, which are then slung down the blade by centrifugal force into cavities in the LP turbine casing at the points of moisture removal, as shown in Figures 5.3 and 5.4. These cavities are located just upstream of the extraction points in the LP turbine. The result is drier steam that continues to expand through subsequent LP turbine blade stages.

Figure 5.5 illustrates that the analysis of moisture removal stages is similar to the analysis of an MS, where the moisture removal stage is viewed as an MS between two turbines. Moisture removal effectiveness, μ, is defined as follows:

$$\mu = \frac{m_{ex\text{-}m}}{m_{total\text{-}m}} = \frac{m_{ex\text{-}m}}{m_{in}\left(1 - x_{in}\right)} \tag{5.1}$$

where:

$m_{ex\text{-}m}$ = moisture that is extracted
$m_{total\text{-}m}$ = total moisture entering the stage
m_{in} = total steam entering the stage
x_{in} = quality of the steam entering the stage.

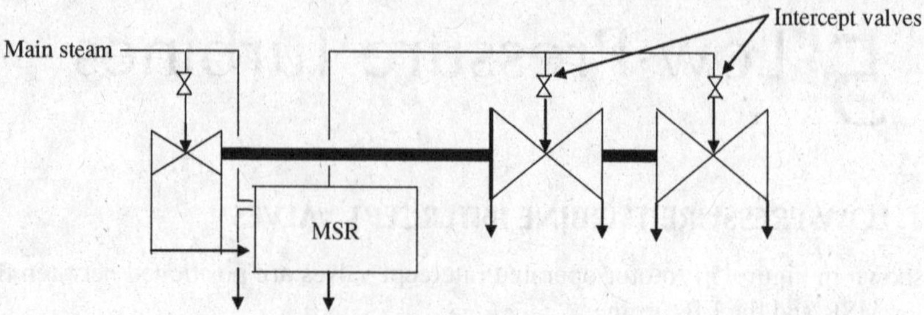

FIGURE 5.1 Intercept valves positioned before LP turbines.

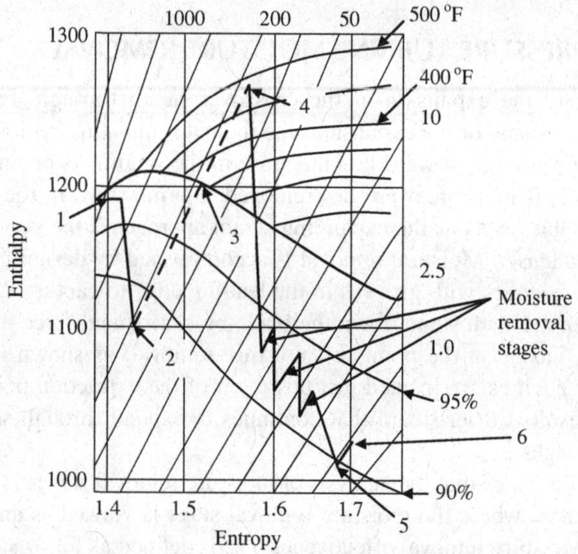

FIGURE 5.2 LP turbine expansion line on the Mollier diagram.

Therefore,

$$m_{ex\text{-}m} = m_{in}\left(1 - x_{in}\right)\mu \tag{5.2}$$

The effectiveness of moisture removal stages is normally provided by the turbine vendor as a curve similar to Figure 5.6.

By knowing the amount of moisture wrung out of the steam entering the stage, one can calculate the enthalpy of the remaining steam as follows:

$$h_{out} = \frac{m_{in}\, h_{in} - m_{ex\text{-}m}\, h_{ex\text{-}m}}{m_{in} - m_{ex\text{-}m}} \tag{5.3}$$

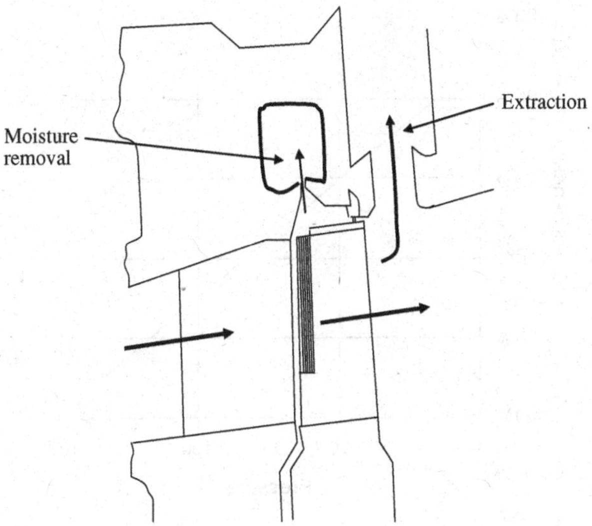

FIGURE 5.3 LP turbine moisture removal stage.

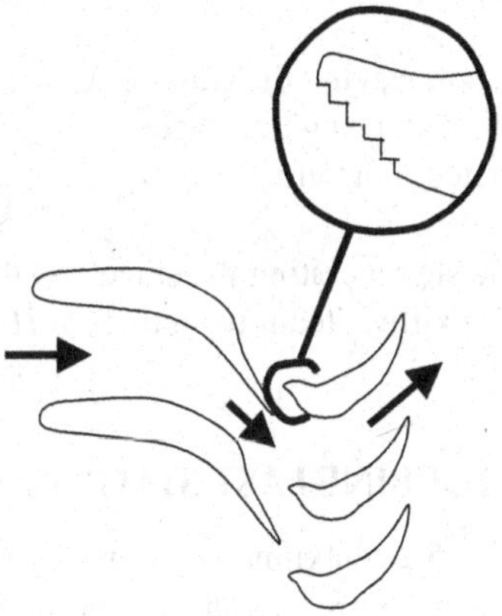

FIGURE 5.4 Cross section of an LP turbine moisture removal stage. (After Steam Turbine Performance Seminar Lecture Notes by Kenneth C. Cotton.)

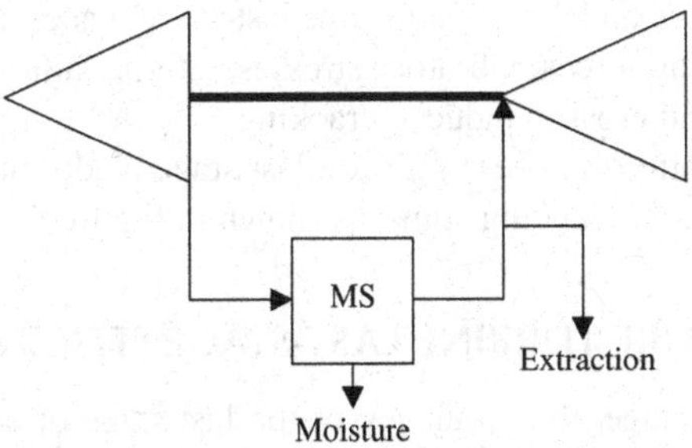

FIGURE 5.5 LP turbine moisture removal stage analysis model.

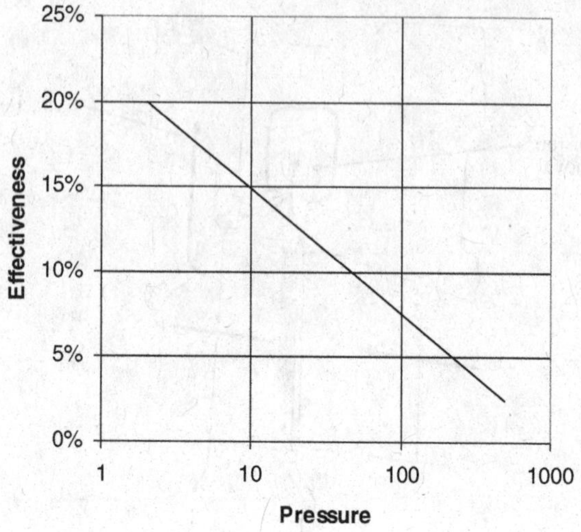

FIGURE 5.6 Internal moisture removal effectiveness.

where:

h_{out} = enthalpy of extraction leaving the stage
h_{in} = enthalpy of extraction entering the stage
$h_{ex\text{-}m}$ = enthalpy of extracted moisture.

Knowing the pressure of the steam exiting the stage and the enthalpy, one may then determine the quality of the exiting steam from the *ASME Steam Tables*.[1,2]

5.3 LOW-PRESSURE TURBINE LAST-STAGE BLADE FAILURES

As one may see from Figure 5.2, maximum power is derived from the LP turbine by expanding the steam to as low a pressure in the main condenser as possible. The resulting high specific volume of the steam along with the high mass flow rate requires very large last stage blades on the LP turbine, as illustrated in Figure 5.7. These long blades are not connected with a shroud as is the case with HP turbine blades (see Figures 3.12 and 3.13). LP turbine last-stage blades are subject to centrifugal and steam bending forces, vibratory stresses, fatigue failures, blade tip failures, blade root failures, and erosion-induced cracking.

In an attempt to minimize these failures, last-stage blades may be designed with a stellite leading edge and lancing stubs, as shown in Figure 5.8.

5.4 LOW-PRESSURE TURBINE LAST-STAGE EFFICIENCY

As with any turbine stage, the efficiency of the last stage of an LP turbine is best when the velocity of the steam passing through the stage is the design velocity, as illustrated in Figure 5.9.

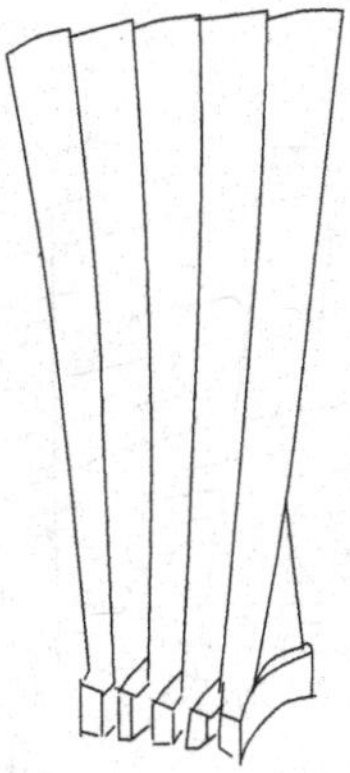

FIGURE 5.7 LP turbine last-stage blades.

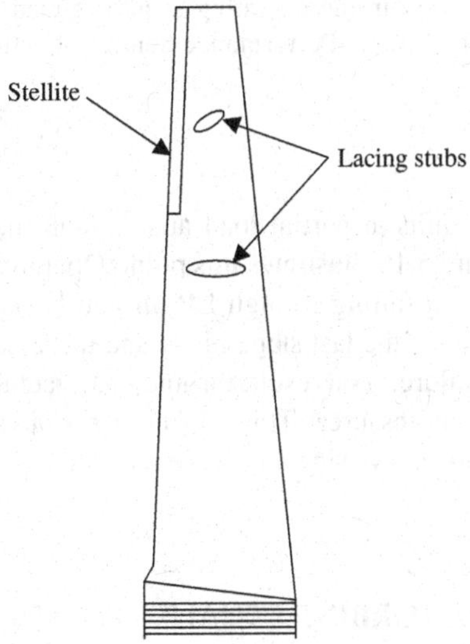

FIGURE 5.8 LP turbine last-stage blade with a stellite leading edge and lancing stubs.

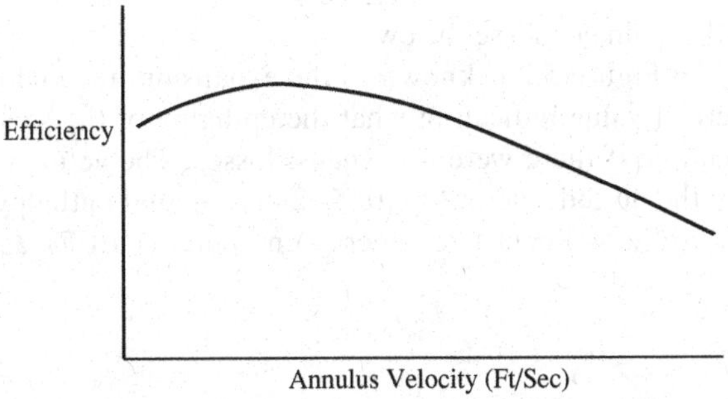

FIGURE 5.9 Efficiency of the LP turbine last stage. (From Steam Turbine Performance Seminar Lecture Notes by Kenneth C. Cotton.)

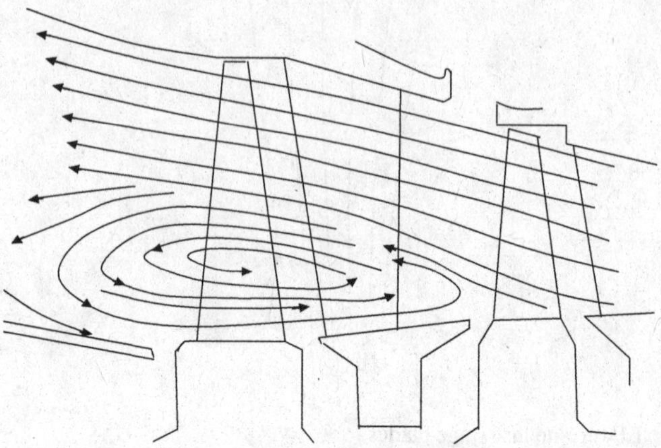

FIGURE 5.10 Low-pressure turbine operating at partial load or high main condenser backpressure. (After Steam Turbine Performance Seminar Lecture Notes by Kenneth C. Cotton.)

Operation of nuclear units at partial load and/or high main condenser pressure can be problematic. Figure 5.10 illustrates this point. Operation at partial load and/or high condenser pressure (resulting in high LP turbine backpressure) causes steam flow separation at the root of the last stage blade and increases vibratory stress contributing to early blade failure. As a result, turbine manufacturers limit the allowable LP turbine operating backpressures. This can be a serious operational limitation, resulting in reduced reactor power in some cases.

5.5 LOW-PRESSURE TURBINE EXHAUST LOSSES

Referring to Figure 5.2, the slight uptick in enthalpy between Stage Points 5 and 6 represents the exhaust loss in the LP turbine. Figure 5.11 illustrates a typical arrangement at the LP turbine exhaust as the steam makes the turn out of the last-stage blades to the main condenser below.

State Point 5 in Figure 5.2 is known as the expansion line end point (ELEP), a purely theoretical value indicating what the enthalpy of the steam exiting the LP turbine would be if there were no exhaust losses. The actual expansion line is illustrated by the dotted line in Figure 5.12. The actual enthalpy of the steam exiting the LP turbine is the utilized energy end point (UEEP), as shown in the figure.

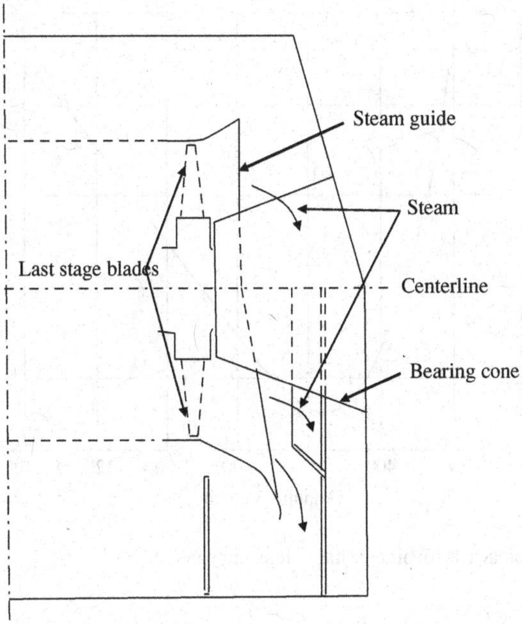

FIGURE 5.11 LP typical turbine exhaust arrangement. (After Steam Turbine Performance Seminar Lecture Notes by Kenneth C. Cotton.)

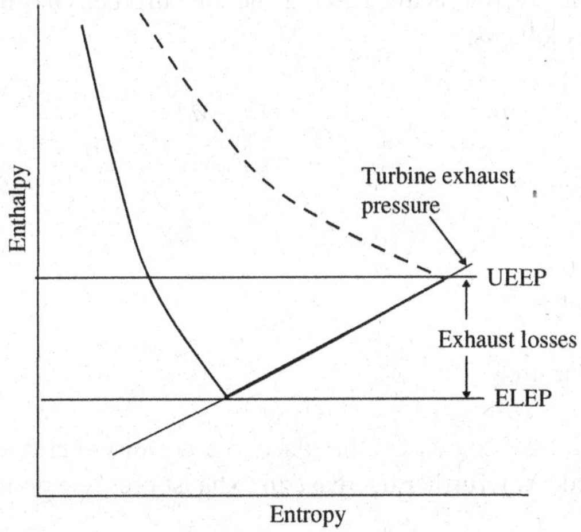

FIGURE 5.12 LP turbine exhaust loss.

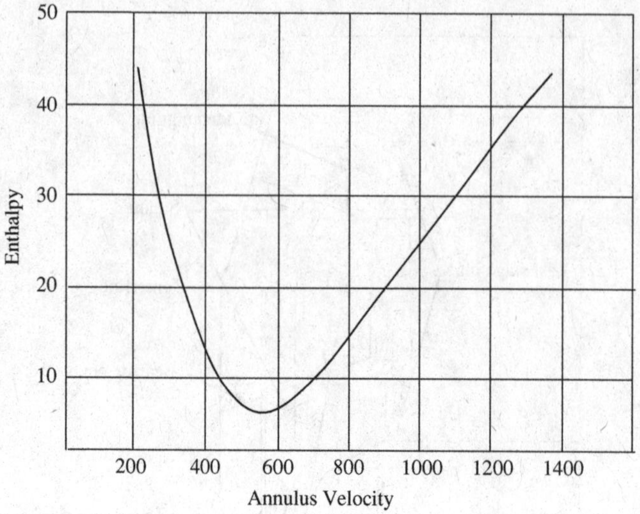

FIGURE 5.13 Typical LP turbine exhaust loss curve.

The difference in the enthalpy between the UEEP and the ELEP is known as the exhaust loss. Figure 5.13 shows a typical exhaust loss curve as provided by the turbine manufacturer.

As may be seen in Figure 5.13, in order to know the exhaust loss, one must know the annulus velocity of the steam leaving the LP turbine. One may calculate the annulus velocity as follows:

$$V_{annulus} = \frac{mv(1-M)}{A_{annulus}} \tag{5.4}$$

where:

m = mass flow rate
v = specific volume
M = moisture
$A_{annulus}$ = annulus area.

As the exhaust pressure decreases, the discharge velocity increases until the sonic velocity is achieved. Any further decrease in exhaust pressure does not increase the discharge velocity.

As illustrated in Figure 5.14, there is an optimum turbine backpressure. Operation above the optimum backpressure reduces electrical output due to the

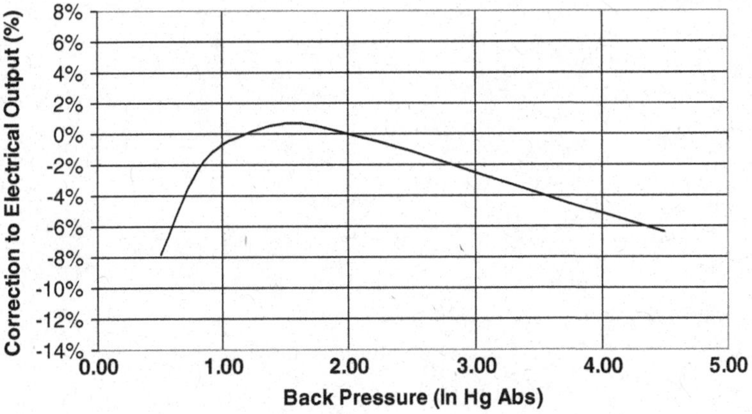

FIGURE 5.14 LP turbine backpressure correction to electrical output.

higher UEEP, whereas operation below the optimum reduces the output due to choked flow (sonic velocity) being achieved in the exhaust, resulting in an increase in the exhaust losses.

REFERENCES

1. Meyer C. A., et al., *ASME Steam Tables*, 6th ed., American Society of Mechanical Engineers, New York, 1993.
2. *ASME Steam Tables Compact Edition*, American Society of Mechanical Engineers, New York, 1993, 2006.

6 Main Condensers

6.1 HEAT REJECTION

Referring back to the discussion of the Rankine cycle in Chapter 1, the work performed by the turbine(s) (the hash-marked area) and the heat that must be rejected through the main condenser are shown in Figure 6.1.

Because the amount of work (i.e., electrical energy produced) and the amount of heat that must be rejected to the environment are a function of the areas under the curves, one may readily see that reducing the main condenser pressure not only increases the amount of useful work produced but also reduces the amount of heat that must be rejected to the environment.

In nuclear plants, the main condenser is a shell-and-tube HX with the steam that is exhausted from the LP turbine being condensed on the shell side by water passing through the tubes. Although the temperature of the water is increased by the transfer of heat from the steam on the shell side, this is an isothermal process on the shell side (i.e., the steam is not cooled); thus, the water is referred to as condenser circulating water (CCW), not condenser cooling water—a common misnomer. Because condensation is an isothermal process, the CCW does not "cool" the condenser.

6.2 MAIN CONDENSER CONFIGURATIONS

Numerous configurations of the main condenser are possible. Figure 6.2 illustrates the simplest and most common main condenser configuration. In this arrangement, the tubes are oriented perpendicular to the LP turbine shaft and CCW passes into and out of water boxes on each end of the tube bundle. The condensed steam, referred to as condensate, passes out of the main condenser and is pumped back to the reactor or SG. Because most nuclear plants are fitted with at least two LP turbines, there would be one main condenser, as illustrated in Figure 6.2, under each LP turbine. Although the pressure in each main condenser should be essentially the same, there is normally a duct tying the condenser shells together at the neck of the condenser between the LP turbine and the tube bundle below to equalize the pressures should one tube bundle be isolated for cleaning.

In the arrangement illustrated in Figure 6.3, the CCW passes through the main condenser shell twice so that it enters and leaves the tube bundle on the same side of the condenser. Note that the coldest water enters the top half of the tube bundle so that the lowest possible LP turbine backpressure is achieved. This arrangement is used occasionally where the arrangement of the turbine building requires shorter tubes and/or shorter tube pulling spaces.

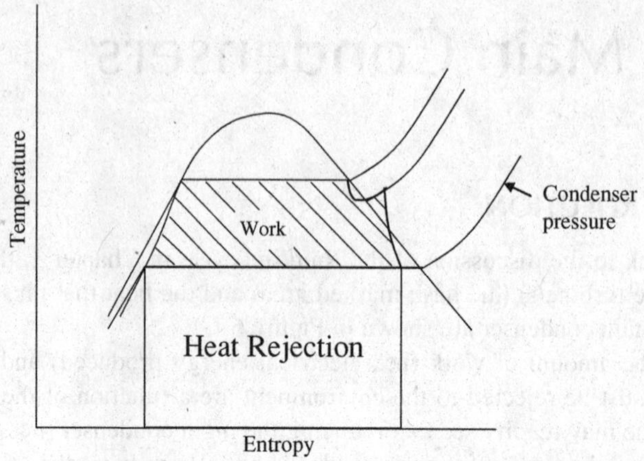

FIGURE 6.1 Work and heat rejection as illustrated on the Rankine cycle.

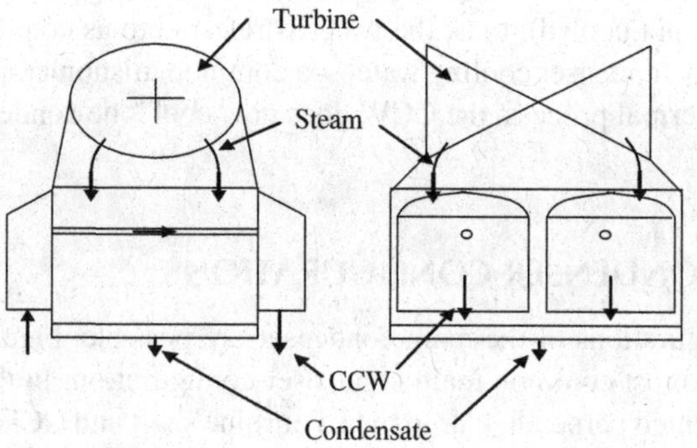

FIGURE 6.2 Side and end views of a single-pass, single-pressure, single-shell main condenser.

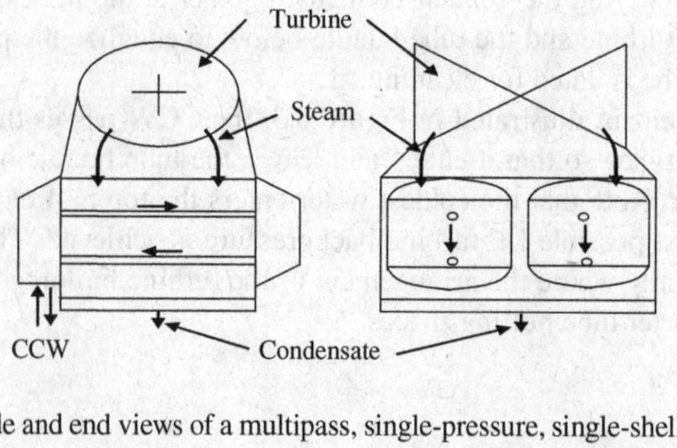

FIGURE 6.3 Side and end views of a multipass, single-pressure, single-shell main condenser.

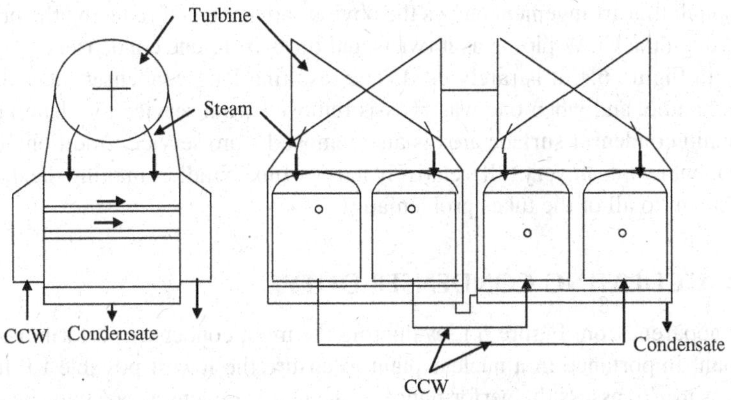

FIGURE 6.4 End and side views of a multipass, multipressure, multishell main condenser.

In the arrangement illustrated in Figure 6.4, the CCW passes through two or more main condenser shells in series so that the pressures in the individual shells are different.

This is a common arrangement when the nuclear station utilizes cooling towers to reject the waste heat into the atmosphere. Multipressure main condensers are common where cooling towers are employed because with cooling towers, the CCW must be pumped up above the cooling tower fill, requiring much higher head CCW pumps that require larger motors that require more auxiliary power. Reducing the CCW flow rate by passing it through the shells in series minimizes this effect. Reducing the CCW flow rate also reduces the cost of the cooling towers. To minimize the impact of hotter water entering the tube bundles in subsequent shells, the surface area of the condenser is increased.

In the arrangement illustrated in Figure 6.5, the CCW passes through two or more main condenser shells in series such that the pressures in the individual shells are different, as in Figure 6.4, but the tubes are oriented parallel rather than perpendicular to the turbine shaft.

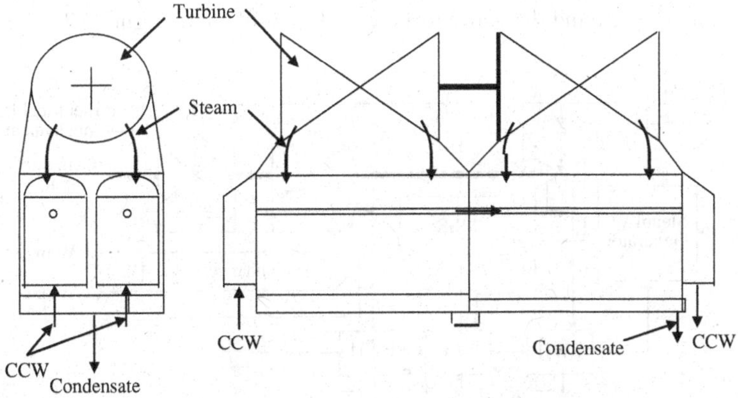

FIGURE 6.5 Side and end views of a single-pass, multipressure, multishell main condenser.

Although this arrangement enjoys the obvious advantage of reducing the pressure drop through the CCW piping as it twists and turns from one condenser shell to the next, as in Figure 6.4, it is rarely used. The resulting long condenser tubes are difficult to handle, and when one waterbox is removed from service for cleaning, half of the main condenser surface area is also removed from service. Additionally, this arrangement results in very tall, relatively narrow tube bundles, making the distribution of steam to all of the tubes problematic.

6.3 CALCULATING CONDENSER DUTY

As it is apparent from Figure 6.1, evaluating the main condenser performance is of paramount importance in a nuclear plant to ensure the lowest possible LP turbine backpressure. To assess the performance of the main condenser, one must be able to determine the amount of heat that is being transferred, commonly referred to as the main condenser duty. Consider a representative turbine cycle shown in Figure 6.6 in which a system boundary is drawn around the main condenser.

Based on the system boundary as defined, one may see that the main condenser duty may be calculated by either of two ways. First, one may calculate the heat added to the condenser as follows:

$$\text{Duty} = m_{exhaust}\left(h_{exhaust} - h_{condensate}\right) + m_{drain}\left(h_{drain} - h_{condensate}\right) \quad (6.1)$$

This is a particularly difficult way to determine the duty, as neither the LP turbine exhaust nor the UEEP can be measured. Second, one may calculate the heat removed by the CCW as follows:

$$\text{Duty} = m_{CCWout}\left(h_{CCWout} - h_{CCWin}\right) = m_{CCWout}\, c_{p\text{-}ccw}\left(t_{CCWout} - t_{CCWin}\right) \quad (6.2)$$

Whereas both the CCW flow rate and the inlet and outlet temperatures may be calculated, the accuracy of these measurements may be subject to question.

A more accurate means of determining the main condenser duty is to place the system boundary around the entire turbine cycle, as shown in Figure 6.7.

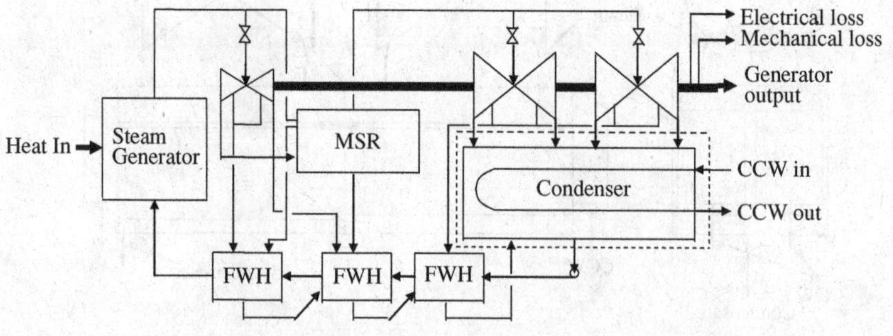

FIGURE 6.6 Representative turbine cycle with system boundary around the main condenser.

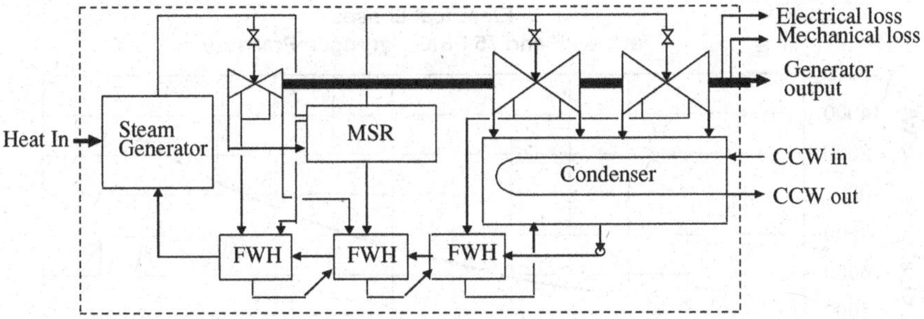

FIGURE 6.7 Representative turbine cycle with system boundary around the entire turbine cycle.

From the first law of thermodynamics,

$$\text{Heat In} + \dot{m}_{CCWin}\, h_{CCWin} = \text{Generator output} + \text{Electrical losses} + \text{Mechanical losses} + \dot{m}_{CCWout}\, h_{CCWout} \tag{6.3}$$

$$\dot{m}_{CCWin} = \dot{m}_{CCWout} \tag{6.4}$$

$$\begin{aligned}\text{Duty} &= \dot{m}_{CCWout}\,(h_{CCWout} - h_{CCWin}) \\ &= \text{Heat In} - (\text{Generator output} + \text{Electrical losses} + \text{Mechanical losses})\end{aligned} \tag{6.5}$$

This approach offers a more accurate means of determining the main condenser duty because the reactor power and electrical output are accurately measured and the electrical and mechanical losses may be estimated from curves such as those shown in Figures 6.8 and 6.9, where PF stands for power factor.

The heat generated by mechanical and electrical losses is typically removed from the station by a separate cooling water system.

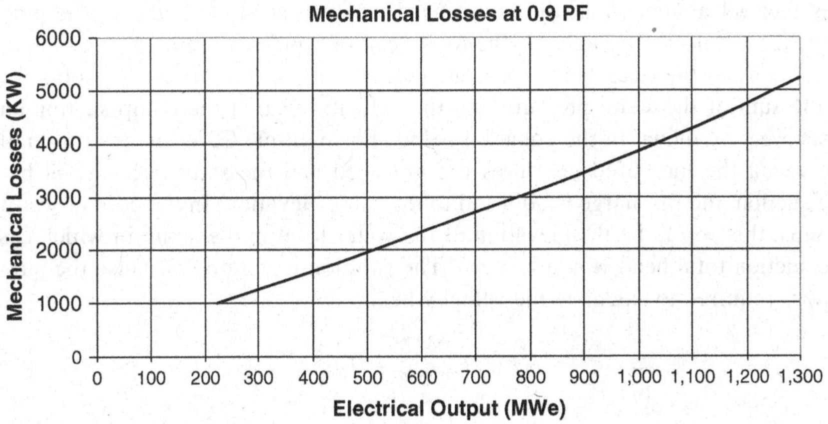

FIGURE 6.8 Typical mechanical loss curve.

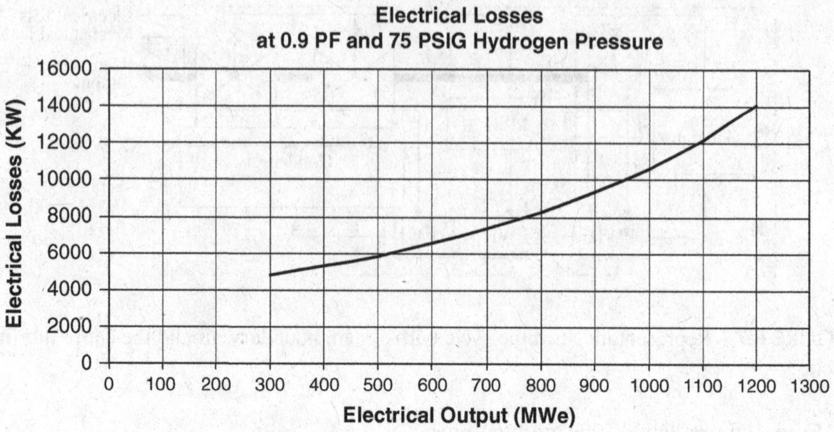

FIGURE 6.9 Typical electrical loss curve.

6.4 CALCULATING CONDENSER CIRCULATING WATER FLOW RATE

Another parameter that is essential in determining the performance of the main condenser is the CCW flow rate. As alluded to previously, directly measuring this flow rate is challenging due to the huge size of the CCW conduits. One of the most accurate methods of flow measurement is by dye dilution. A known concentration of dye is injected quickly into the CCW at a precise flow rate. The diluted concentration of fluorescence is measured downstream by a highly accurate instrument, and the dilution factor is used to accurately measure the flow rate. Although measuring the CCW flow by dye dilution on a routine basis is impractical, this method is useful in establishing a baseline for monitoring changes in CCW flow by calibrating the head loss through a point of transition such as the main condenser outlet waterbox.[1]

Another method of determining the CCW flow rate is to determine the CCW pump(s) total head and read the flow rate from the pump head curve. (Of course, this method assumes that the pump head curve is accurate and that the pump impeller has not deteriorated.) The total head of a pump is defined as the difference between the total discharge head and the total suction head. The total head is the sum of the static pressure and the velocity head at the pump suction and discharge, as shown in the equation below. For multiple CCW pumps (normally the case), the measured pressures are averaged and the static pressure at both the suction and discharge is adjusted to the same elevation. In the case of wet-pit pumps, this would be the elevation of the water level in the sump in which case the suction total head would be zero. The solution is iterative because the pump flow is required to calculate the velocity head.

$$TH = H_{discharge} - H_{suction} \tag{6.6}$$

$$TH = \left(H_{static} + H_{velocity}\right)_{discharge} - \left(H_{static} + H_{velocity}\right)_{suction} \tag{6.7}$$

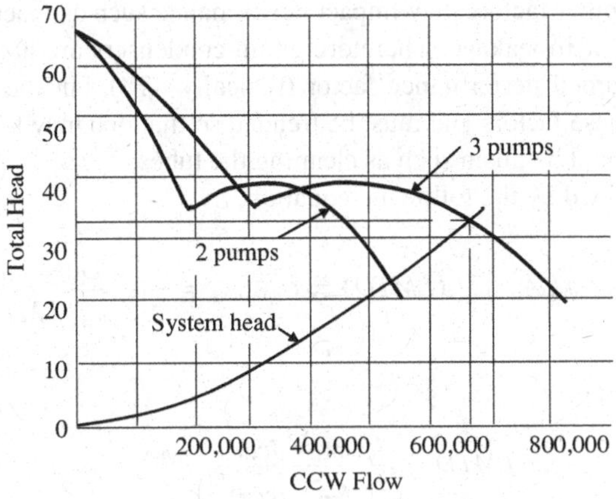

FIGURE 6.10 Determining the CCW flow rate from the pump total head curve.

$$TH = \left(H_{static} + \frac{V^2}{2g_c}\right)_{discharge} - \left(H_{static} + \frac{V^2}{2g_c}\right)_{suction} \tag{6.8}$$

Figure 6.10 illustrates how the CCW flow rate could then be estimated from the CCW pump curve where two CCW pumps are operating in parallel. Assuming that an accurate measure of the CCW inlet and outlet temperatures is known, perhaps the most accurate method for determining the CCW flow rate is to simply divide the duty by the temperature rise as follows:

$$\text{Duty} = m_{CCW}\, c_{p\text{-}ccw}\left(t_{CCWout} - t_{CCWin}\right) \tag{6.9}$$

$$m_{CCW} = \frac{\text{Duty}}{c_{p\text{-}ccw}\left(t_{CCWout} - t_{CCWin}\right)} \tag{6.10}$$

6.5 CALCULATING THE MAIN CONDENSER PERFORMANCE

In monitoring the performance of the main condenser, the most important parameter to monitor is the performance factor (also known as the cleanliness factor or fouling factor). The performance factor is defined as the ratio of the actual overall heat transfer coefficient, *U*, in operation divided by the theoretical overall heat transfer coefficient if the condenser was perfectly clean and operating as designed, as shown below.

$$PF = \frac{U_{operating}}{U_{clean}} \tag{6.11}$$

Immediately after being placed into service, the performance of the main condenser begins to deteriorate, primarily due to microfouling of the inside diameter of the tubes. However, other factors may impact performance such as macrofouling blocking the tubes or air in-leakage. Therefore, main condensers are normally specified based on an assumed performance factor (typically 85%), but the performance is variable due to these factors and must be trended so that one may know when a corrective action should be taken such as cleaning the tubes.

$U_{operating}$ is defined by the following equation:

$$\text{Duty} = U_{operating}\, A_{condenser}(LMTD) \Rightarrow U_{operating} = \frac{\text{Duty}}{A_{condenser}(LMTD)} \tag{6.12}$$

where

$$LMTD = \frac{t_{CCWout} - t_{CCWin}}{\ln\left(\dfrac{t_{sat} - t_{CCWin}}{t_{sat} - t_{CCWout}}\right)} \tag{6.13}$$

where the saturation temperature is determined from the *ASME Steam Tables*[2] for the condenser pressure.

6.6 MAIN CONDENSER EFFECTIVENESS

The effectiveness, P, of a condenser is defined as the temperature rise through the condenser divided by the maximum amount that the temperature could rise (i.e., the CCW outlet temperature would be equal to the saturation temperature).

$$P = \frac{t_{CCWout} - t_{CCWin}}{t_{sat} - t_{CCWin}} \tag{6.14}$$

Therefore, if the value of P is known, one may know the saturation temperature and therefore the condenser pressure by consulting the *ASME Steam Tables*. Solving for t_{sat},

$$t_{sat} = \frac{t_{CCWout} - t_{CCWin}(1-P)}{P} \tag{6.15}$$

Fortunately, the formula for P for a condenser is quite simple

$$P = 1 - e^{-NTU} \tag{6.16}$$

where

$$NTU = \frac{UA}{m_{ccw} c_{p\text{-}ccw}} \tag{6.17}$$

Therefore, if the value of U is known, one may calculate the condenser pressure for a given CCW inlet temperature and flow rate by consulting the *ASME Steam Tables*.

6.7 MAIN CONDENSER OVERALL HEAT TRANSFER COEFFICIENT

The Heat Exchange Institute (HEI), an association of HX manufacturers, has provided a method for calculating the value of U in their *Standards for Steam Surface Condensers* as follows[3]:

$$U = C\sqrt{V}\,F_t\,F_m\,F_f \tag{6.18}$$

where:
C = constant based on the tube diameter
V = tube velocity
F_t = temperature correction
F_m = tube material correction
F_f = fouling factor (performance factor).

It follows that for a clean main condenser operating as designed

$$U_{clean} = C\sqrt{V}\,F_t F_m \tag{6.19}$$

The HEI provides the values for C, F_t, and F_m in Reference 3.

The HEI values are empirical in nature based on operating experience. As such, they have changed over the years as more experience has been gained. Therefore, they may be subject to change.

A more rigorous, analytical method for determining the value of U_{clean} is the additive resistance method (ARM), which is endorsed by the ASME Performance Test Code committee, PTC 12.1, Steam Surface Condensers.[4] U_{clean} is defined as follows:

$$U_{clean} = \frac{1}{r_{shell} + r_{wall} + \left(\dfrac{d_o}{d_i}\right) r_{tube}} \tag{6.20}$$

where the reference area is the shell side of the tubes
r_{shell} = shell-side resistance to convection heat transfer
r_{wall} = tube wall resistance to heat transfer and
r_{tube} = tube-side resistance to convection heat transfer

In each case, the resistance to heat transfer is the inverse of the coefficient of heat transfer, h. The tube-side coefficient of heat transfer is

$$h_{tube} = \frac{k_{ccw}}{d_i}\,\mathrm{Nu}_t \tag{6.21}$$

where:
k_{ccw} = thermal conductivity of the CCW
Nu = Nusselt number.

There are several equations for determining the Nusselt number for water flowing in a round tube. The oldest and most commonly used is the Colburn[5] analogy as follows:

$$\mathrm{Nu}_t = 0.023\,\mathrm{Re}_t^{0.8}\,\mathrm{Pr}_t\left(\mu_s/\mu_t\right)^{0.14} \tag{6.22}$$

where Re_t and Pr_t are the tube-side Reynolds and Prandtl numbers, respectively, and μ_s and μ_t are the shell-side and tube-side viscosities, respectively, and where

$$\mathrm{Re}_t = \left(\frac{m_t}{a_t}\right)\left(\frac{d_i}{\mu_t}\right) \tag{6.23}$$

and

$$\mathrm{Pr}_t = \frac{\mu_t c_{p\text{-}ccw}}{k_{ccw}} \tag{6.24}$$

A more accurate equation for determining the Nusselt number is the Petukhov equation[5] as follows:

$$\mathrm{Nu} = \frac{\left(f/2\right)\mathrm{Re}_t\,\mathrm{Pr}_t}{1.07 + 12.7\sqrt{f/2}\left(\mathrm{Pr}_t^{2/3} - 1\right)} \tag{6.25}$$

where

$$f = \left(1.58\ln\mathrm{Re} - 3.28\right)^{-2} \tag{6.26}$$

The resistance to heat transfer through the tube wall is as follows:

$$r_{wall} = \frac{d_o}{2\,k_t}\ln\left(\frac{d_o}{d_i}\right) \tag{6.27}$$

where k_t is the thermal conductivity of the tube material.

The resistance to heat transfer on the shell side of a condenser is a function of the design of the tube bundle and is not easily determined directly. Therefore, this value is back-calculated from the design overall heat transfer coefficient, $U_{clean\text{-}design}$, taken from the manufacturer's datasheet as follows[5]:

$$r_{shell\text{-}design} = \frac{1}{U_{clean\text{-}design}} - r_{wall\text{-}design} - \left(\frac{d_o}{d_i}\right) r_{tube\text{-}design} \tag{6.28}$$

However, when the condenser is operating at conditions other than design, a correction must be made to r_{shell} as follows. One estimate of the shell-side coefficient of heat transfer for a condenser is as follows[5]:

$$h_{shell} = 0.729\left[\frac{k_f^3\,\rho_f^2\,g\,h_{fg}}{\mu_f\left(t_{sat} - t_{CCWin}\right)d_o}\right]^{1/4} \tag{6.29}$$

therefore,

$$r_{shell} = \frac{1}{h_{shell\text{-}design}} \frac{\left[\dfrac{k_f^3\,\rho_f^2\,h_{fg}}{\mu_f\left(t_{sat}-t_{CCWin}\right)}\right]_{design}^{1/4}}{\left[\dfrac{k_f^3\,\rho_f^2\,h_{fg}}{\mu_f\left(t_{sat}-t_{CCWin}\right)}\right]^{1/4}} \tag{6.30}$$

where:

ρ_f = density of the steam
h_{fg} = latent heat of steam
μ_f = viscosity of the condensate film
k_f = shell-side thermal conductivity of the condensate film
and the condensate film temperature is $\left(t_{sat} + t_{CCWin}\right)/2$.

The difference between the resistance to heat transfer in a clean main condenser and one that is in service is generally referred to as fouling resistance, although, as indicated herein, poor performance could also be due to air binding, etc. The fouling resistance is as follows:

$$r_{fouling} = \frac{1}{U_{clean}} - \frac{1}{U_{operating}} \tag{6.31}$$

where $U_{\text{operating}}$ is

$$U_{operating} = \frac{\text{Duty}}{A_{condenser}(LMTD)} \tag{6.32}$$

and U_{clean} may be determined by either the HEI method or the ARM.

Figure 6.11 shows the results of a comparison between the PF of the main condenser at a nuclear station with an on-line tube cleaning system when calculated over a period of one year by the HEI method and the ARM.

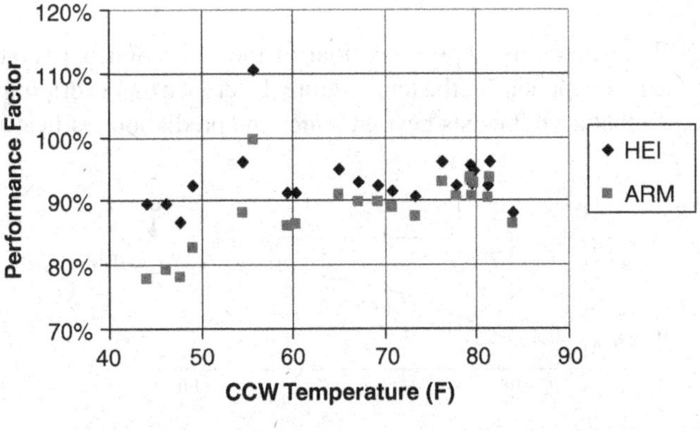

FIGURE 6.11 Main condenser performance factor as calculated by HEI and ARM methods.

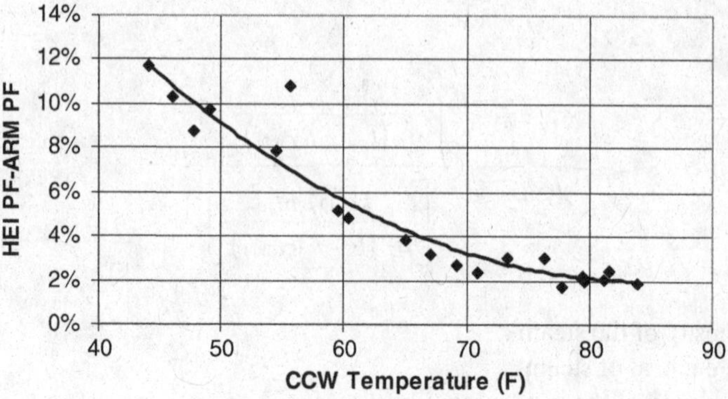

FIGURE 6.12 Difference in the main condenser performance factor as calculated by HEI and ARM methods.

Figure 6.12 presents an illustration showing the difference between the PF for the main condenser as determined by the HEI method and the ARM method. As one may see, there is fairly good agreement at higher CCW inlet temperatures but not at lower temperatures, with the HEI method being much more optimistic. As a result, one may be lulled into thinking that the main condenser is performing well during the winter months and not so well in the summer when in fact the condenser performance was poor during the winter as well.

6.8 MULTIPRESSURE CONDENSERS

As previously stated, where cooling towers are employed, reducing the CCW flow rate is desirable to reduce the auxiliary power consumed by CCW pumps due to the higher head required to pump the CCW up over the cooling tower fill. This is achieved by passing the CCW through the main condenser shells in series. The result is a multipressure condenser, as discussed in Section 6.2. To compensate for this reduced flow rate and higher temperature, the surface area of the main condenser is increased. Consider Figure 6.13, showing the temperature rise through a single pressure main condenser.

As the CCW temperature approaches that of the saturation temperature on the shell side of the main condenser, the temperature difference that is driving heat transfer is reduced so that a limit exists beyond which the predictions of heat transfer are

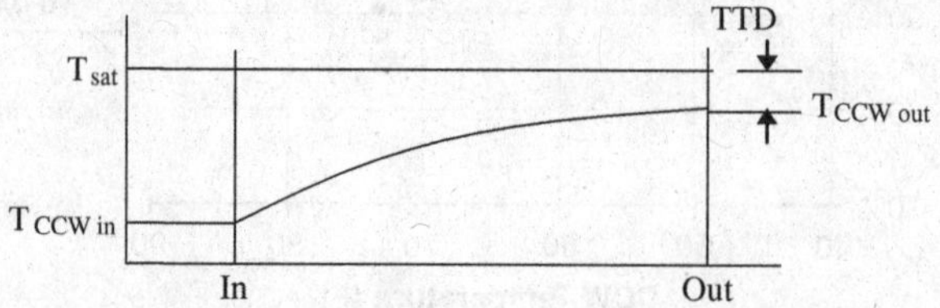

FIGURE 6.13 CCW temperature rise through a single-pressure main condenser.

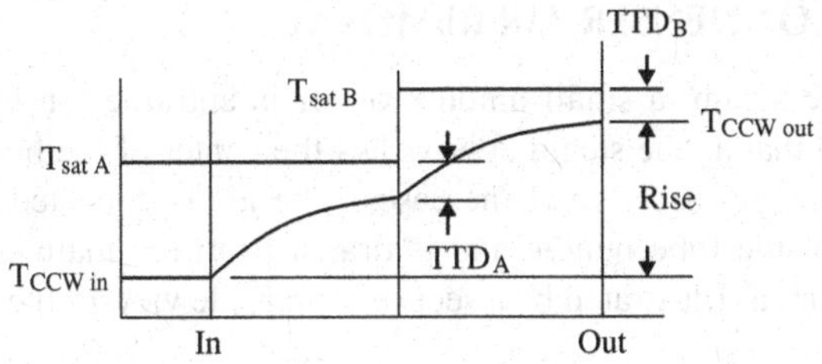

FIGURE 6.14 CCW temperature rise through a multipressure main condenser.

no longer applicable and longer tubes are not effective. The HEI recommends that the TTD be no less than 5°F. The solution to this problem is multipressure condensers. Figure 6.14 shows the CCW temperature rise through a multipressure condenser.

As the CCW temperature approaches the saturation temperature of the first main condenser shell, the CCW flow passes to a second or even a third shell where the saturation temperature is higher so that heat transfer continues unimpeded.

6.9 MAIN CONDENSER TUBE BUNDLE DESIGN

Consider Figure 6.15, which illustrates the cross section of a main condenser tube bundle.

In order for all of the main condenser surface areas to be effective, steam from the LP turbine must penetrate the tube bundle so that steam is condensed on all or nearly all of the tubes. Main condenser manufacturers have devised various designs for tube sheets to promote steam flow into the tube bundle. When designing a main condenser for a cooling tower application, the designer is tempted to increase the height of the tube bundle to increase the surface area. The result may be a poor performing condenser.

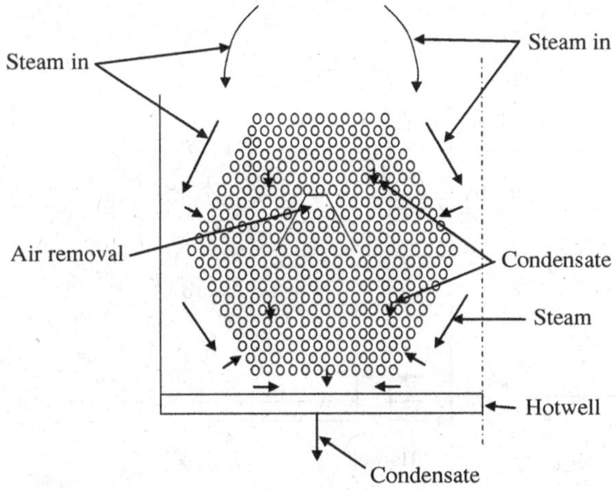

FIGURE 6.15 Cross section of a main condenser tube bundle.

6.10 MAIN CONDENSER AIR REMOVAL

In addition to the steam, a small amount of air is entrained in the steam that is not condensed so that as the steam approaches the center of the bundle the volume of air as a percentage of the total increases. The air is collected under a shroud near the center of the tube bundle and is drawn from the main condenser by air removal equipment, as illustrated by a sectional and side view of the main condenser in Figure 6.16.

The Standards for Steam Surface Condensers[3] shows the HEI-recommended air removal equipment capacity for each shell of the main condenser.

The following assumptions are made in sizing the air removal equipment:

- The air that is removed is saturated with water vapor.
- The pressure at the air/vapor outlet is the main condenser pressure.
- The temperature of the mixture is 7.5°F less than saturation.
- The main condenser pressure is 1.0 in HgA or design pressure, whichever is less.

Sizing the air removal equipment is based on the Gibbs–Dalton Law, and the definition of absolute humidity, ω, is as follows

$$p_t = p_a + p_v \tag{6.33}$$

$$\omega = \frac{m_v}{m_a} = \frac{M_v}{M_a} \times \frac{p_v}{p_a} \tag{6.34}$$

$$\omega = \left(\frac{18}{29}\right)\frac{p_v}{p_a} \tag{6.35}$$

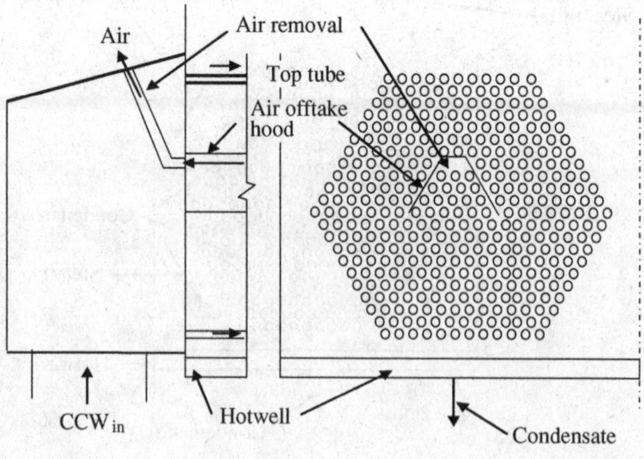

FIGURE 6.16 Cross section and side view of a main condenser tube bundle.

$$\omega = 0.62 \frac{p_v}{p_t - p_v} \tag{6.36}$$

where:

p_t = total pressure
p_a = partial pressure of the air
p_v = partial pressure of the water vapor
m_a = mass flow rate of the air
m_v = mass flow rate of the water vapor
M_a = molecular weight of air
M_v = molecular weight of water vapor.

Therefore,

$$m_t = m_a + \dot{m}_v$$

$$\omega = \frac{m_v}{m_a} = 0.62 \frac{p_v}{p_t - p_v} \tag{6.37}$$

$$m_v = m_a \left(0.62 \frac{p_v}{p_t - p_v} \right) \tag{6.38}$$

$$m_t = m_a + m_a \left(0.62 \frac{p_v}{p_t - p_v} \right) \tag{6.39}$$

$$m_t = m_a \left(1 + \left(0.62 \frac{p_v}{p_t - p_v} \right) \right) \tag{6.40}$$

6.11 CCW SYSTEM VACUUM PRIMING

Most nuclear stations are sited near rivers, lakes, or the ocean to provide the required CCW. An initial siphon of the CCW through the main condenser is established by throttling the CCW pumps at the main condenser discharge valves so that the CCW pumps need only to overcome the friction losses through the system. However, because air comes out of solution from the CCW when the water is heated, a vacuum priming system is provided to remove the air to maintain the siphon. As illustrated in Figure 6.17, the vacuum priming connection is on the outlet waterbox. Any air that comes out of solution in the inlet waterbox travels to the outlet waterbox through the condenser top tube.

Nuclear stations using cooling towers operating as a closed system do not normally require a vacuum priming system because the main condenser waterboxes are below the hydraulic gradient and are pressurized.

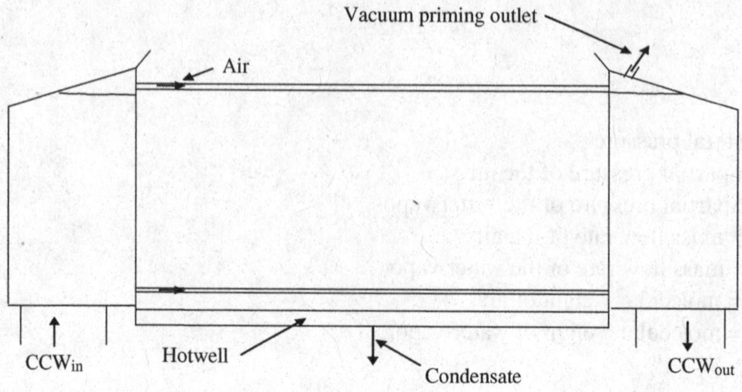

FIGURE 6.17 Main condenser vacuum priming.

6.12 MAIN CONDENSER PERFORMANCE PROBLEMS

In addition to tube microfouling and macrofouling, which obviously impede heat transfer to the CCW in the main condenser, other issues can also play a part in reduced performance. These issues include air binding, condensate subcooling, inadequate vacuum priming, and tube bundle design.

6.12.1 AIR BINDING

If more air enters the turbine cycle than can be removed, the condenser is said to be air-bound. Figure 6.18 illustrates how air binding can effectively remove a portion of the surface area of the main condenser from service, increasing the main condenser pressure.

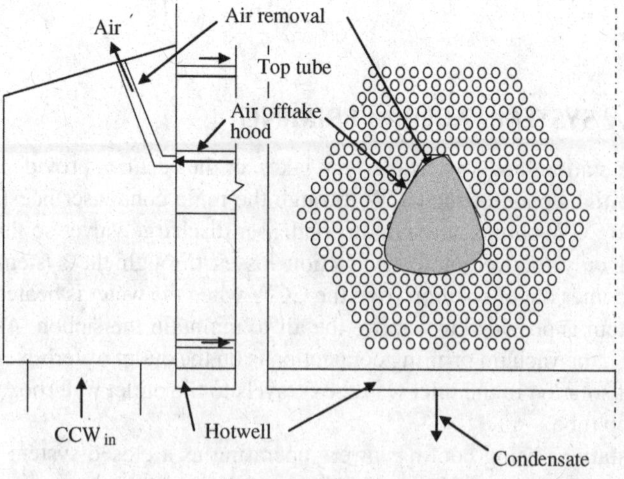

FIGURE 6.18 Cross section and side view of a main condenser tube bundle with air binding.

Sources of main condenser air in-leakage include the following:

- Turbine shaft seals
- LP turbine/condenser expansion joints
- Valve stem seals
- Condenser isolation valve leaks
- Feedwater heater operating vents
- LP turbine rupture disk leaks.

Methods for detecting air binding include the following:

- Bringing an additional vacuum pump into service. If the main condenser pressure goes down, air binding is occurring.
- As unit load drops, the main condenser pressure should drop in a predictable manner. If not, air binding is occurring.
- Bleeding air into the main condenser. If the condenser pressure rises, air binding is occurring.

6.12.2 Condensate Subcooling

Another potential source of poor main condenser performance is condensate subcooling. As Figure 6.19 illustrates, the condensate that drips from the main condenser tube may have a tendency to be subcooled.

The temperature of the condensate coming from a well-designed main condenser should be approximately equal to the saturation temperature (within 1.0 °F). Figure 6.20 illustrates that this is accomplished by directing steam down below the tube bundle.

Colder condensate requires more extraction steam to reheat the condensate in the first stage of feedwater heating, thus reducing the steam flow through the last stages of the LP turbine. Therefore, reheating the condensate before it reaches the hotwell is important.

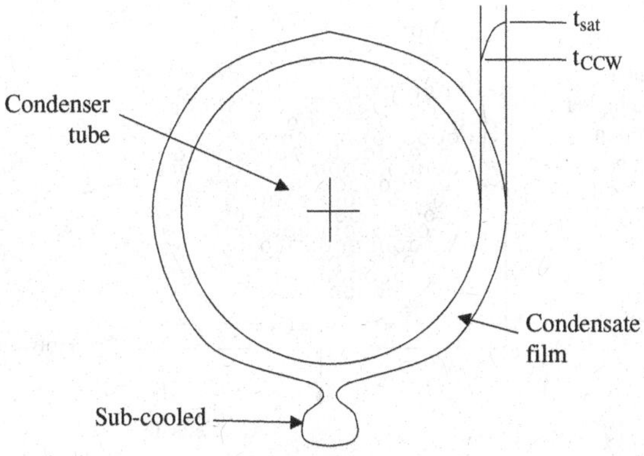

FIGURE 6.19 Condensate dripping from a main condenser tube.

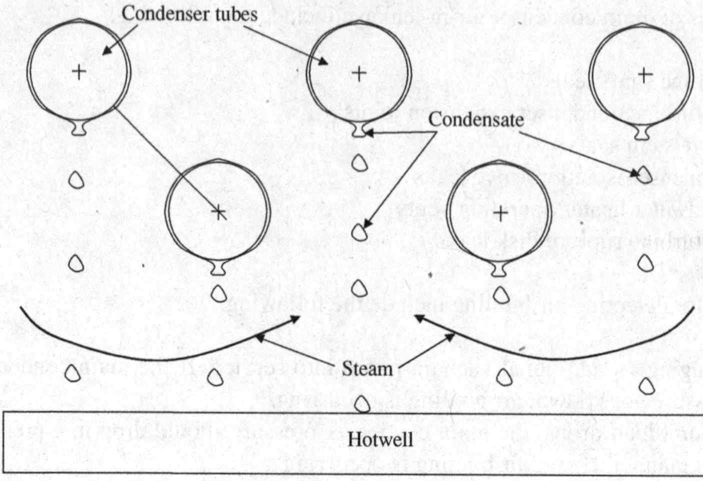

FIGURE 6.20 Steam reheating condensate dripping from a main condenser tube.

6.12.3 Inadequate Vacuum Priming

Inadequate CCW vacuum priming is illustrated in Figure 6.21. In addition to reducing the main condenser effective surface area, inadequate CCW vacuum priming reduces the CCW flow with an associated increase in the CCW outlet temperature. In order to monitor the CCW vacuum, main condensers are fitted with sight glasses on the outlet waterbox to monitor the water level.

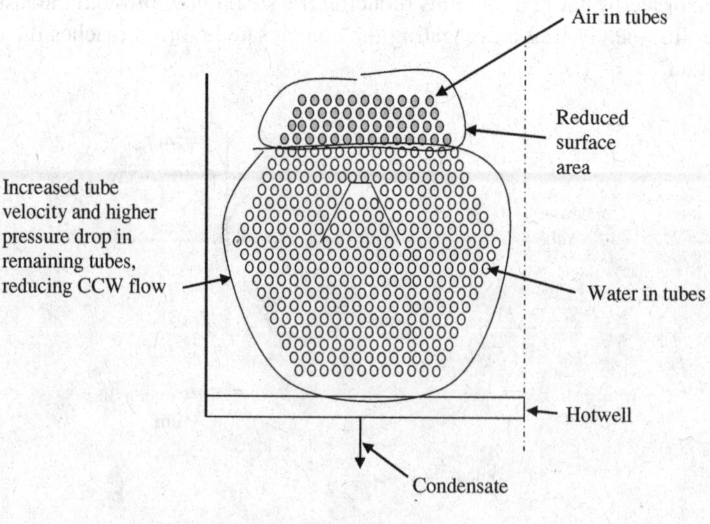

FIGURE 6.21 Inadequate CCW vacuum priming.

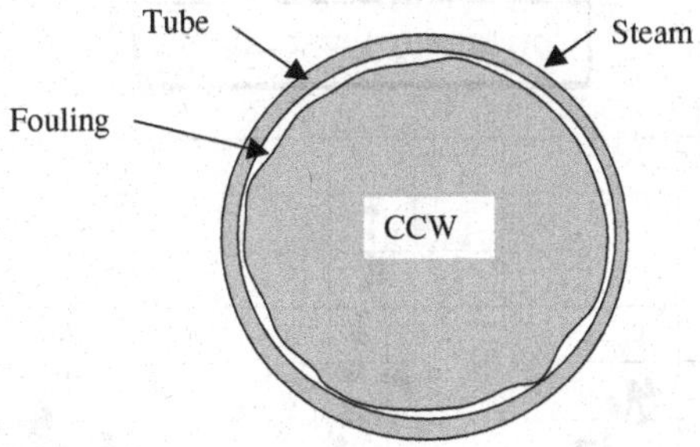

FIGURE 6.22 Main condenser tube microfouling.

6.12.4 Tube Microfouling

The inevitable condenser tube microfouling illustrated in Figure 6.22 is the most common source of poor main condenser performance.

Tube-side microfouling increases the resistance to heat transfer, resulting in a higher main condenser pressure and may be damaging to the tube material. However, it does not normally significantly reduce the CCW flow rate, as it normally presents as a slime bacteria that is slick. On-line chemical treatment or condenser tube cleaning systems such as the Taprogge ball cleaning system can significantly improve the performance factor when maintained. However, these systems are often assigned a low maintenance priority, especially during the winter months, because they may do little to increase the electrical output at that time. The factors that affect tube-side microfouling include the following:

- Tube velocity—Low velocity promotes increased slime formation.
- Tube material—Copper-based tube materials foul less than ferrous materials.
- CCW temperature—Slime forms more easily at higher temperatures.
- CCW water quality—The cleaner the water, the better.
- Presence of silt—Silt can collect in the bottom of tubes at low velocities.

Figure 6.23 presents a case study of microfouling in a nuclear station. After the main condenser was retubed with stainless steel tubes, performance was essentially perfect. However, as the station proceeded to operate through the winter, performance declined until a chlorination treatment was initiated. During an outage, the tubes were cleaned and performance was restored somewhat but not to the original condition. After the outage, chlorination was discontinued, resulting in rapid deterioration in performance as before. When chlorination was initiated again, the decline was staunched, but improvement in performance was only gradual. It can be concluded that when the nuclear station decides to let main condenser performance slide during the winter months, the subsequent recovery of performance can be problematic.

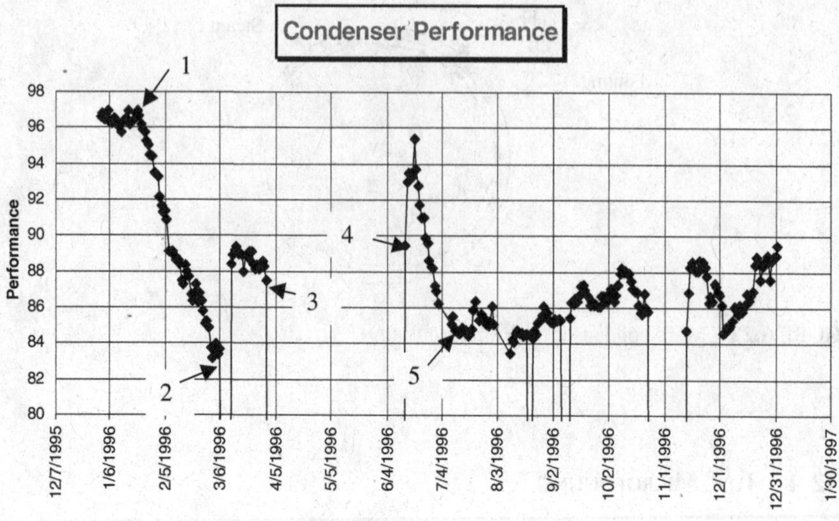

1 Retubed condenser with stainless steel
2 Initiated chlorination
3 Tubes cleaned during outage
4 Unit restarted—no chlorination
5 Initiated chlorination

FIGURE 6.23 Case study of condenser performance in a nuclear station.

Where cooling towers are employed in a closed cycle, a scale can form on the tubes, especially near the outlet of the condenser. This is because these systems are normally operated at higher concentrations of dissolved solids and because these systems operate at higher temperatures where calcium is more prone to plate out on the tubes. Calcium deposits on main condenser tubes are very difficult to remove, often requiring cutting tools or chemical treatment.

6.12.5 Tube Macrofouling

Main condenser tube macrofouling illustrated in Figure 6.24 is a frequent source of poor main condenser performance, especially at nuclear stations that draw their CCW from sources containing active aquatic life.

Tube-side macrofouling in the main condenser partly or completely blocks individual tubes. Macrofouling increases the pressure drop through the other tubes, reducing CCW flow and increasing the main condenser outlet temperature. Blocking the tubes reduces the main condenser effective surface area while increasing the tube velocity in the unblocked tubes. The net effect is to increase the main condenser pressure.

Sources of macrofouling include debris, Asiatic clams, mollusks, zebra mussels, fish, seagrass, plastic bags, mud, and rocks. Some nuclear stations have found that installing a Taprogge debris filter at the end of the CCW inlet piping helps to control macrofouling.

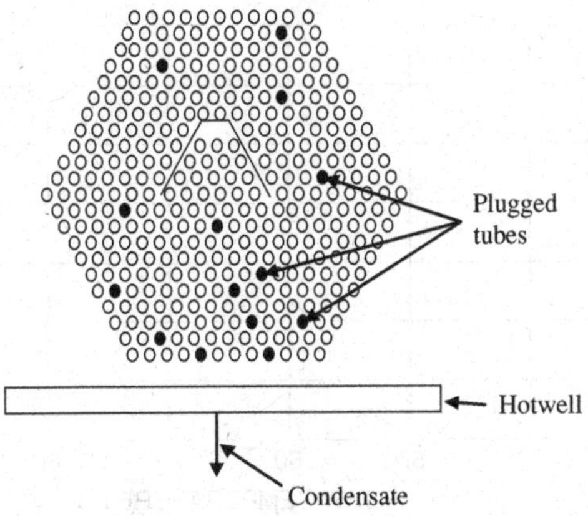

FIGURE 6.24 Main condenser tube macrofouling.

6.13 IMPACT OF MAIN CONDENSER PERFORMANCE ON NUCLEAR STATION ELECTRICAL OUTPUT

Figures 6.25 and 6.26 illustrate the impact of main condenser performance on the LP turbine backpressure and electrical output, respectively, for a typical nuclear station. As one may see, although main condenser performance impacts the LP turbine backpressure during the cold months, it does not translate into lost electrical output until the warmer months. Thus, the tendency is to not maintain performance in the winter months.

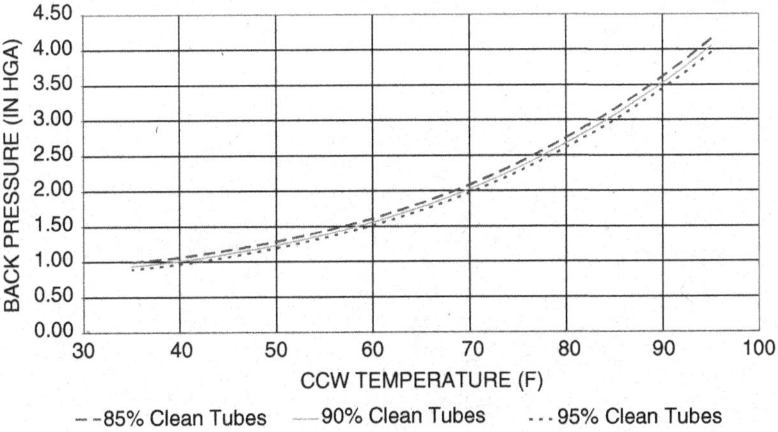

FIGURE 6.25 Impact of main condenser performance on LP turbine backpressure.

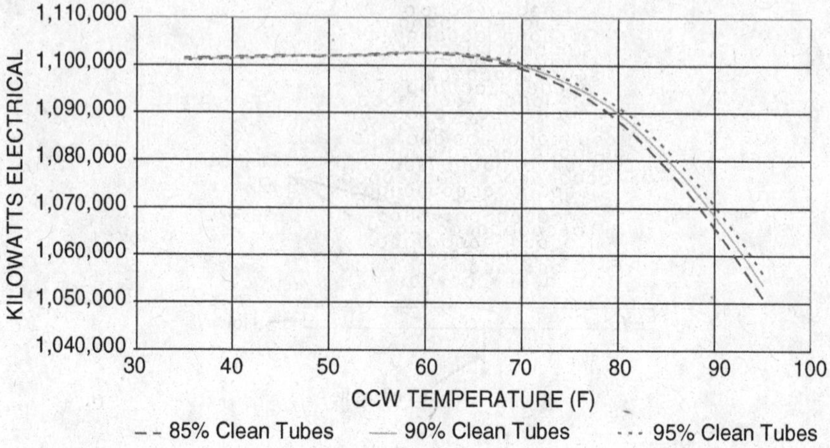

FIGURE 6.26 Impact of main condenser performance on net turbine electrical output.

REFERENCES

1. March, P. A. and C. W. Almquist, New Techniques for Monitoring Condenser Flow Rate and Fouling, *Power*, March 1989.
2. Meyer, C. A., et al., *ASME Steam Tables*, 6th ed., American Society of Mechanical Engineers, New York, 1993.
3. *Standards for Steam Surface Condensers*, 11th ed., Heat Exchange Institute, Cleveland, OH, 2012.
4. Burns, J. M., et al., *Improved Test Methods, Modern Instrumentation and Rational Heat Transfer Analysis Proposed in Revised ASME Surface Condenser Test Code PTC 12.5, 91-JPGC-PTC-8*, American Society of Mechanical Engineers, New York, 1991.
5. Thomas, L. C., *Heat Transfer Professional Version*, 2nd ed., Capstone Publishing Corporation, Tulsa, OK, 1999.

7 Feedwater Heaters

7.1 CONDENSATE HEATING DEVICES

Referring to the simplified turbine cycle illustrated in Figure 7.1, the piping system delivering the water from the main condenser to the main feedwater pump (MFP) is referred to as the condensate system and the piping system that delivers water from the discharge of the MFP system to the SG (or reactor in the case of a BWR) is referred to as the feedwater system.

The water leaving the main condenser is essentially a saturated liquid at the main condenser pressure. The condensate and feedwater systems heat the water prior to entering the nuclear boiler. As a result, less thermal energy coming from the reactor core is required to heat the feedwater to saturation conditions before being converted to steam, and the thermodynamic efficiency of the turbine cycle is thereby increased. The vast majority of this heating occurs in the low-pressure and high-pressure feedwater heaters (FWHs). However, any amount of condensate heating that can be achieved prior to entering the lowest pressure FWH is beneficial, as it reduces the demand for extraction steam, and this steam passes through the last stage(s) of the LP turbine where it produces additional power. Condensate heating devices include gland steam condensers (also called steam packing exhausters), steam jet air ejectors, off-gas condensers, SG blowdown HX, and main feed pump turbine condensers.

Gland Steam Condenser: A portion of the steam in an HP turbine leaks down the turbine shaft past the glands to either the steam seal regulator where it is injected into the LP turbine glands or the gland steam condenser. Likewise, a portion of the steam that is injected into the LP turbine glands also leaks off to the gland steam condenser where both streams are condensed by condensate from the main condenser, thereby slightly increasing the temperature of the condensate.

Steam Jet Air Ejectors: In some nuclear stations, the vacuum in the main condenser is maintained by the use of steam jet air ejectors that draw steam from the turbine cycle and pass it through a restriction that creates a low-pressure region based on the Bernoulli principle. Air and steam from the main condenser are drawn into the region of low pressure, and the steam is condensed by condensate from the main condenser.

Off-Gas Condensers: In a BWR nuclear station, the mixture of water vapor and air removed from the main condenser is radioactive. Off-gas condensers reduce the moisture content of the stream prior to entering the charcoal filter columns by lowering the dew-point of the mixture.

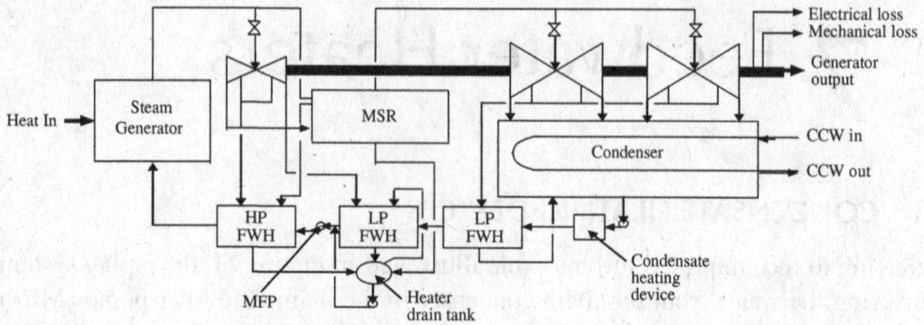

FIGURE 7.1 Simplified turbine cycle.

Steam Generator Blowdown Heat Exchangers: As discussed in Chapter 2, PWR nuclear stations discharge a relatively small quantity of blowdown from the SG in order to maintain proper water chemistry. SG blowdown HX extracts the available energy from this stream by cooling it with condensate from the main condenser.

Main Feed Pump Turbine Condensers: For nuclear stations employing turbine-driven MFP(s), the exhaust from each MFP is normally either discharged into the main condenser through a large duct or discharged into a separate MFP turbine condenser. In the latter case, the steam is normally condensed by diverting a portion of the CCW from the main condenser. However, in a few nuclear stations, this steam is condensed by condensate from the main condenser.

7.2 FEEDWATER AND CONDENSATE EXTRACTION POINTS

Referring back to the discussion of the Rankine cycle in Chapter 1, a series of FWHs extract a portion of the steam passing through the HP and LP turbines. "Extraction" is the proper term because the steam is extracted or pulled from the turbine by the condensation taking place in the FWH. However, although the flow rate of the extracted steam is determined by the FWH's capacity to condense the steam, the operating pressure of the FWH is set by the pressure in the turbine at the extraction point minus the pressure drop through the extraction piping (generally less than 5%).

The simplified turbine cycle illustrated in Figure 7.1 shows FWHs extracting steam from the HP turbine, the cold reheat piping between the HP and LP turbines, and the LP turbine. Although only three extraction points are illustrated on the Rankine cycle in Figure 7.2, in practice there may be as many as seven extraction points in a typical turbine cycle with most taken from the LP turbine.

Figure 7.3 shows an indication of how much the efficiency of the turbine cycle may be improved as a function of the number of FWHs for an array of possible temperature rises. The maximum temperature rise is the difference between the inlet temperature and the saturation temperature of the FWH. Heat rate is an inverse

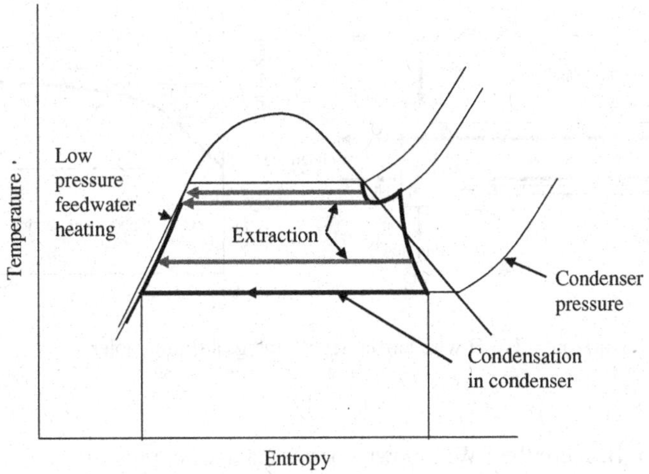

FIGURE 7.2 Rankine cycle for nuclear station with feedwater heating.

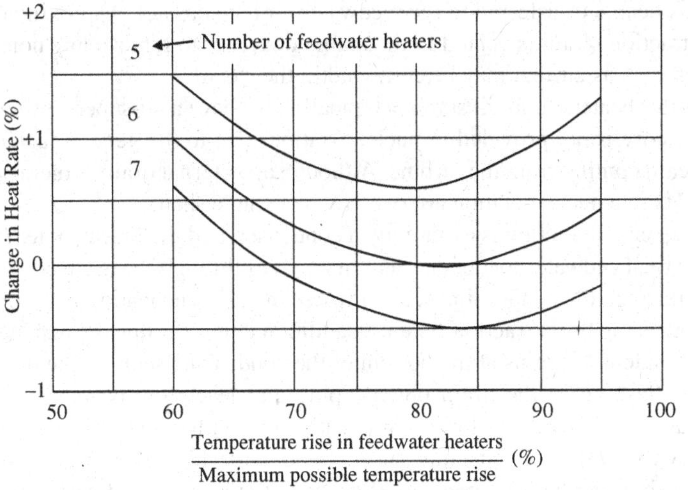

FIGURE 7.3 Change in turbine cycle heat rate for an array of the number of FWHs. (From Spencer, R.C. and Booth, J.A., Heat Rate Performance on Nuclear Steam Turbine-Generators, *Proceedings of the American Power Conference*, 1968.)

function of the turbine cycle efficiency, so a lower heat rate of, say, 0.5% results in an increase in thermodynamic efficiency by that amount. Therefore, the actual number of FWHs is an economic decision.

Figure 7.4 shows a simplified longitudinal section of a typical horizontal FWH in a nuclear station with a partial length integral drain cooler. Assuming one or more FWHs operating at a higher pressure, the drain from that FWH is cascaded into the

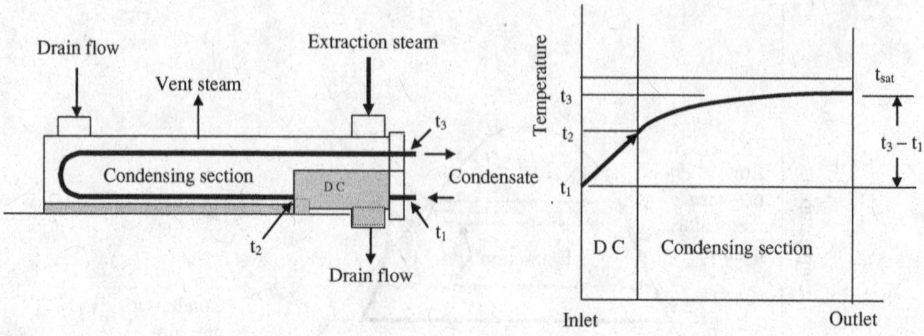

FIGURE 7.4 Horizontal FWH with partial length integral drain cooler.

FWH shown. Because the FWH shown operates at a lower pressure, a portion of the drain flow flashes into steam, supplementing the extraction steam in the condensing section. Then, the heat is transferred to the feedwater (or condensate), raising the temperature before the steam is condensed in an isothermal process before entering the drain cooling section where additional heat is removed as the temperature is reduced as heat is transferred to the feedwater (or condensate). Approximately 0.5% of the extraction steam is vented to the main condenser to remove any noncondensable gases such as air that may be entrained in the steam.

Feedwater heaters in fossil stations typically also have a de-superheating section, but this feature is not provided in nuclear stations due to the general lack of superheated steam coming from the turbine. Although an external drain cooler is possible, most FWHs in a nuclear station are two HX in a single shell.

The condensing section is essentially a condenser, and the drain cooler is a shell-and-tube HX. Feedwater (or condensate) first passes through the drain cooler section and then through the condensing section where most of the heating is accomplished while condensing the extraction steam. In addition to recovering any residual energy from the condensed drains at the bottom of the condensing section, the drain cooler eliminates flashing in the drain outlet piping and level control valve, permitting smooth cascading to the next lower FWH without water hammer. The drain cooler is enclosed within a shroud and a tube sheet or endplate where the tubes transition into the condensing section. The condensed water in the bottom of the condensing section passes into the drain cooler through a snorkel pipe. Flow within the drain cooler is baffled to promote flow perpendicular to the tubes. Control of the condensed water level in the bottom of the condensing section is very important because one must keep steam out of the drain cooler to prevent damage to the tubes and baffles. Otherwise, with the loss of the water seal, steam could flood into the drain cooler and eliminate all subcooling. Damage to the drain cooler may also occur by steam leaking between the outlet tube sheet and the tube. This is especially true for tubes that have been plugged, as no condensing is occurring on the tube that would otherwise obstruct steam flow down the tube. Therefore, the tube sheet should be sufficiently thick to minimize steam flow down the tube. An external drain cooler could eliminate many of these problems but would be more costly in terms of building space and piping.

Causes of poor FWH performance include the following:

- Shell-side fouling due to corrosion products or soluble materials
- Tube-side fouling
- Tube plugging
- Restriction or a leak in the extraction piping resulting in lower FWH pressure
- Air binding of the tube surface area due to inadequate venting
- Channel-side bypassing due to a damaged pass-partition plate
- Tube leaks
- Improper level control.

The FFW temperature is particularly important because it establishes the FFW flow for the same nuclear power level. The condition of lower pressure FWHs as well as the HP turbine extraction pressure should be examined in troubleshooting a low or high FFW temperature. The HP turbine extraction pressure often runs higher than the design HB predicts, resulting in a higher FFW temperature and thus a lower FFW flow rate.

7.3 FWH ANALYSES AT REFERENCE CONDITIONS

Referring to Figure 7.4, the FWH parameters that are commonly monitored are the heat transfer rate, Q, the TTD, and the drain cooler approach, DCA, and are defined as follows.

$$Q = m\, c_p \left(t_3 - t_1\right) \tag{7.1}$$

$$TTD = T_{sat} - t_3 \tag{7.2}$$

$$DCA = t_{DC} - t_1 \tag{7.3}$$

It should be noted that poor FWH performance may be the result of a problem in another component. For example, a high TTD may be the result of poor performance in a lower pressure FWH as the FWH compensates for a lower inlet temperature.

The FWH extraction steam flow is set by the steam demand of the FWH, which is determined by first law analysis around the FWH as shown below, where the vent steam flow equals 0.5% of the extraction steam flow.

$$\begin{aligned} & m_{drain\ in}\, h_{drain\ in} + m_{extraction}\, h_{extraction} + m_{condensate}\, h_{condensate\ in} \\ & = m_{drain\ out}\, h_{drain\ out} + m_{vent}\, h_{vent} + m_{condensate}\, h_{condensate\ out} \end{aligned} \tag{7.4}$$

$$m_{drain\ out} = m_{drain\ in} + m_{extraction} - m_{vent}\ m_{vent} = 0.005\ m_{extraction} \tag{7.5}$$

$$m_{drain\ out} = m_{drain\ in} + m_{extraction}\left(1 - 0.005\right) \tag{7.6}$$

$$m_{extraction}\left(h_{extraction} - 0.995\, h_{drain\ out} - 0.005\, h_{vent}\right)$$
$$= m_{condensate}\left(h_{condensate\ out} - h_{condensate\ in}\right) - m_{drain\ in}\left(h_{drain\ in} - h_{drain\ out}\right) \tag{7.7}$$

$$m_{extraction} = \frac{m_{condensate}\left(h_{condensate\ out} - h_{condensate\ in}\right) - m_{drain\ in}\left(h_{drain\ in} - h_{drain\ out}\right)}{h_{extraction} - 0.995\, h_{drain\ out} - 0.005\, h_{vent}} \tag{7.8}$$

The nuclear industry's practice of accurately measuring feedwater flow affords an opportunity to monitor the thermal performance of FWHs. In analyzing an FWH, the following assumptions are made:

- Condensate in and out and drains in and out are saturated liquids at the indicated temperatures.
- For FWHs in the feedwater system (downstream of the main feed pump), there is a slight increase in enthalpy over saturation.
- Extraction enthalpy is taken from the turbine expansion line.
- Vent steam enthalpy is saturated vapor at the heater shell pressure.

In some turbine cycles, some of the FWHs do not have a drain cooler but rather discharge the condensate to a heater drain tank that collects the drains from the higher pressure heater. Sometimes, the belly drains from the MS and the drains from the higher FWH are pumped forward into the condensate system, rather than allowing them to be cascaded to the next lower pressure FWH. Figure 7.5 shows a typical arrangement.

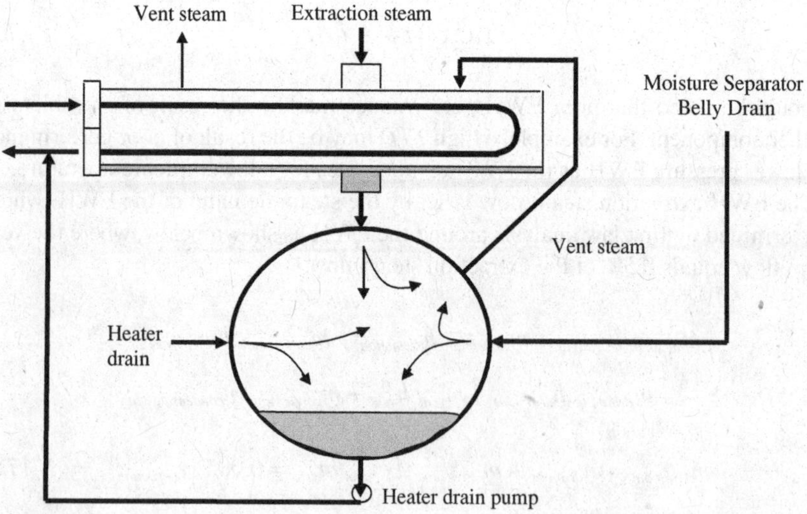

FIGURE 7.5 Feedwater heater with heater drain tank with drains pumped forward.

In this case, the FWH extraction steam flow is determined by first law analysis around the FWH and the drain tank as shown below.

$$m_{dr\ in}h_{dr\ in} + m_{extraction}h_{extraction} + m_{cond}h_{cond\ in} + m_{MS\ drain}h_{MS\ drain}$$
$$= m_{dr\ out}h_{dr\ out} + m_{vent}h_{vent} + m_{cond}h_{cond\ out} \tag{7.9}$$

$$m_{dr\ out} = m_{dr\ in} + m_{extraction}\left(1 - 0.005\right) + m_{MS\ drain} \tag{7.10}$$

$$m_{extraction} = \frac{m_{cond}\ (h_{cond\ out} - h_{cond\ in}) - m_{MS\ drain}(h_{MS\ drain} - h_{dr\ out}) - m_{dr\ in}\ (h_{dr\ in} - h_{dr\ out})}{h_{extraction} - 0.995\ h_{drain\ out} - 0.005\ h_{vent}} \tag{7.11}$$

The condensate in-flow is equal to the condensate out-flow minus heater drain flow. An initial value of condensate flow must be assumed. The final value is determined by iteration. The heater drain tank pressure is taken as the FWH pressure (i.e., the vent line pressure drop is normally negligible).

The only way to increase the extraction steam flow, thus increasing the thermodynamic efficiency of the nuclear turbine cycle, is to increase the effectiveness (surface area) of the FWHs or to increase the number of FWHs in the cycle.

The resistances to heat transfer in an FWH listed on the HX manufacturer's datasheets are referenced to the outside tube diameter, A_s. Therefore, the inverse of the sum of the resistances is equal to the calculated heat transfer coefficient without correction for the different surface areas. That is,

$$U = \frac{1}{r_{sc} + r_{sf} + r_w + r_{tf} + r_{tc}} = \frac{Q}{A_s\ EMTD} \tag{7.12}$$

where:

r_{sc} = shell-side convection resistance
r_{sf} = shell-side fouling resistance
r_w = tube wall resistance
r_{tf} = tube-side fouling resistance
r_{tc} = tube-side convection resistance
$EMTD$ = effective mean temperature difference.

The performance of an FWH is based on the effectiveness method as defined by the following

$$Q = C_t(t_o - t_i) = C_t P(T_{sat} - t_i) \tag{7.13}$$

$$P = \frac{t_o - t_i}{T_{sat} - t_i} \tag{7.14}$$

when referenced to the tube side.

The effectiveness as defined here is referenced to the cold stream (tube side). Equations for the effectiveness of the subcooling and condensing sections of the FWHs are as follows:

$$P_1 = \frac{1 - e^{[-NTU_1(1-R_1)]}}{1 - R_1\, e^{[-NTU_1(1-R_1)]}} \tag{7.15}$$

$$P_2 = 1 - e^{-NTU_2} \tag{7.16}$$

where the subcooling section is assumed to be a pure counter-flow HX and the condensing section is isothermal.

$$NTU_1 = \frac{U_1\, A_{s,1}}{C_t} \tag{7.17}$$

$$NTU_2 = \frac{U_2\, A_{s,2}}{C_t} \tag{7.18}$$

The variables are

$$R_1 = \frac{C_t}{C_{s,1}} \tag{7.19}$$

and

$$C_t = m_t\, c_{p-t} \tag{7.20}$$

Because the tube-side temperature exiting the subcooling zone cannot be measured, the physical properties of the fluid (specific heat, thermal conductivity, and viscosity) on the tube side are based on the average tube-side temperature. Because the temperature of the shell-side drains in the subcooling zone is very close to that of the tube-side fluid, the properties of the shell-side flow are assumed equal to those of the tube side. Therefore,

$$C_{s,1} = m_{d,o}\, c_{p-t} \tag{7.21}$$

where

$$m_{d,o} = m_e + m_{d,i} \tag{7.22}$$

If the overall coefficients of heat transfer for the subcooling and condensing zones are known, then the heat transferred in the drain cooling section may be calculated from

$$Q_1 = C_t\, P_1\, (T_{sat} - t_1) \tag{7.23}$$

and the tube-side temperature leaving the drain cooler is

$$t_2 = t_1 + \frac{Q_1}{C_t} \tag{7.24}$$

Similarly, the heat transferred in the condensing section, Q_2, is

$$Q_2 = C_t\, P_2\, (T_{sat} - t_2) \tag{7.25}$$

the outlet temperature is

$$t_3 = t_2 + \frac{Q_2}{C_t} \tag{7.26}$$

and the *TTD* is

$$TTD_{calc} = T_{sat} - t_3 \tag{7.27}$$

The shell-side outlet temperature may be calculated from

$$T_o = T_{sat} - \frac{Q_1}{C_{s,1}} \tag{7.28}$$

and the drain cooler outlet approach is

$$DOA_{calc} = T_o - t_1 \tag{7.29}$$

A check on the methods of FWH analysis is to confirm the following equations:

$$Q = Q_1 + Q_2 \tag{7.30}$$

$$C_t(t_3 - t_1) = C_t\, P_1(T_{sat} - t_1) + C_t\, P_2\, (T_{sat} - t_2) \tag{7.31}$$

where the values for P_1 and P_2 are calculated as shown above.

This procedure should yield results that agree fairly well with the vendor data at reference or design conditions.

7.4 FWH ANALYSES AT OPERATING CONDITIONS

The analysis described above illustrates that if the overall coefficient of heat transfer is known, the thermal performance of the FWH may be characterized reasonably well at reference conditions by the effectiveness method. However, in order to use this approach at other than reference conditions during normal operation, the coefficients of heat transfer for the tube-side and shell-side convection resistances must be calculated.

The overall coefficient of heat transfer for the subcooling zone and the condensing zone may be calculated by developing an equation for the individual resistances to heat transfer for the tube side and shell side because tube-side and shell-side fouling resistances and the tube wall resistance are assumed to be constant.

The tube-side convection coefficient for both the subcooling and condensing zones may be determined as follows:

First, calculate the tube-side flow area, A_t

$$d_i = d_0 - 2t_W \tag{7.32}$$

$$A_t = 2N\frac{\pi d_i^2}{4} \tag{7.33}$$

where:

d_i = tube inside diameter
d_o = tube outside diameter
t_W = tube wall thickness
N = number of u-tubes.

Then, calculate the tube-side mass flux, G_t

$$G_t = \frac{m_t}{a_t} \tag{7.34}$$

Calculate the tube-side Reynolds number, Re_t, and the Prandtl number, Pr_t

$$\text{Re}_t = \frac{G_t d_i}{\mu_t} \tag{7.35}$$

$$\text{Pr}_t = \frac{\mu_t C_{p-t}}{k_t} \tag{7.36}$$

The Petukhov correlation is used to calculate the tube-side Nusselt number. First, calculate the Fanning friction number

$$f = (1.58 \ln \text{Re}_t - 3.28)^{-2} \tag{7.37}$$

then calculate the Nusselt number as follows:

$$\text{Nu} = \frac{\left(\frac{f}{2}\right)\text{Re}_t\,\text{Pr}_t}{1.07 + 12.7\left(\frac{f}{2}\right)^{\frac{1}{2}}\left(\text{Pr}_t^{\frac{1}{3}} - 1\right)} \tag{7.38}$$

Finally, calculate the tube-side convection coefficient

$$h_t = \text{Nu}_t\frac{k_t}{d_i} \tag{7.39}$$

The tube-side convection resistance is the inverse of the convection coefficient referenced to the shell-side area

$$r_{tc} = \left(d_o / d_i \right) \frac{1}{h_t} \tag{7.40}$$

The procedure for explicitly calculating the shell-side convection coefficient for the subcooling zones of the FWHs is complex and requires extensive knowledge of the internal design of the HX. However, the value used to arrive at the overall coefficient of heat transfer specified by the HX manufacturer may be determined by the back-calculation method, and the corresponding resistance to heat transfer may be compared with that specified in the HX datasheet at reference conditions as follows:

$$h_{sc,1} = \frac{1}{1/U_1 - [\, r_{tc} + r_{tf} + r_w + r_{sf} \,]} \tag{7.41}$$

$$r_{sc,1} = \frac{1}{h_{sc,1}} \tag{7.42}$$

This procedure amounts to a check on the adequacy of the equation for predicting the tube-side convection coefficient as all of the other variables are taken from the HX datasheet.

For the condensing zone, the shell-side convection coefficient may be computed directly from the following equation from the Bleed Heater Manufacturers Association expressed in English units:

$$h_{sc,2} = \frac{1}{0.06834\, T_{sat}^{-0.8912}} \tag{7.43}$$

However, if the resulting shell-side convection coefficient exceeds 2,500 Btu/(hr ft^2 °F), this value is to be used. The corresponding resistance to heat transfer is as follows:

$$r_{sc,2} = \frac{1}{h_{sc,2}} \tag{7.44}$$

7.5 ANALYSIS OF FWH TEST RESULTS USING "APPARENT" FOULING RESISTANCE

The approach in evaluating FWH test results is to determine an "apparent" fouling resistance that will yield the measured total heat transfer. This "apparent" fouling resistance may be due to fouling on the tube side or the shell side of the tube or may be due to some condition other than fouling that may result in poor heat transfer. Possible problems with FWH performance in addition to tube-side or shell-side fouling include air binding, bypassing around the pass partition plate in the channel head, plugged tubes, and improper level control. The "apparent" fouling resistance is expressed as a fouling ratio, *FR*, which is a multiple of the sum of the design fouling resistances on the tube and shell sides such that

$$U = \frac{1}{r_{sc} + r_w + r_{tc} + FR(r_{sf} + r_{tf})_{design}} \tag{7.45}$$

For a given test condition, there is a fouling ratio that would yield a value for the overall heat transfer coefficient when inserted in the following equation that would result in

$$Q_{test} = Q_{calc.} \tag{7.46}$$

where

$$Q_{test} = C_t(t_3 - t_1) \tag{7.47}$$

$$Q_{calc} = Q_1 + Q_2 = C_t P_1(T_{sat} - t_1) + C_t P_2(T_{sat} - t_2) \tag{7.48}$$

Because P_1 and P_2 are functions of U_1 and U_2 as discussed above, one may select a value for *FR* that, when incorporated into the equation for U above with the appropriate expression for r_{sc} and r_{tc}, will satisfy the requirement that the calculated heat transfer rate must equal the measured value. Because for FWHs with an integral subcooling zone, the temperature leaving the subcooling zone cannot be measured, the value of *FR* must be determined by trial and error. The accuracy of the proposed approach may be evaluated by computing the required fouling ratio at baseline conditions where the fouling ratio should be 1.0.

7.6 FWH CASE STUDY

An acceptance test was conducted on the turbine cycle of a nuclear plant shortly after its commissioning, permitting the "apparent" fouling of the five stages of feedwater heating plus an external drain cooler to be calculated in the essentially new condition and comparing those values with the specified *FR* where *FR* = 1.0. As one may see from Figure 7.6, the calculated *FR* for four of the six HX was less than 1.0 in the essentially new condition. The two exceptions are the external drain cooler HX0 and HX 5, located downstream of a deaerator (an open FWH). Note that, whereas the calculated *FR* for HX 5 was only slightly higher than the design value, that of HX0 was almost four times the design *FR*.

Approximately 16 years later, a subsequent test of the turbine cycle was conducted. The results as shown in Figure 7.6 indicated that the calculated *FR* had increased dramatically, resulting in poorer thermal performance for the turbine cycle, with the greatest increase in fouling occurring in the FWHs closest to the main condenser.

Subsequent investigation revealed a manufacturing defect in HX0 in which the portions of the baffle plates designed to be attached to the shell of the HX to force

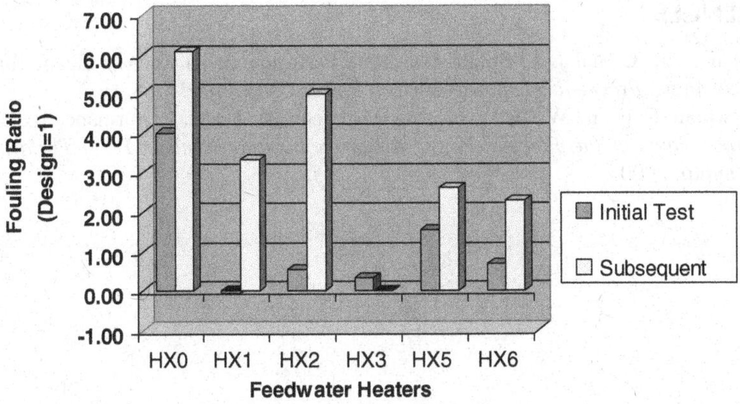

FIGURE 7.6 Feedwater heater fouling test results.

the condensate flow across the tube bundle were not installed and, as a result, the condensate was passing between (above) the tube bundle and the shell of the HX and not being reheated as intended.

A review of the operating history of the nuclear station revealed that an intrusion of LP turbine bearing seal oil into the main condenser had occurred, resulting in the seal oil being mixed with the condensate in the main condenser hotwell where it passed to the FWHs and coated the interior of the tubes in the FWHs.

This actual case study[2] illustrates how the use of the apparent fouling resistance method of FWH analysis may be utilized to identify deficiencies in FWH performance that would not otherwise be detected.

7.7 FWH DRAINS

The physical arrangement of the FWH drain systems varies greatly among nuclear stations. The arrangement illustrated in Figure 7.1 (where the drains from the FWH resulting from condensing the steam from the HP turbine exhaust are pumped forward) is but one of several possible arrangements. Some nuclear stations do not pump any of the drains forward but rather have them cascade from one FWH to another and finally drain into the main condenser hotwell. Another common arrangement is to drain the water from the lowest pressure FWH into a drain tank and pump it forward, as shown in Figure 7.1, for the higher pressure FWH. In that case, none of the drains from the FWHs are discharged to the main condenser. All of the FWHs are normally provided with bypass drains to the main condenser to be used during start-up and emergency situations.

The FWH drain valves that control the level in the FWH are normally located below the FWH to minimize flashing in the valves. For those drains that are pumped forward, the heater drain pumps are sized for the maximum flow and head required to ensure that the proper level is maintained in the heater drain tank.

REFERENCES

1. Spencer, R. C. and J. A. Booth, Heat Rate Performance on Nuclear Steam Turbine-Generators, *Proceedings of the American Power Conference*, 1968.
2. Bowman, C. F. and W. Cichowlas, Nuclear Feedwater Heater Performance Indicators, *Proceedings of the Electric Power Research Institute Nuclear Plant Performance Seminar*, 2000.

8 Cycle Isolation and the Mass Balance

8.1 CYCLE ISOLATION

Because in a nuclear turbine cycle electrical power is produced by expanding steam through turbines to the main condenser, any leakage of steam out of that steam path results in less power output. Leakages may include both steam and high-temperature water used to replace extraction steam such as FWH drains. These leakages may be around a turbine to the main condenser or into the environment through leaking pressure relief valves or rupture disks. Some potential excessive leakage paths are necessary, such as turbine drains. Some paths are orifices designed to limit flow during normal operation. Other paths are normally open during the start-up of the turbine and close when operating temperatures are reached. Examples of these are low-point drains such as the following:

- Main steam and bypass piping to the main condenser
- Before the HP turbine stop valves
- Between the stop valves and the first stage nozzles
- HP and LP turbine steam seals
- HP and LP turbine casings
- Main feed pump turbine steam supply piping
- Emergency belly drains from the moisture separator
- Emergency FWH drains
- Before the extraction non-return valves.

Extraction non-return valves are essential to the detection, isolation, and disposal of any accumulation of water to prevent water induction into the turbine.

In addition to reducing electrical output, poor cycle isolation can result in steam cutting of valves and orifices, leading to more steam leaks. Leakages into the atmosphere often result in an observable steam plume coming from the relief valves, etc. Excessive steam leakage to the main condenser may often be discovered by monitoring the temperature downstream of cycle isolation valves or by using an acoustic valve leakage analyzer.

8.2 LOSS OF ELECTRICAL OUTPUT FROM POOR CYCLE ISOLATION

Electrical power output loss from poor cycle isolation may be estimated as follows:

$$\Delta P = m_{leakage}\left(h_{leakage} - h_{UEEP}\right) \tag{8.1}$$

Of course, the difficult part is estimating the mass flow rate of the leaking steam or water. The following is one method of estimating leakage steam flow in English units.

Steam as an ideal gas has the following properties where P^* and T^* denote properties in the discharge plain of the pipe and P_o and T_o denote properties just upstream of the point of discharge:

$$k = 1.33 \text{ (for steam)} \tag{8.2}$$

$$P^*/P_o = 0.540 \tag{8.3}$$

$$T^*/T_o = 0.859 \tag{8.4}$$

From the ideal gas law, the velocity in the exit plain is

$$k = V^2 / g_c RT^* \tag{8.5}$$

$$V = \sqrt{k\, g_c\, R\, T^*} = \sqrt{(1.33)(32.17)(85.76)(557.1)} = 1{,}430 \text{ (ft/s)} \tag{8.6}$$

where $R = 85.76$ (ft-lbf/lbm-°R) for steam and T^* is in °R.

Note that T^* is not the temperature of the reservoir but is defined by

$$T^* = 0.859\, T_o \tag{8.7}$$

The mass flow rate may then be calculated from

$$m = V \rho A \tag{8.8}$$

where A is the cross-sectional area of the discharge pipe and ρ, the density of the steam in the discharge plain, is a function of P^*, the pressure of the low-pressure reservoir.

An alternate solution may be found in the *ASME Steam Tables*[1] as follows.

Use Figure 14 in the *ASME Steam Tables* entitled "Critical, (Choking), Mass Flow Rate for Isentropic Process and Equilibrium Conditions." This figure provides the mass flow rate per square inch of flow area per lbf/in^2 of pressure, m', as a function of upstream enthalpy and pressure. Based on the critical pressure ratio, the pressure upstream of the diffuser is

$$P_o = \frac{P^*}{0.54} \tag{8.9}$$

The value of m' may be found from Figure 14.

Therefore,

$$m = m' A P_o \tag{8.10}$$

8.3 TURBINE SHAFT SEALING STEAM

Figure 8.1 shows a diagram of a typical turbine shaft sealing steam system in a nuclear station. High-pressure steam leaking from the HP turbine stop/control valves and from the HP turbine seals is supplemented as required by main steam discharges into the steam seal header. From there, sealing steam is injected into the LP turbine seals. Figure 8.2 shows the typical LP turbine seals. Any excess steam in the sealing steam header is discharged to the main condenser. The low-pressure leak-off is collected in a header and is discharged to the gland steam condenser

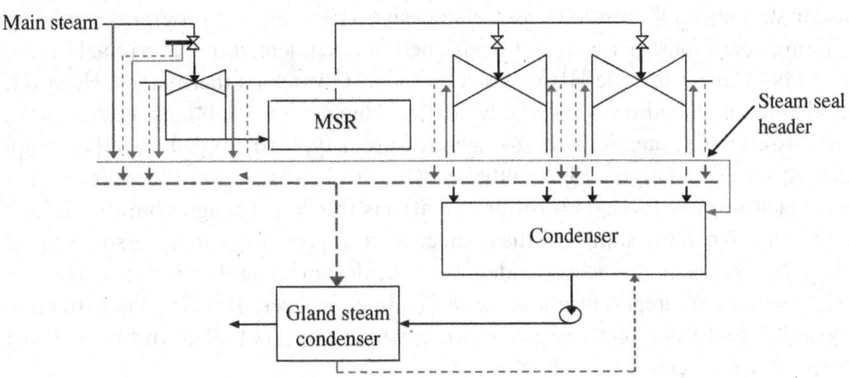

FIGURE 8.1 Turbine shaft sealing steam system.

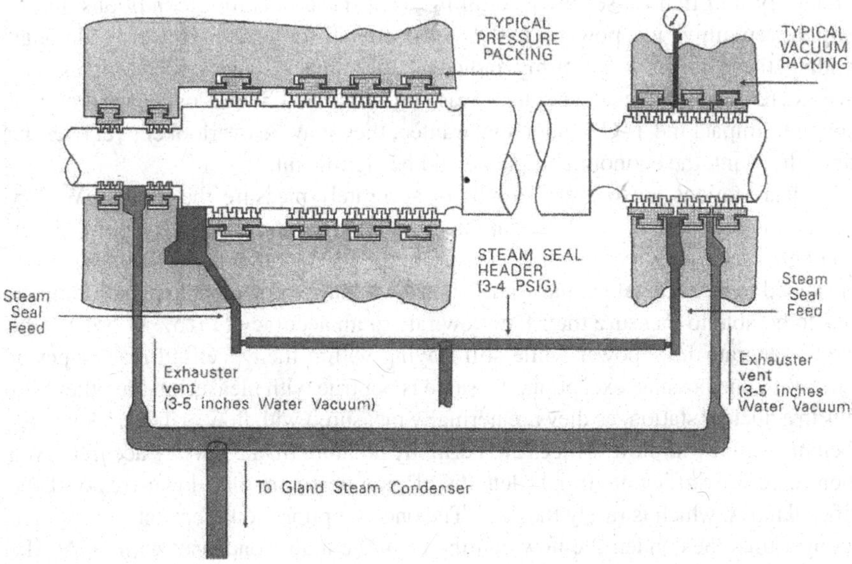

FIGURE 8.2 LP turbine shaft seals.

(also referred to as the steam packing exhauster). Condensate from the main condenser condenses the steam, and condensate from the gland steam condenser drains to the main condenser.

8.4 MASS BALANCE

A mass balance is helpful in identifying cycle isolation problems and equipment deficiencies. Design HB drawings showing the thermodynamic state points of the steam and water in the turbine cycle, including the mass flow rates, for an array of power levels are normally provided by the turbine–generator (T–G) vendor. Some nuclear stations may even have available design HB computer programs modeling operating conditions that were not envisioned or documented in the original design HB. In both instances, the HB is based on the T–G vendor's thermal kit. However, actual operating conditions invariably deviate from design conditions. For example, as discussed in Chapter 3, the first-stage pressure is typically higher than the design value, resulting in a higher FFW enthalpy, which results in a lower FFW flow for the same reactor power level. This difference affects the flow through both the HP and LP turbines. Another example is the main condenser pressure with the associated LP turbine backpressure that varies widely from the design value. The result is higher or lower condensate temperature leaving the main condenser, affecting the extraction flow to the first stage of feedwater heating, which impacts the LP turbine exhaust flow and thus the condensate flow.

In practice, the turbine cycle is rarely aligned as envisioned in the original design. Some cycle isolation issues are frequently known to exist such as leaking valves and heater drains being dumped to the main condenser. Normal practice is to attempt to quantify and trend lost power resulting from these operating conditions and to lump the remaining lost power into the "unaccounted-for losses." However, although trending these losses is certainly indicated, in some instances these losses may have existed when the nuclear station was commissioned. If the losses are not great enough to impact the T–G vendor's guarantee, they may be overlooked, yet over the life of the plant, the economic impact could be significant.

Nuclear stations go to great lengths to accurately measure the final FFW flow rate, as this flow rate is a critical parameter in determining the reactor power level. Although an accuracy of 2% when measuring the FFW with a flow nozzle was once considered to be acceptable, many nuclear stations have implemented plant modifications to be able to measure the FFW flow rate to an accuracy of 0.5% in order to be able to generate 1.5% power while still staying within their licensed reactor power limit. With one possible exception, the same is not true with measuring the other flow rates in a nuclear station, as they are normally measured with flow orifices. Flow rates when measured with flow orifices are normally no more than 2% to 5% accurate even when there are sufficient straight lengths of pipe upstream and downstream of the orifice flanges, which is rarely the case. The one exception would be a nuclear station that measures the condensate flow coming from the main condenser with an ASME flow nozzle. ASME flow nozzles were commonly used during T-G acceptance tests where accurate measurement of the condensate flow was important, but many nuclear stations remove these flow nozzles after completion of the tests to reduce the pressure drop in the condensate system. Therefore, in general, the only flow measurement that may be considered accurate in a nuclear station is the FFW flow. In conducting a mass balance, one must occasionally rely upon less accurate flow measurements such as the blowdown flow rate from a PWR SG. Some minor flows such as HP turbine control valve stem leak-off must be taken from the design HB. However, any inaccuracies in these flow rates do not result in significant errors in the overall mass balance. Measured FWH drain and condensate flows should normally be ignored in favor of calculated values based on an energy balance around the MSR and FWHs, as discussed in Chapters 4 and 7. It should be noted that, as discussed in Chapter 4, either confirmation of the MS effectiveness or an accurate measurement of the MS belly drain flows is highly desirable for PWR nuclear stations.

In addition to the FFW flow, the condensate flow, extraction enthalpies, and feedwater/condensate system temperatures must be known in order to accurately calculate extraction steam flows. As one may see from Figures 3.4 and 5.2, the extraction enthalpy is a function of the HP or LP turbine stage pressure at the extraction point. In nuclear turbines, the expansion line at or very near full reactor power remains constant as that reflected in the 100% power design HB because normal wear in the turbine blades is minimal. Therefore, because the operating pressure in the FWHs is normally known, the extraction pressure and thus the extraction enthalpy may be determined by adding to the FWH pressure through the extraction piping. When performing a design HB, a common practice is to assume a 5% pressure drop through the extraction piping. Therefore, the extraction enthalpy shown on the 100% power design HB may be assumed in calculating the extraction steam flow rate at or very nearly at 100% power. Feedwater/condensate system temperatures between stages of feedwater heating are generally available. Because these temperature sensors are normally periodically calibrated, they are generally quite accurate. In the case of nuclear stations employing FWH drain designs that are pumped forward so that the condensate temperature differs from the FFW flow, the condensate flow must be determined by iteration.

The mass balance may be calculated as follows:

- Begin with the FFW flow rate.
- Subtract the blowdown flow (PWR) or add the control rod drive flow (BWR) to arrive at the main steam flow rate.
- Subtract the MSR reheat flow based on 102% of measured drain flow (if applicable—see Chapter 4), any main steam flow to the MFP turbine, steam jet air ejector, etc., and HP turbine stop valve leakages to arrive at the HP turbine throttle flow.
- Subtract the extraction steam to any high-pressure FWHs using the calculation method described in Chapter 7 and the HP turbine gland leak-offs to arrive at the HP turbine exhaust flow.
- From the HP turbine exhaust flow, subtract the steam flow going to the feedwater heater to arrive at the flow entering the MS.
- Subtract the calculated or measured MS belly drain flow (see Chapter 4) from the flow entering the MS to arrive at the hot reheat flow going to the LP turbine(s).

- Subtract the calculated extraction steam flows from the LP turbine(s) to the low-pressure FWHs (see Chapter 7) and moisture that is removed from the last few LP turbine stages to arrive at the LP turbine exhaust flow.
- Add to the LP turbine exhaust flow the MFP turbine exhaust flow (or MFP turbine condenser drain flow), the gland steam condenser drain flow, the FWH operating vents and drain flows, miscellaneous leakage flows, and the blowdown flow (PWR) (or subtract the control rod drive flow [BWR]) to arrive at the condensate flow.

For nuclear stations that pump the FWH drains forward, the condensate flow plus the heater drain tank flow must equal the measured final FWH flow rate. Because the extraction flows to the LP FWHs are a function of the condensate flow rate, iteration would be required in this case. One should assume the design condensate flow value as an initial estimate and iterate on the calculated condensate flow until the assumed condensate flow plus the heater drain tank flow equals the measured final FWH flow rate.

8.5 MASS BALANCE SPREADSHEET

A common practice is to trend electrical output as a function of CCW inlet temperature, as shown in Figure 6.25. However, with a mass balance, much more can be done. The mass balance may be computed in an Excel Workbook. Plant data along with the appropriate enthalpies may be downloaded from the plant computer, and miscellaneous flow rates may be entered from the design HB drawing to perform the calculations, as shown in Chapters 2–7. The plant data having been collected and massaged, key turbine cycle equipment parameters such as reactor power, MSR TTD's, FWH TTD's, and DCA's, and the main condenser performance factor may be conveniently calculated and trended based on actual plant performance. Adverse trends may then be addressed in a timely manner without waiting for them to show up in lost electrical power. Because nuclear stations primarily deal with saturated steam and water, many parameters that would require laborious lookups from the *ASME Steam Tables*[1,2] may be reduced to polynomial regressions such as enthalpy as a function of saturation temperature, saturation temperature as a function of pressure, etc. Bowman[3] developed a Turbine Cycle Equipment Evaluation (TCEE) Excel workbook for the Electric Power Research Institute (EPRI) to facilitate a comparison between plant data and the state point values predicted by design HB and to key turbine cycle equipment parameters that are thermal performance indicators.

REFERENCES

1. Meyer, C. A., et al., *ASME Steam Tables*, 6th ed., American Society of Mechanical Engineers, New York, 1993.
2. *ASME Steam Tables, Compact Edition*, American Society of Mechanical Engineers, New York, 1993, 2006.
3. Bowman, C. F., *Turbine Cycle Equipment Evaluation (TCEE) Workbook*, Electric Power Research Institute, Product ID: 3002005344, 2015.

9 Heat Rejection Systems

9.1 SCOPE OF THE HEAT REJECTION SYSTEM

The heat rejection system (HRS) is a system that rejects the waste heat from the main and MFP turbine condensers into the environment, thereby elevating the temperature of the CCW that is pumped through the condensers. Because the design of the HRS is dictated by the nuclear station site considerations, HRS configurations vary widely. HRS configurations may be classified as open, helper, or closed systems or combinations thereof and are defined as follows:

- *Open*: CCW is drawn from a river, lake, or ocean adjacent to the station and returned to the same at an elevated temperature.
- *Helper*: CCW is drawn from a river, lake, or ocean and then passed through a heat dissipation device such as a cooling tower before being returned to its original source.
- *Closed*: CCW is drawn from a heat dissipation device such as a cooling tower and returned to the device.

Most of the existing nuclear stations employ an open HRS. Others have been backfitted to become helper systems due to environmental regulations. These same environmental considerations have dictated that more recent and future HRS will be closed systems. Although more expensive in both capital and operating costs, the financial impact of having to provide a closed HRS may be minimized by proper design (see Chapter 6), and this option expands the potential future nuclear station site locations, as they would no longer be married to sites adjacent to a river, lake, or ocean.

The basic elements of an open HRS include CCW pumps, piping, and valves to and from the condensers on the tube side of the condensers. Helper and closed HRS also include the device that dissipates the heat into the atmosphere and may include mechanical draft cooling towers (MDCTs), natural draft cooling towers (NDCTs), spray systems, and dedicated cooling lakes. These devices are covered in detail in subsequent chapters. Helper HRS often include cooling tower lift pumps, valves, and piping to deliver the CCW to a cooling tower. Some open or helper HRS employ underwater diffusers that promote the mixing of the CCW into the river, lake, or ocean. A closed HRS includes means to provide makeup to the CCW to replace that which is evaporated and lost as drift and the blowdown that must be discharged. Some nuclear stations also employ online condenser ball cleaning systems and debris filters and chemical treatment systems to increase the main condenser performance (see Chapter 6).

Determining the size of the HRS is a classical tradeoff of capital cost against future gains in electrical output as determined by economic policy and forecasting the value and future demand for electrical energy. Higher CCW flow, a larger main condenser and cooling tower, etc. result in lower LP turbine backpressures, causing more T-G terminal output. However, larger CCW pumps and more MDCT cells would mean a higher station load. Do the economics dictate an HRS of sufficient size to support full reactor power under all operating conditions, including extreme environmental conditions that may only infrequently occur? These are issues that must be addressed when planning a new nuclear station.

9.2 CCW PUMPS, VALVES, AND PIPING

Due to the huge volumes of CCW required to remove the waste heat from a nuclear station, the CCW pumps constitute one of the largest auxiliary power (station) loads. Due to the practical size limitation of electric motors, nuclear stations typically employ multiple (at least three) CCW pumps. For nuclear stations that employ a closed HRS with cooling towers, the required CCW pump head is up to three times as great as one employing an open or helper HRS, so more CCW pumps are normally required. Multiple CCW pumps may also offer the advantage of permitting the station to continue to operate at full power with one or more CCW pump out of service.

Occasionally, nuclear stations with closed HRS use dry-pit pumps, but vertical wet-pit CCW pumps are more commonly employed with each pump located in a separate sump fitted with a stop log and mounted on the open deck of the CCW pumping station. The sumps should be designed according to the recommendations of the Hydraulic Institute Standards[1] and the motors should have weather-protected NEMA Type II enclosures. Trash racks and traveling water screens are normally provided to protect the CCW pumps and the condensers from the intrusion of debris. Each CCW pump is fitted with a large single-speed, electric motor-operated butterfly valve on the discharge, and the pump is started against a closed valve that is opened slowly upon pump start to maintain a downward thrust on the pump foundation and to avoid water hammer in the system.

CCW piping may be cast-in-place concrete square conduits (often used inside the turbine building) or concrete pressure pipes. Carbon steel pipes are not normally used except for short sections at the main condenser water box entrance where bolted, flanged butterfly valves are employed to permit the isolation of a water box for cleaning. Rubber expansion joints of single-arch spool-type are located between the main condenser isolation valves and the bottom of the water box to accommodate the main condenser tube to shell expansion. If an online tube cleaning system is employed, sponge rubber balls are injected into the spool piece at the inlet and collected in the spool piece at the discharge of the main condenser. The CCW piping and water boxes should be designed for the maximum operating pressure created by the CCW pumps.

Figure 9.1 shows a typical hydraulic gradient in an open HRS located on a lake. Note that the impeller of the vertical wet-pit pump is located sufficiently below the lowest lake level to provide sufficient net positive suction head (NPSH) for the pump. Because the CCW pump is required to provide only sufficient hydraulic pressure to overcome the pressure drop through the CCW piping and valves and the main condenser, the hydraulic gradient falls below the elevation of the main condenser tubing

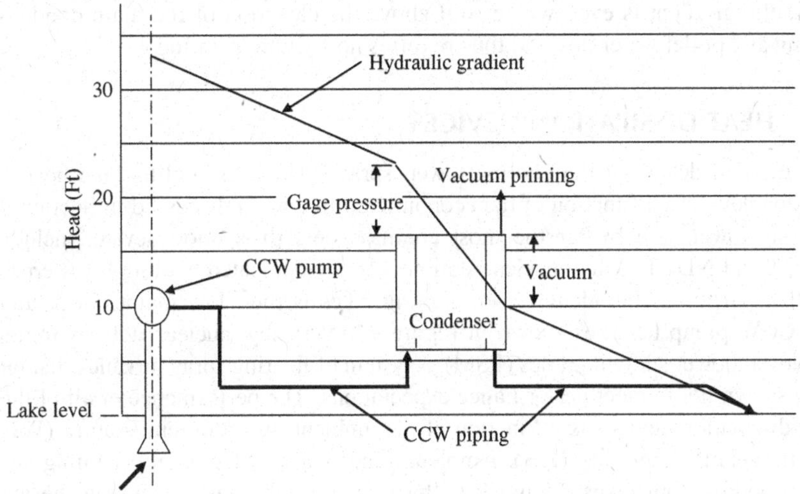

FIGURE 9.1 Typical hydraulic gradient in an open HRS located on a lake.

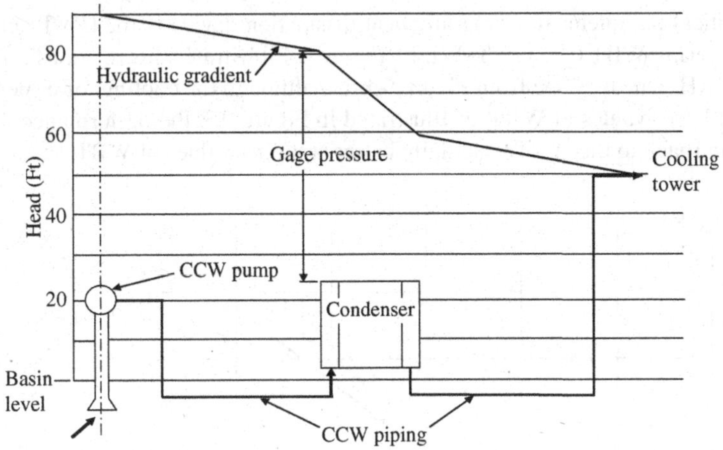

FIGURE 9.2 Typical hydraulic gradient in a closed HRS with a cooling tower.

as it passes through the main condenser tubes, necessitating vacuum priming at the outlet water box, as discussed in Chapter 6.

Figure 9.2 shows a typical hydraulic gradient in a closed HRS with a cooling tower. Note that the impeller of the vertical wet-pit pump is located sufficiently below the lowest cooling tower basin level to provide sufficient NPSH for the pump. Because the CCW pump is required to provide enough hydraulic pressure to not only overcome the pressure drop through the CCW piping and valves and the main condenser but also raise the CCW to the cooling tower hot water basin, the hydraulic gradient is everywhere well above the elevation of the main condenser tubing and outlet water box and thus requires no vacuum priming.

9.3 HEAT DISSIPATION DEVICES

The thermal design of the cooling tower, spray system, or cooling lake heat dissipation device is a function of the economic optimization discussed in Section 9.1. Cooling towers are by far the most common heat dissipation device, including MDCT and NDCT. A few nuclear stations employ dedicated cooling lakes created for that purpose. The advantage of cooling lakes is their low maintenance and a low CCW pump head, as shown in Figure 9.1. Very few nuclear stations initially employed power spray modules (PSM), a system of floating spray modules that have been shown not to meet performance expectations. The performance of all of these heat dissipation devices is a function of the ambient wet-bulb temperature (WBT) and the relative humidity (RH). Ambient wind is also a factor in designing spray ponds and cooling lakes. Figure 9.3 illustrates the relationships among the cold-water temperature (CWT) leaving a typical cooling tower and the WBT and RH. One may see from Figure 9.3 that the cooling tower performs slightly better at lower values of WBT, as one might expect. This general relationship is applicable to any evaporative heat dissipation device.

A critical parameter in sizing any heat dissipation device is the CWT approach to the entering WBT (i.e., CWT–WBT). Figure 9.4 illustrates this relationship for an array of RH. One may see from Figure 9.4 that although the cooling tower performs better at lower values of WBT, as illustrated in Figure 9.3, the performance relative to the approach to the WBT is actually better at higher values of WBT.

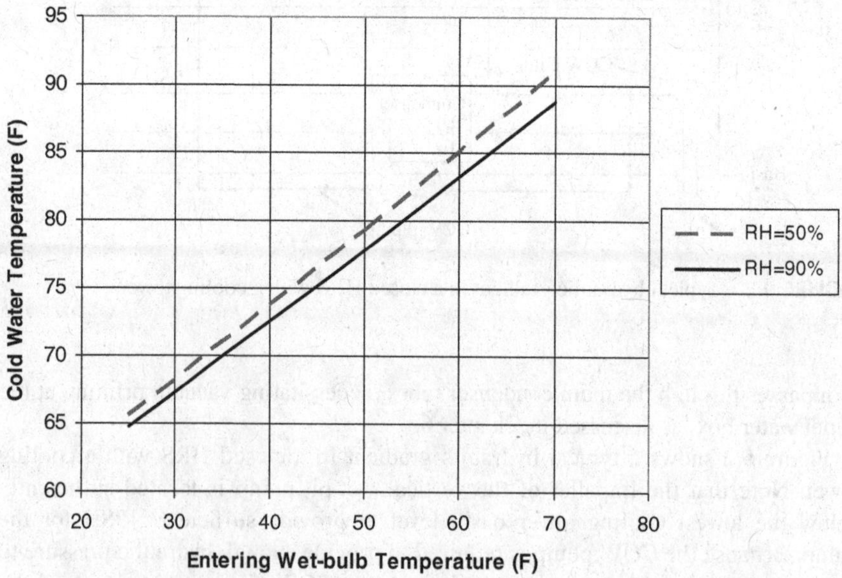

FIGURE 9.3 CWT from a cooling tower vs. entering WBT for an array of RH.

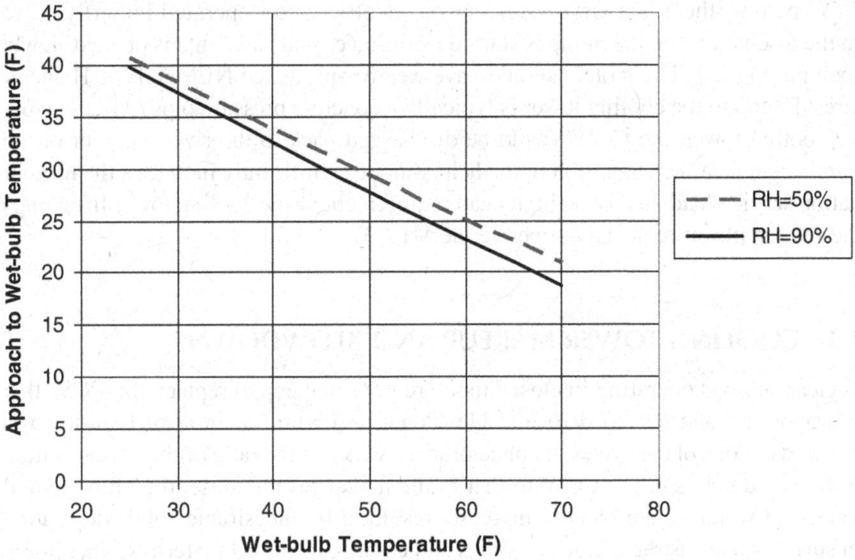

FIGURE 9.4 CWT from a cooling tower approach to WBT vs. entering WBT.

The performance of the heat dissipation device is normally specified at one meteorological point (i.e., WBT, RH, and wind speed) with the vendor providing curves similar to Figure 9.3 to predict performance at other operating conditions. The guaranteed design point relative to these parameters is a matter of personal judgment. A common practice is to specify the guaranteed performance at some extreme value such as the WBT that is exceeded only 1% of the time. However, consideration should be given to specifying the annual average meteorology, selected because the reliance on correction curves is reduced. Normally, any acceptance test would be performed during the first year after commissioning. The opportunity to conduct an acceptance test at extreme conditions would likely occur only once, and if the test is conducted at some other time, reliance on the correction curve is increased. If the guarantee is specified at the annual average meteorology, the opportunity to conduct the test presents itself twice in the first year, and the actual test is more likely to be performed nearer to the guarantee point.

9.4 COOLING TOWER LIFT PUMPS

Nuclear stations with helper HRS normally require a cooling tower lift pumping system where the CCW is pumped from the CCW discharge canal up to the top of one or more cooling towers. The designs of these lift pumping stations are frequently very challenging, as the flow in the discharge channel is normally quite high and not conducive to proper pumping station sump design, in accordance with the recommendations of the Hydraulic Institute Standards. Cooling tower lift pumps are almost invariably of the wet-pit design similar to the CCW pumping station but without the intake screens, as the CCW has already passed through the condensers. As with the

CCW pumps, there is a large single-speed, electric motor-operated butterfly valve on the discharge, and the pump is started against a closed valve that is opened slowly upon pump start. The motors should have weather-protected NEMA Type II enclosures. Piping to the cooling tower is typically a concrete pressure pipe. After leaving the cooling tower, the CCW would be discharged back to the river, lake, or ocean from which it came. Operation in the helper mode significantly increases the nuclear station service load due to the high head requirement of the cooling tower lift pumps and the fan power required to operate the MDCT.

9.5 COOLING TOWER MAKEUP AND BLOWDOWN

Nuclear stations operating in closed mode require makeup to replace the CCW that is evaporated and lost to drift and blowdown. A certain amount of blowdown is required to control the cycles of concentration, which is the ratio of the concentration of dissolved solids in the CCW to that in the makeup. Otherwise, the ratio of solid volume to water volume would increase, resulting in undesirable solids deposition on surfaces such as the condenser tubes and cooling tower fill materials. Therefore,

$$C = \frac{M}{B} \tag{9.1}$$

and

$$B = \frac{(E+D)}{(C-1)} \tag{9.2}$$

where:

C = cycles of concentration
M = makeup
B = blowdown
E = evaporation
D = drift.

Some nuclear stations located at sites where makeup is scarce may run very high cycles of concentrations and control scaling by adding acid to the system. Nuclear stations with an abundance of available makeup normally maintain a fairly low number of cycles of concentrations so that the corrosive and scaling tendencies of the CCW are balanced. The Ryznar stability index is commonly used for determining the corrosive or scale-forming tendencies of CCW. In order to calculate this index, the following data on the makeup is required as inputs:

- Calcium hardness
- Total alkalinity
- Total dissolved solids
- Temperature
- pH.

The Ryznar stability index should be maintained between values of 6.0 and 7.0.

Cooling tower makeup can come from a variety of sources including a river or lake, groundwater, or the discharge from a sewage treatment plant. Straining and chemical treatment to eliminate invasion of live species such as mollusks in the CCW is highly desirable.

9.6 CHEMICAL TREATMENT OF CCW

Where permitted by environmental regulations, nuclear stations often employ chemical treatment regimens to improve the performance of the main condenser by controlling the buildup of microfouling (see Chapter 6). The most common chemical used are an oxidizing biocide such as chlorine. Even in open or helper HRS, low levels of this treatment may be permitted for an hour or two each day as the chemical demand of an oxidizing biocide is quickly satisfied when released into a large body of water, making it more benign. In closed HRS employing cooling towers, oxidizing biocides are not as effective as nonoxidizing biocides because they are quickly scrubbed out of the CCW as it passes over the cooling tower fill. Nonoxidizing biocides are proprietary complex organic chemicals such as "Clamtrol" that (unlike oxidizing biocides) are nonvolatile and remain in the CCW for long periods. Nonoxidizing biocides are both very expensive and highly toxic, and any release into the environment through blowdown requires extensive pre-release treatment. Nonoxidizing biocides are often used in conjunction with a comprehensive chemical treatment regimen including a surfactant, etc. to maintain complete chemical control of the CCW.

9.7 ONLINE MAIN CONDENSER TUBE CLEANING SYSTEMS AND DEBRIS FILTERS

Figure 9.5 illustrates the online main condenser tube cleaning systems (top figure) and the debris filters (bottom figure) manufactured by Taprogge. In the tube cleaning system, sponge rubber balls that are slightly larger than the inside diameter of the main condenser tube are injected into the riser pipe entering the main condenser water box. The balls are randomly propelled through the tubes by the pressure drop across the tube bundle and collected at a screen in the discharge piping below the outlet water box. The balls are then pumped through a ball collector where worn balls are collected as they pass through a screen and periodically replaced with a charge of new balls. The sponge rubber balls interrupt the formation of microfouling on the main condenser tubes, resulting in cleaner tubes and better main condenser performance. In the event that a substantial slime film has built up prior to the operation of the system, a load of abrasive balls may be used to remove the film. A separate tube cleaning system is normally installed for each water box. The debris filter as shown in Figure 9.5 is not required to operate a tube cleaning system.

Some nuclear stations that have experienced substantial main condenser tube macrofouling have installed debris filters (bottom figure) that collect objects such as mollusks, fish, sticks, rocks, etc. that manage to pass through the CCW pumps and flush them directly into the discharge pipe. Although this device reduces the amount

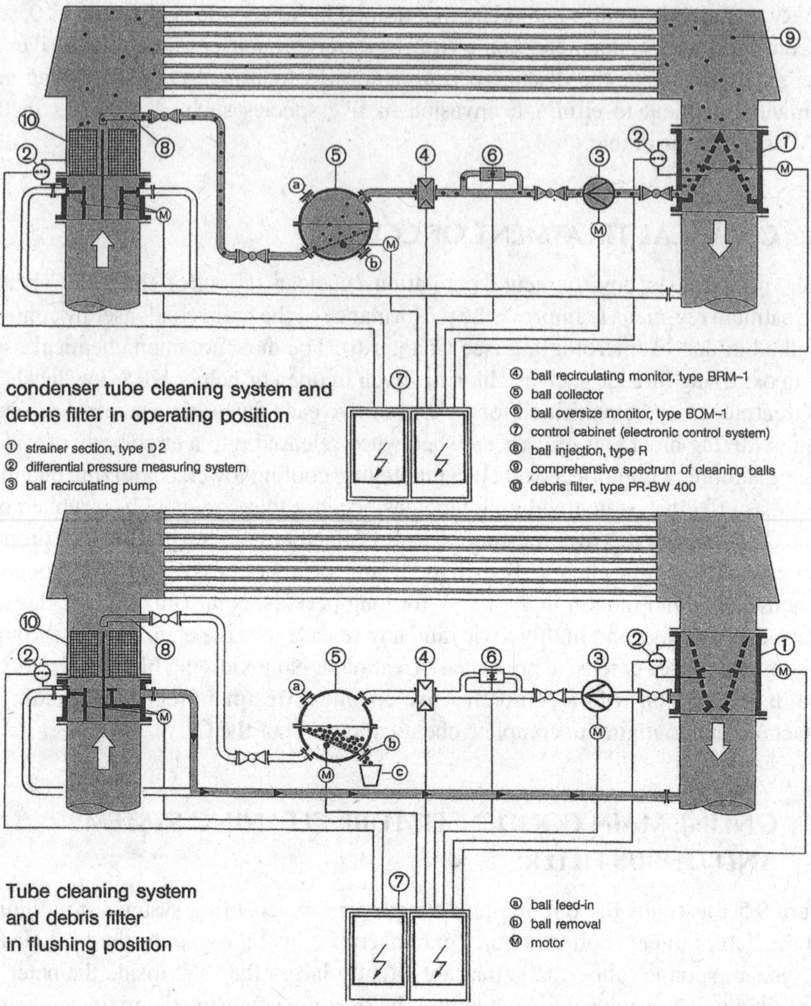

FIGURE 9.5 Taprogge tube cleaning system and debris filter. (Courtesy of Taprogge.)

of main condenser tube plugging due to macrofouling, it also diverts approximately 10% of the CCW flow around the main condenser that is then unavailable to reduce the main condenser pressure. When used, debris filters are typically only employed on the last water box riser pipe in a line, as that is typically where debris collects. The tube cleaning system as shown in Figure 9.5 is not required in order to operate a debris filter.

REFERENCE

1. *Hydraulic Institute Standards for Centrifugal, Rotary & Reciprocating Pumps*, 14th ed., Hydraulic Institute, Piqua, OH, 1983.

10 Cooling Towers

10.1 TYPES OF COOLING TOWERS

10.1.1 Cross-Flow Mechanical Draft Cooling Tower

Figure 10.1 Illustrates a common type of cross-flow MDCT. A cross-flow MDCT consists of one or more cells, each of which consists of a large fan affixed on the top of a structure containing splash-type fill material. CCW that has been heated in the nuclear station condenser(s) is pumped up to the top of the structure and distributed in the hot water flume atop the structure. It then passes through specially designed nozzles embedded in the floor of the hot water flume. The nozzles are fitted with splash plates to evenly distribute the CCW over the splash-type fill below. The motor-driven fan atop each cell of the MDCT operates to draw air into the side of the structure through louvers. The air passes across the water as it is continually separated into fine drops by the splash fill falling to the cold-water basin below. After interacting with and cooling the falling CCW droplets, the air passes through drift eliminators and leaves through the fan at the top of the MDCT. The drift eliminators are designed to minimize the number of fine drops of CCW carried out of the fill section of the MDCT and passed through the fan. A fan stack surrounds each fan to increase the efficiency of the fan by minimizing recirculation of the air. The cold CCW is collected from all of the MDCT cells in a single cold-water basin located below all of the MDCT cells and then returned to the nuclear station.

Some more recent MDCT designs configure two cells back-to-back with a fan above each such that the air enters from one side only. This arrangement requires less land area while concentrating more fan power with two motors rather than one large motor.

Although most MDCTs consist of cells aligned in a straight line, some designs take the back-to-back concept one step farther by arranging a collection of one of the back-to-back cells in a circle rather than in a line.

10.1.2 Cross-Flow Natural Draft Cooling Tower

Figure 10.2 illustrates a cross-flow NDCT. A cross-flow NDCT operates in a manner similar to a cross-flow MDCT except that the airflow is created by the chimney effect of a tall veil. Even when there is no CCW flow to the NDCT, airflow is induced by the lower density of the air at the top of the veil. When the nuclear station is not operating but there is still CCW flow to the NDCT, there is less dense saturated air entering the veil because the molecular weight of water vapor (18) is less than that of air (29). Depending on the water loading on the fill section, this may result in an increase in airflow as the lighter air tends to rise. When the nuclear station

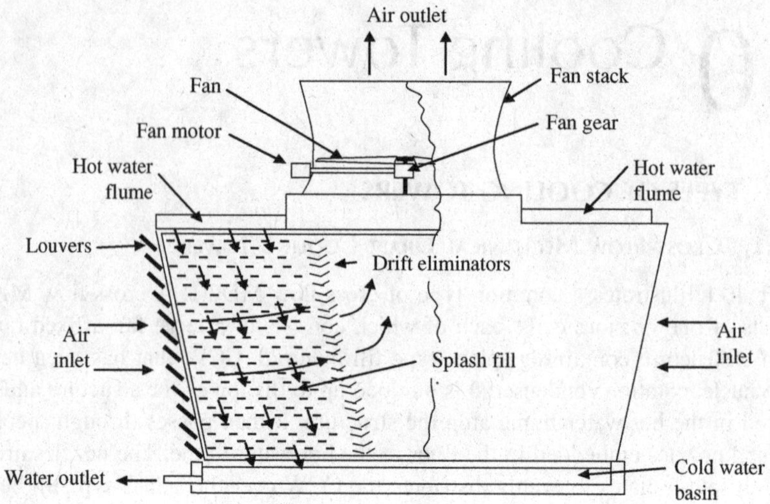

FIGURE 10.1 Cross-flow mechanical draft cooling tower.

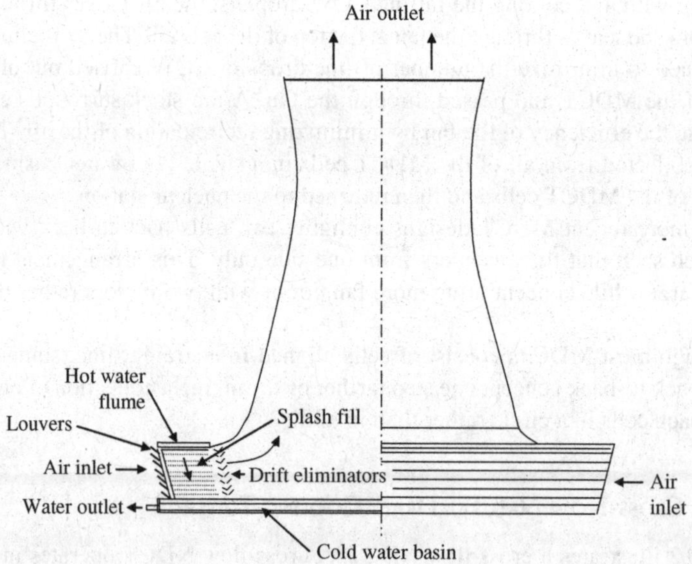

FIGURE 10.2 Cross-flow natural draft cooling tower.

is operating, the heat added to the CCW results in even greater airflow, as warm saturated air is even less dense. A canopy between the donut-shaped heat transfer section around the veil containing the splash-type fill material and the veil encloses the interior of the NDCT so that no air can bypass the fill section.

As an alternative to the expensive veil, some more recent MDCT designs locate the fans in and above the space in the interior of the circle created by the heat

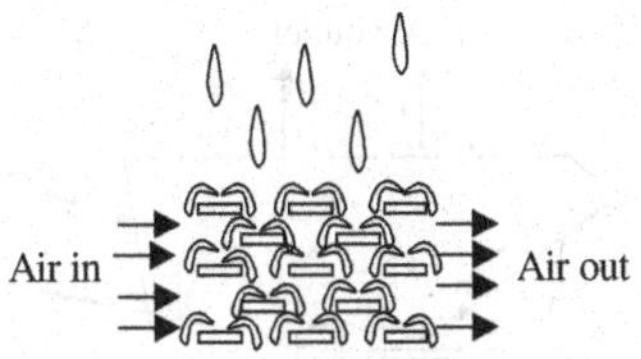

FIGURE 10.3 Splash-type cooling tower fill material.

transfer section. This design has the added benefit of concentrating the plume from the MDCT, making it more buoyant and less likely to recirculate back to the inlet of the fill section.

10.1.3 Splash-Type Cooling Tower Fill Material

Figure 10.3 illustrates a splash-type fill that is commonly used with cross-flow MDCT and NDCT. A splash-type fill material may be constructed from a variety of materials such as wood slats but is most commonly made of plastic in large cross-flow cooling towers serving nuclear stations. The plastic slats may be oriented either parallel to or perpendicular to the airflow. (The parallel orientation results in less resistance to airflow.) The slats, normally supported by a system of plastic or coated wire hangers, are frequently perforated and come in a variety of shapes for strength.

One of the major drawbacks with cross-flow cooling towers, both MDCT and NDCT, is the propensity for ice to build up on the fill material in cold weather. Various schemes have been employed to prevent ice buildup such as diverting a larger portion of the CCW to the perimeter of an NDCT or shutting off or reversing the fan flow in MDCT, but all such schemes require diligent observation to identify when ice buildup has begun. As a result, cross-flow cooling towers are in somewhat disfavor where there is the potential for ice formation.

10.1.4 Counter-Flow Mechanical Draft Cooling Tower

Figure 10.4 illustrates a counter-flow MDCT. As is the case with a cross-flow MDCT, each cell in a counter-flow MDCT consists of a large fan affixed on top of a structure, but the fill material is of a thin-film variety. The heated CCW is pumped through a piping system located above the fill section and passes through specially designed nozzles located at the bottom of pipes fitted with splash plates to evenly distribute the CCW over the thin-film fill material. As with a cross-flow MDCT, the motor-driven fan atop each cell operates to draw air into the side of the structure. The air passes through the fill material where the CCW is running down the fill material in thin sheets before it falls to the cold-water basin below. After interacting with and cooling the falling CCW, the air passes through drift eliminators and leaves through the fan on top of the MDCT. The cold CCW is collected from all of the MDCT cells in a single cold-water basin located below all of the MDCT cells and returned to the nuclear station.

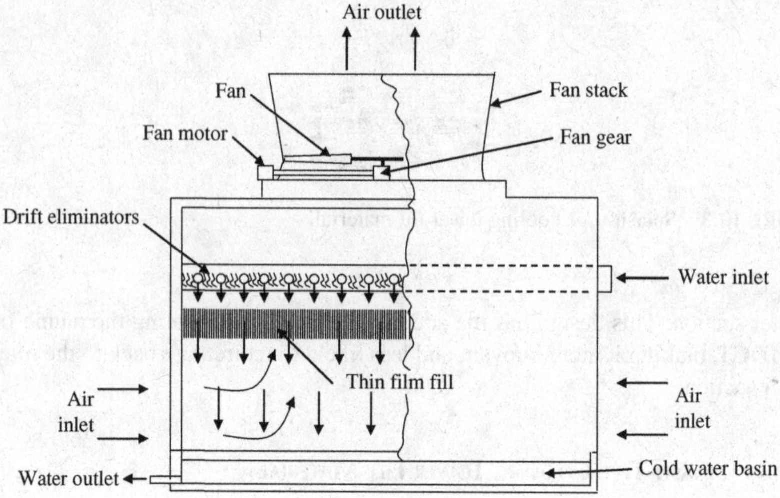

FIGURE 10.4 Counter-flow mechanical draft cooling tower.

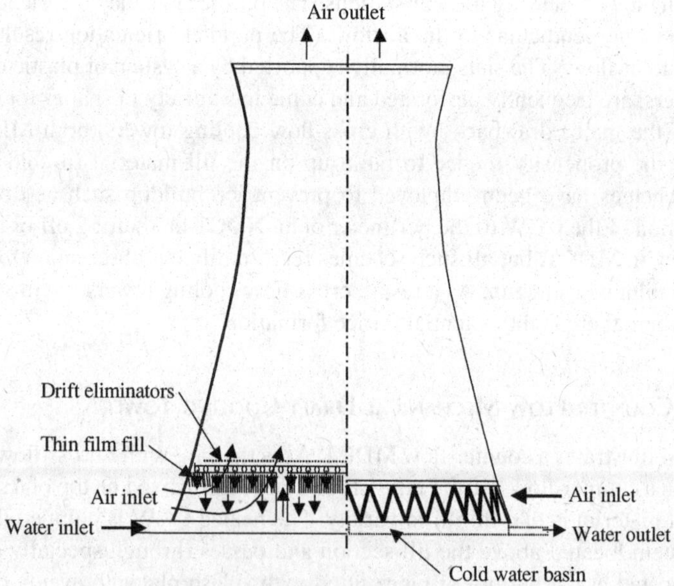

FIGURE 10.5 Counter-flow natural draft cooling tower.

10.1.5 Counter-Flow Natural Draft Cooling Tower

Figure 10.5 illustrates a counter-flow NDCT. The heated CCW is pumped up through concrete standpipes that distribute the CCW through a piping system located above the fill section. It then passes through specially designed nozzles located at the bottom of the pipes fitted with splash plates to evenly distribute the CCW over the thin-film fill material. A counter-flow NDCT operates in a manner similar to a counter-flow MDCT, except that, as in the case of a cross-flow NDCT, the airflow is created by the chimney effect of a tall veil. As with a cross-flow NDCT, even when there is no CCW flow to the NDCT, the airflow is induced by the difference in the density of the air at the top of the veil, which is less than that at the ground level. Airflow is also increased when the nuclear station is operating because the heat added to the CCW results in warm saturated air that is even less dense. Counter-flow NDCTs are by far more common in the world than cross-flow NDCTs as they are aggressively and successfully marketed by Hamon Cooling Tower Company in cooperation with Research-Cottrell Corporation. Counter-flow NDCTs have proven to be much more resistant to ice damage due to the much more durable fill material and the fact that damage is normally limited to the outer edges of the HX section. Some NDCTs have been converted from cross-flow to counter-flow for this reason.

10.1.6 Thin-Film-Type Cooling Tower Fill Material

Figure 10.6 illustrates a thin-film-type fill commonly used with counter-flow MDCT and NDCT. The thin-film-type fill consists of several layers of very thin vertical sheets hung from structural members such that when the CCW lands on the top of the top sheet, it tends to dribble down the sheet in a thin film before falling to the basin below. Originally, the material for these sheets was asbestos cement, but due to the concern about the health risks with asbestos, they are now made of plastic. Some vendors offer sheets with an enhanced surface texture that is intended to promote better heat transfer.

There are a number of other less common types of cooling towers such as dry cooling towers that rely on direct contact with air to condense the steam from the LP

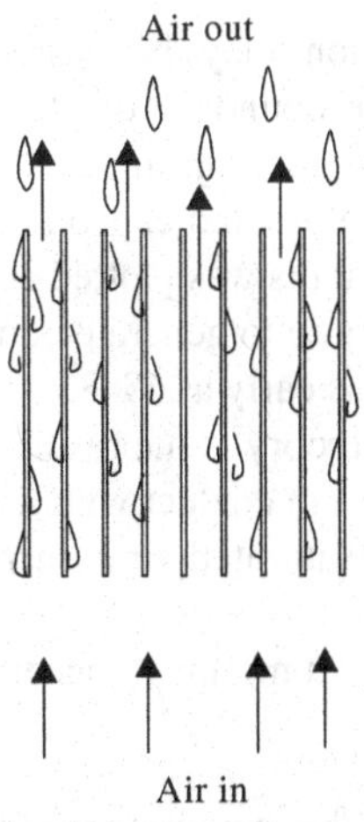

FIGURE 10.6 Thin-film-type cooling tower fill material.

turbine(s) and wet-dry cooling towers that enhance the heat transfer from the steam to the air by spraying water in the HX during periods of high ambient temperatures (see Chapter 17). Additionally, there are fan-assisted NDCTs that reduce the height of the NDCT veil by locating fans around the base of the cooling tower to promote airflow.

10.2 MERKEL EQUATION

The first known attempt at modeling the processes taking place in an evaporative cooling tower was made by Dr. Fredrick Merkel[1] in 1925 when he proposed a theory relating the heat transferred from the air to a theoretical film by convection to the heat transferred from the film to the ambient air by evaporation. This theory based on counter-flow contact of water and air was suited, but not limited, to various types of cooling towers.

Merkel made several simplifying assumptions that reduce the governing relationships for a counter-flow tower to a single separable ordinary differential equation as follows:

$$\mu\, dt_w/\left(h_s - h_a\right) = KaV/L \tag{10.1}$$

where:

t_w = water temperature, °F
h_s = enthalpy of saturated air at the temperature of the water, Btu/lbm
h_a = enthalpy of air, Btu/lbm
K = mass transfer coefficient, lbm/hr-ft^2
a = interface area per unit of volume, ft^2/ft^3
V = cooling volume, ft^3
L = water flow rate, lbm/hr.

KaV/L is a dimensionless performance characteristic that is often referred to as the number of transfer units (NTU).

The left side of this equation may be easily integrated by the four-point Tchebycheff method because the boundary conditions are known. The right-hand side of this equation is a dimensionless group, also known as the number of diffusion units, which relates the variables and capabilities of a particular cooling tower design. The number of diffusion units was plotted as a function of the cooling tower liquid (water)-to-gas (air) (L/G) ratio to generate cooling tower "demand curves" by the Foster Wheeler Corporation as early as 1943.

The simplicity of the Merkel theory is the result of the simplifying assumptions. Merkel posited that the bulk water in contact with a stream of air is surrounded by a film of saturated air and that the saturated air is surrounded by a bulk stream of air, as shown in Figure 10.7.

Merkel made the following assumptions concerning the heat transfer between water and air:

1. The saturated air film is at the temperature of the bulk water.
2. The saturated air film offers no resistance to heat transfer.

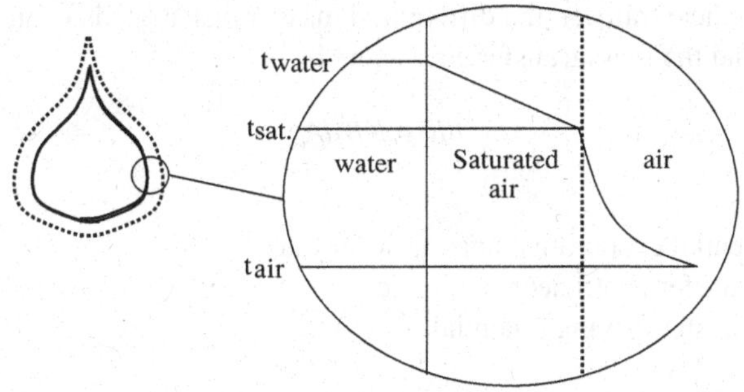

FIGURE 10.7 Merkel evaporative cooling model.

3. The vapor content, or absolute humidity, of the air is proportional to the partial pressure of the water vapor (i.e., it acts like an ideal gas).
4. The heat transferred from the air to the film by convection is proportional to the heat transferred from the film to the ambient air by evaporation. This assumption results in a Lewis factor of one.
5. The specific heat of the air–water vapor mixture and the heat of vaporization are constant.
6. The loss of water by evaporation is neglected.
7. The force driving heat transfer is the differential enthalpy between the saturated air and the bulk air.

10.3 DERIVATION OF THE MERKEL EQUATION

It is well understood that the total heat transfer rate in a cooling tower may be expressed as

$$dQ_t = dQ_s + dQ_e \tag{10.2}$$

where:

dQ_t = differential heat transfer rate
dQ_s = differential sensible heat transfer rate
dQ_e = differential latent heat transfer rate.

For sensible heat transfer,

$$dQ_s = H\left(t_W - t_{db}\right) dA_i \tag{10.3}$$

where:

H = local heat transfer coefficient
t_W = local water temperature
t_{db} = local air dry-bulb temperature
dA_i = differential interface area.

For evaporative heat transfer, the differential mass transfer rate is related to the driving potential and the mass transfer coefficient by

$$dE = KB\,dA_i \tag{10.4}$$

where:
dE = differential evaporative mass transfer rate
K = mass transfer coefficient
B = mass transfer driving potential.

and

$$B = x_s - x \tag{10.5}$$

where:
x_s = mass fraction of the water in the saturated air at the water surface temperature
x = mass fraction of the water in the air in the bulk air stream.

and

$$x_s = T_s/(1+T) \tag{10.6}$$

$$x = T/(1+T) \tag{10.7}$$

where:
T = absolute humidity

so that

$$B = (T_S - T)/(1+T) \tag{10.8}$$

The differential heat transfer rate is related to the differential mass transfer rate through the enthalpy of saturated water vapor, h_g, by

$$dQ_e = h_g dE \tag{10.9}$$

where:
h_g = enthalpy of saturated water vapor.

Therefore, substituting into the equation for dQ_t,

$$dQ_t = H\left(t_W - t_{db}\right)dA_i + h_g K\left(x_s - x\right)dA_i \tag{10.10}$$

Let

$$dA_i = a\,dV \tag{10.11}$$

where:
a = interface area per unit of volume
V = cooling volume.

The equation for dQ_t may be written in the form

$$dQ_t = Ka\,dV\left[\left(H/Kc_{p(a)}\right)c_{p(a)}\left(t_W - t_{db}\right) + h_g\left(x_s - x\right)\right] \qquad (10.12)$$

where:

$c_{p(a)}$ = specific heat of moist air.

Merkel simplified this relationship by suggesting that for a film of saturated air at the air/water interface, an equilibrium condition can be expressed by equating heat transferred from air to the film by convection to the heat transferred from the film to the ambient air by evaporation:

$$H\left(t_W - t_{db}\right) = Kh_g\left(x_s - x\right) \qquad (10.13)$$

By multiplying both sides of this equation by $c_{p(a)}$ and collecting terms, the result is the Lewis factor.

$$Le = H/K\,c_{p(a)} = h_g\left(x_s - x\right)/c_{p(a)}\left(t_W - t_{db}\right) \qquad (10.14)$$

If one assumes a value of $Le = 1$, then dQ_t reduces to

$$dQ_t = Ka\,dV\left[c_{p(a)}\left(t_W - t_{db}\right) + h_g\left(x_s - x\right)\right] \qquad (10.15)$$

Noting that the heat transferred to the air is lost from the water,

$$dQ_t = d\left(L\,c_{p(w)}t_W\right) \qquad (10.16)$$

where:

L = water flow rate

and also noting that the differential change in enthalpy of moist air is

$$dh = c_{p(a)}dt + h_g dx \qquad (10.17)$$

By assuming constant average values of $c_{p(a)}$ and h_g, the equation for enthalpy becomes

$$h = c_{p(a)}t + h_g x \qquad (10.18)$$

Combining the equations above yields

$$d\left(Lc_{p(w)}t_W\right) = Ka\left(h_s - h_a\right)dV \qquad (10.19)$$

where:

h_s = enthalpy of saturated air at the temperature of the water

h_a = enthalpy of air.

If one assumes that L and $c_{p(w)}$ are constant, where $c_{p(w)} = 1$, then

$$\mu\, dt/\left(h_s - h_a\right) = Ka\,dV/L \tag{10.20}$$

Thus, the well-known Merkel equation is obtained, which was the standard method of cooling tower performance analysis for many years. However, as shown above, this approach requires several significant simplifying assumptions.

The right side of the Merkel equation is a measure of the depth of the fill required in a counter-flow cooling tower for a given airflow.

10.4 DETERMINING THE VALUE OF *KaV/L*

Figure 10.8 shows a representation of the heat transfer occuring in a counter-flow cooling tower on the psychrometric chart showing the properties of air at standard atmospheric pressure. One may see that the saturated air (which is at the temperature of the water) moves down the saturation curve as it is cooled due to the cooler air that is passing over it and the air is heated as it passes over the fill sheets, resulting in heat transfer from the water to the air with the driving potential of $h_w - h_a$.

The right side of Figure 10.9 graphically illustrates the integration of the left side of the Merkel equation, which is equal to the *KaV/L* or the NTU of the cooling tower.

Several methods of integration are available to determine the value of NTU. The simple Tchebycheff method[2] is as follows:

$$\frac{KaV}{L} = \int_{t_{cold}}^{t_{hot}} \frac{dt}{h_S - h_a} = \frac{t_{S-in} - t_{S-out}}{4}\left(\frac{1}{\Delta h_1} + \frac{1}{\Delta h_2} + \frac{1}{\Delta h_3} + \frac{1}{\Delta h_4}\right) \tag{10.21}$$

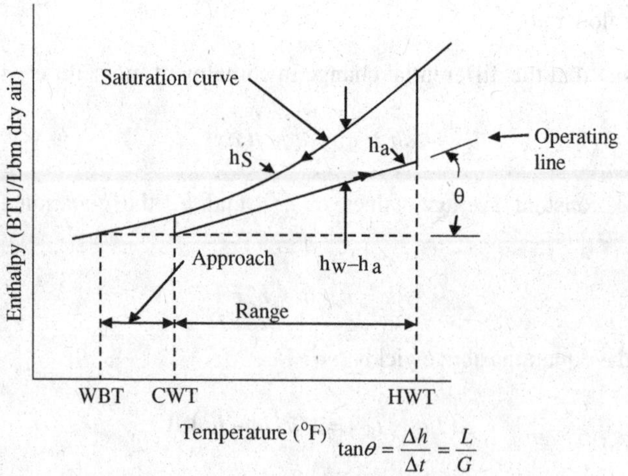

FIGURE 10.8 Psychrometric chart showing the driving potential for heat transfer in a counter-flow cooling tower.

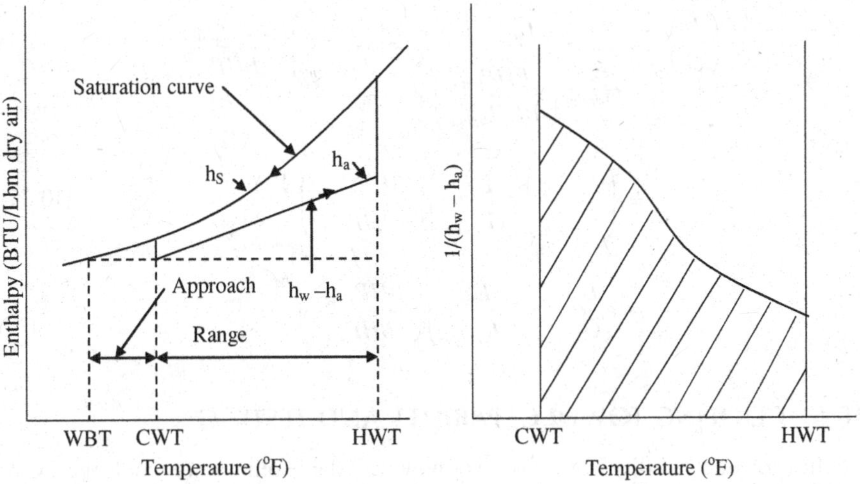

FIGURE 10.9 Integration of the inverse of the driving potential as a function of temperature equals NTU.

10.5 LIQUID-TO-GAS RATIO

Referring to Figure 10.8, the tangent of the ratio of the increase in the enthalpy of the air as it passes through the tower divided by the increase in the temperature of the water is said to be the liquid-to-gas (*L/G*) ratio, a very important parameter. The *L/G* ratio may be determined by dividing the *L* (i.e., CCW flow rate) in gal/min by the *G* (i.e., gas – saturated air) in ft³/min to give 116.5 as shown below

$$L/G = \frac{\dfrac{L\left(\text{gal}/\text{min}\right)60\left(\text{min}/\text{hr}\right)62.0\left(\text{lbm}/\text{ft}^3\right)}{7.48\left(\text{gal}/\text{ft}^3\right)}}{G\left(\text{ft}^3/\text{min}\right)60\left(\text{min}/\text{hr}\right)0.071\left(\text{lbm}/\text{ft}^3\right)} = \frac{L\left(\text{gal}/\text{min}\right)}{G\left(\text{ft}^3/\text{min}\right)} = 116.5 \qquad (10.22)$$

For an MDCT, when comparing test conditions to design conditions, the airflow measured during a test must be corrected for the actual water loading and for the fan horsepower measured during the test according to the fan laws

$$\frac{G_{corrected}}{G_{design}} = \left(\frac{BHP_{test}}{BHP_{design}}\right)^{1/3} \qquad (10.23)$$

$$G_{corrected} = G_{design}\left(\frac{BHP_{test}}{BHP_{design}}\right)^{1/3} \qquad (10.24)$$

$$\left(\frac{L}{G}\right)_{test} = \frac{L_{test}}{G_{design}\left(\frac{BHP_{test}}{BHP_{design}}\right)^{1/3}} = \left(\frac{L_{test}}{G_{design}}\right)\left(\frac{BHP_{design}}{BHP_{test}}\right)^{1/3}$$

$$= \left(\frac{L_{design}}{L_{design}}\right)\left(\frac{L_{test}}{G_{design}}\right)\left(\frac{BHP_{design}}{BHP_{test}}\right)^{1/3} \tag{10.25}$$

$$= \left(\frac{L}{G}\right)_{design}\left(\frac{L_{test}}{L_{design}}\right)\left(\frac{BHP_{design}}{BHP_{test}}\right)^{1/3}$$

10.6 COOLING TOWER CAPABILITY AND TESTING

Cooling tower capability is a measure of how well the cooling tower cools the CCW with respect to the guaranteed performance. An approximate definition for cooling tower capability is the percentage of CCW flow that the cooling tower can cool to the specified CWT at the specified WBT. However, this definition neglects the water loading impact on the airflow rate. A more accurate definition of cooling tower capability is as follows:

$$\text{Tower capability} = \frac{\left(L/G\right)_{test\ corrected}}{\left(L/G\right)_{design}} \tag{10.26}$$

Figure 10.10 illustrates how one determines the design *L/G* of a cooling tower that is designed to achieve a specified approach to the design WBT. The straight diagonal line provided by the cooling tower vendor shows the amount of *KaV/L* (NTU) that a given cooling tower fill material is capable of achieving as a function of *L/G* of the

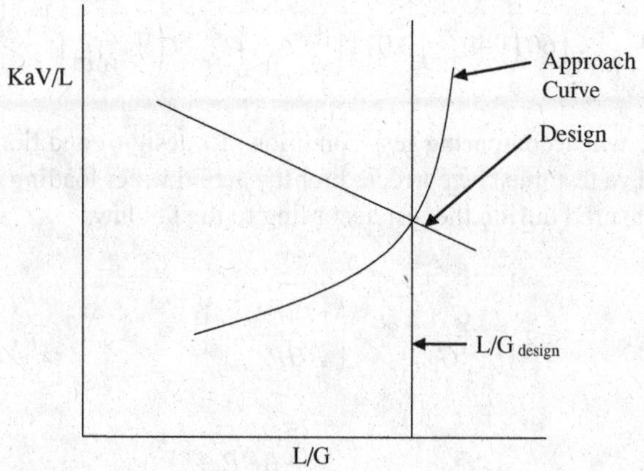

FIGURE 10.10 Design *L/G* for a cooling tower.

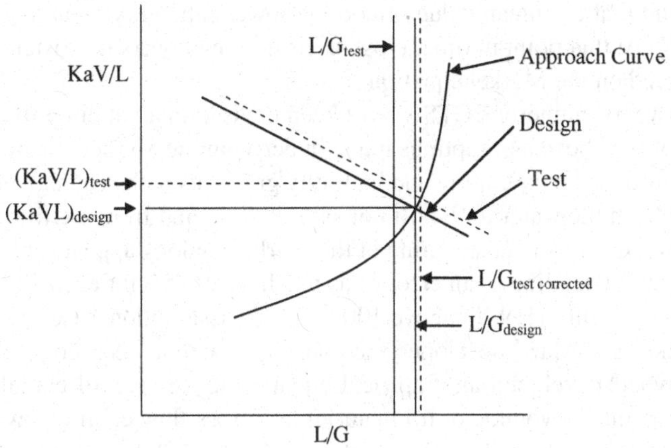

$$Tower\ Capability = \frac{\left(\frac{L}{G}\right)_{test\ corrected}}{\left(\frac{L}{G}\right)_{design}}$$

FIGURE 10.11 Results of cooling tower test.

cooling tower. Note that a lower *L/G* (i.e., less water loading) would result in a higher NTU but would require a larger cooling tower to cool the same amount of CCW. The curved line provided by the Cooling Tower Institute (CTI) shows the estimated *L/G* required to achieve the indicated NTU for a cooling tower that is designed to achieve a specified approach to the design WBT. The intersection of these two lines yields the design *L/G* to achieve the specified approach to the design WBT. However, cooling tower tests are almost never conducted at design conditions. Cooling towers are tested in accordance with the CTI Code ATC-105, Acceptance Test Code for Water-Cooling Towers.

Consider a cooling tower test in which the value of *KaV/L* is determined by measuring the inlet and outlet CCW temperatures. The test results establish a different point on the *KaV/L* vs. *L/G* plot, as shown in Figure 10.11.

The procedure is to draw a second straight line parallel to the vendor-supplied line through the intersection of KaV/L_{test} and L/G_{test}. One may see that because the line representing the test results (dotted) is above the line representing the vendor's design, the test line intersects the approach curve slightly to the right of that representing the design value of *L/G*. The results of the test indicate an *L/G* slightly greater than the design value. The ratio of these two values represents the cooling tower capability relative to the design value.

10.7 COOLING TOWER SIMULATION ALGORITHM (CTSA) DEVELOPMENT

For many years most, if not all, of the heat transfer calculations related to cooling tower design and performance calculations were based on the Merkel equation. A survey of 39 MDCTs and 32 NDCTs conducted by Tennessee Valley Authority (TVA[3]) in 1983 found that the mean value of cooling tower capability demonstrated by the test was 85%. At that point in time, most of the cooling towers in existence had been designed based on the Merkel equation.

Over the years, numerous CTSA have been devised in an attempt to compensate for several of the above assumptions made in deriving the Merkel equation. In 1949, Mickley[4] introduced temperature and humidity gradients with heat and mass transfer coefficients from the water to the film of saturated air and from the film to the bulk stream of air. One major shortcoming of the Merkel analogy appears at high temperatures, potentially leading to an error of as much as 2%–5% for each 10° increase in hot water temperature (HWT) above 100°F due to assumption 1 (see Section 10.2). In 1952, Baker and Mart[5] developed the concept of a "hot water correction factor". In 1955, Snyder[6] developed an empirical equation for an overall enthalpy transfer coefficient per unit of volume of fill material in a cross-flow cooling tower based on tests that he conducted. In 1956, Zivi and Brand[7] extended the analysis of Merkel to cross-flow cooling towers. In 1961, Lowe and Christie[8] performed laboratory studies on several types of counter-flow fill, employing the Merkel analogy in data reduction. In 1975, Hallett[9] presented the concept of a cooling tower characteristic curve of the general form

$$KaV/L = C\left(L/G\right)^{-n} \tag{10.27}$$

where C and n are empirically derived constants.

In 1976, Kelly[10] used the model of Zivi and Brand along with laboratory data to produce a volume of cross-flow cooling tower characteristic curves and demand curves to be used in graphical solutions and design calculations. In 1979, Penney and Spalding[11] introduced a model for NDCT using a finite difference method. In 1981, Majumdar and Singhal[12] extended the model to MDCT.

More recent research has demonstrated that the Lewis factor is not constant but dependent upon the nature of the boundary layer near the exchange surface and the thermodynamic state of the mixture. Numerous attempts have been made to compensate for these factors, but the Lewis factors for wet cooling towers are in a range from 0.5 to 1.3. In 1983, Bourillot[13], conducting research for EPRI, concluded that the Lewis factor for wet cooling towers is approximately 0.92. In 1989, Bell[14] reported on tests conducted by EPRI of eight cross-flow and eight counter-flow fill materials. Based on this test data, Feltzin and Benton[15] concluded in 1991 that for counter-flow fills a Lewis factor of 1.25 is appropriate.

In 1983, several papers on computer codes to predict cooling tower performance were published. Johnson[16] proposed a model based on the NTU-effectiveness approach used for HX. Bourillot[17] developed the TEFERI model based on heat and mass transfer equations similar to Zivi and Brand. The TEFERI model for counter-flow towers is a one-dimensional code in which water and air temperatures and flow rates at the inlet are assumed to be uniform. However, the code calculates the loss of water due to evaporation, so the water flow rate does not remain uniform as it passes through the cooling tower. Benton[18] developed the Fast Analysis Cooling Tower Simulator (FACTS) model that employs an integral formulation of the equations for the conservation of the mass of air and water vapor, conservation of energy, and the Bernoulli equation to arrive at a numerical solution apart from the Merkel analogy. FACTS is a steady-state, steady-flow model that is more sophisticated than a one-dimensional model but contains simplifications that prevent it from being classified as a true two-dimensional code. FACTS can accommodate variable inlet water and air temperatures and hybrid fills. However, the assumption that was made by Merkel that the loss of water due to evaporation is neglected is also incorporated into the FACTS model. Figure 10.12 shows the FACTS model agreement with a cross-flow MDCT Browns Ferry Nuclear (BFN), a cross-flow NDCT Sequoyah Nuclear (SQN), and a counter-flow NDCT Paradise Fossil Plant.

FACTS is widely used by utilities to model cooling tower performance.

In 1984, Lefevre[19] revisited the derivation of the Merkel equation and evaluated the errors of several of the assumptions made. He also revisited the original basis for the Merkel equation that is the energy balance

$$L\,c_{p(w)}dt_w + c_{p(w)}t_w dL = G\,dh_a = K\left(h_s - h_a\right)a\,dV \tag{10.28}$$

This equation simply states that the heat loss from the water, which is the water flow rate times the specific heat of water times the change in the water temperature plus the heat lost by evaporation, is equal to the airflow rate time the change in the air enthalpy, which is equal to the mass transfer coefficient times the enthalpy difference

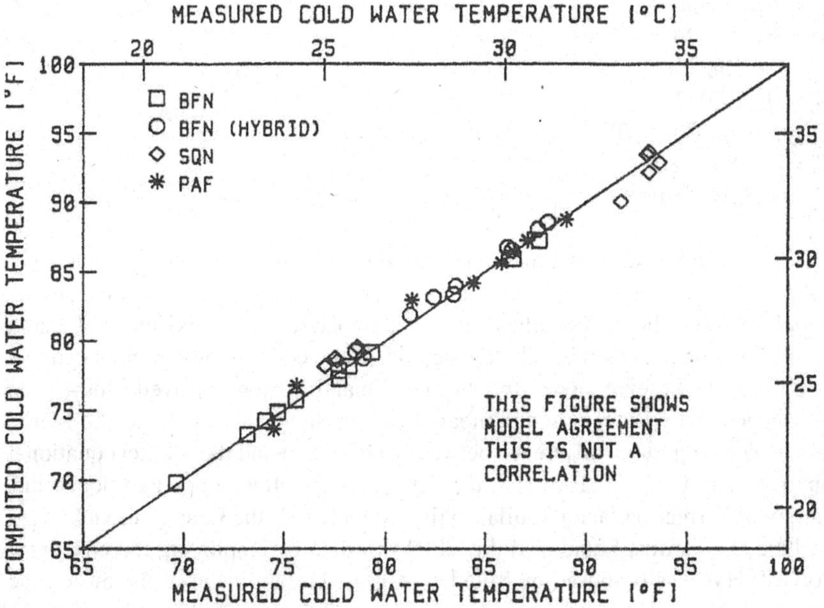

FIGURE 10.12 FACTS verification.

times the interface area per unit of volume times the incremental cooling volume. Merkel used the left and right sides of this equation and arrived at his equation by assuming L to be constant and dL to be zero. Lefevre pointed out that one may just as easily use the middle and right-hand portion of this equation and by dividing by the inlet water flow, L_{in}, (with units of lbm) to arrive at

$$\left(G/L_{in}\right)\mu\, dh_a/\left(h_s-h_a\right)=Ka\,V/L_{in} \tag{10.29}$$

Lefevre pointed out that one may as easily integrate the right side of this equation as the Merkel equation and not have to assume L to be constant and dL to be zero. However, this model continues to suffer from the other deficiencies associated with the other approximations made by Merkel, including its shortcomings at higher water temperatures. Therefore, Lefevre applied a dimensionless Merkel correction factor, M, such that

$$M\left(G/L_{in}\right)\mu dh_a/\left(h_s-h_a\right)=Ka\,V/L_{in} \tag{10.30}$$

where

$$M=1+\exp\left\{C_0+t_w\left[C_1+t_w\left(C_5+C_6t_w\right)\right]+t_a\left(C_3+C_2t_w+C_4t_a\right)\right\} \tag{10.31}$$

and

$C_0 = -7.845656$
$C_1 = 0.0730229$
$C_2 = -6.29829 \times 10^{-5}$
$C_3 = 0.0112047$
$C_4 = -7.646384 \times 10^{-7}$
$C_5 = -0.00038804$
$C_6 = 1.26972 \times 10^{-6}$.

Lefevre also pointed out that a more exact model could be employed by not neglecting evaporation so that the driving potential could be expressed as an air enthalpy potential. However, he did not advocate this approach due to the existence of a large body of data generated by the enthalpy method. Many cooling tower vendors employ some form of the Lefevre model. In 1991, Feltzin and Benton[20] derived a more exact model suggested by Lefevre and compared the results of this model to the Merkel equation. They found the difference between their model and the Merkel equation to be on the order of 1%–3%. However, the Feltzin and Benton model does not include an empirical correction factor similar to that contained in the Lefevre model.

In 1992, Desjardins[21] analyzed the EPRI test data by employing the concept of an "offset" HWT as proposed by Mickley in 1949. Desjardins used the more exact model suggested by Lefevre and the Gaussian quadrature method of integration and reported that the resulting model results in an improvement in correlation with the EPRI data.

10.8 TECHNICAL BASIS FOR FACTS

The following are the other major assumptions made in FACTS:

1. The flow of air is two-dimensional in the fill region of a cross-flow tower and one-dimensional in the fill region of a counterflow tower.
2. WBT is equivalent to the adiabatic saturation temperature.
3. The cooling tower is externally adiabatic.
4. The atmosphere around the NDCT is isentropic.
5. The water flows vertically downward inside the tower.
6. Evaporation loss is neglected in the water mass balance.

The FACTS model is based on the conservation of the mass of air and water vapor as well as the conservation of energy for the gas phase and energy for the water phase. These conservation equations in conjunction with the Bernoulli equation constitute the set of equations that are solved by FACTS to simulate cooling tower performance. The form of the Bernoulli equation used is

$$p_1 + \Delta_1 v^2{}_1 / 2g_c + \Delta_1 g y_1 / g_c = p_2 + \Delta_2 v^2{}_2 / 2g_c + \Delta_2 g y_2 / g_c + \text{losses} \tag{10.32}$$

where the subscripts 1 and 2 represent two locations along a streamline and
p = pressure
Δ = density
V = total velocity
g_c = Newton's constant
g = acceleration of gravity.

From Section 10.2,

$$dQ_s = H\left(t_W - t_{db}\right) dA_i \tag{10.33}$$

$$dE = K\,B\,dA_i \tag{10.34}$$

$$dQ_e = h_g dE \tag{10.35}$$

It is convenient to express these equations in terms of the absolute humidity, T, and to let

$$dV = dx\,dy\,dz \tag{10.36}$$

so that the three equations of interest are then

$$dQ_s = Ha\left(t_W - t_{db}\right) dx\,dy\,dz \tag{10.37}$$

$$dE = Ka[(T_S - T)/(1+T)]\,dx\,dy\,dz \tag{10.38}$$

$$dQ_e = h_g\, Ka\ [(T_S - T)/(1+T)]dx\, dy\, dz \tag{10.39}$$

These equations are applied in their steady-state, steady-flow form. The independent variables are the horizontal distance (x), vertical distance (y), total mass flow rate of water, inlet HWT, and the ambient wet- and dry-bulb temperatures. The dependent variables in the conservation equations are air velocity, absolute humidity, the enthalpy of the air–water vapor mixture, the water temperature, and pressure. The WBT and dry-bulb temperature are determined using the following thermodynamic relationships for air–water vapor mixtures from computed values of t, h_a, and p. The interrelationship among the dependent and independent variables is evident from the formulation of the conservation equations that follow. The conservation of mass for the water vapor within a control volume (CV) is expressed as follows:

$$IIIKa[(T_S - T)/(1+T)]dx\, dy\, dz = II[T\Delta/(1+T)]V * dA \tag{10.40}$$

where:

$V*dA$ = dot product of the two vectors V and dA.

For conservation of energy for the air within a CV is as follows:

$$III\{h_g Ka[(T_S - T)/(1+T)] + Ha\left(t_W - t_{db}\right)\}\, dx\, dy\, dz = II[h_a\Delta/(1+T)]V * dA \tag{10.41}$$

Finally, the conservation of energy for the water within a CV is as follows:

$$Lc_{pw}dt_w = -III\{h_g Ka[(T_S - T)/(1+T)] + Ha\left(t_W - t_{db}\right)\}\, dx\, dy\, dz \tag{10.42}$$

Simulation of the mass, momentum, and heat transfer processes in the cooling tower requires that the tower be discretized, or divided into computational cells. Each cell is treated as a CV, and the governing equations are applied to each. At each cell, the computed dependent variables from the adjacent upstream cells are utilized. These variables are defined at nodes located at the midpoints of the cell boundaries. The use of boundary nodes assures the conservation of mass and energy from cell to cell. Applying the Bernoulli equation and conservation equations to each cell results in a set of nonlinear simultaneous equations relating to the dependent variables. These implicit nonlinear simultaneous equations are solved using the Gauss–Seidel method.

For counter-flow towers, the air is assumed to flow between collinear hyperboloid pathlines. The fraction of air mass flow between each pathline is computed and reflects flow resistance in both the fill and the rain zones. The pressure drop and transfer characteristics of the fill are integrated in the radial direction to obtain average values. These are weighted by the velocity head, airflow, and water flow. These average values are used with the one-dimensional integral conservation equations.

For cross-flow towers, the airflow distribution is evaluated using the Bernoulli equation (with head loss) and the conservation of mass for air. These equations are applied to each computational cell.

The specified inlet conditions of both air and water (temperatures and flows) can vary across the inlet plane. FACTS requires as input a sensible heat transfer coefficient, *H*, and a mass transfer coefficient, *K*, which are a function of the fill characteristics as input. FACTS can model towers containing hybrid fills or fills that have voids or obstructions. FACTS allows for the input of separate correlations for spray and rain regions in counter-flow towers.

REFERENCES

1. Merkel, F. *Verduftungskuhlung*. VDI Forschungsarbeiten, Berlin, Germany, no. 275, 1925.
2. Chebyshev, P. L. Théorie des mécanismes connus sous le nom de parallélogrammes, *Mémoires des Savants étrangers présentés à l'Académie de Saint-Pétersbourg*, vol **7**, 1854, pp. 539–586.
3. Boroughs, R. D., and J. E. Terrell, *A Survey of Utility Cooling Towers*, Tennessee Valley Authority, Chattanooga, TN, April 1983.
4. Mickley, H. S., Design of Forced Draft Air Conditioning Equipment, *Chemical Engineering Progress*, vol. 45, 1949, p. 739.
5. Baker, R. and L. T. Mart, The Merkel Equation Revisited, *Refrigeration Engineering*, 1952, p. 965.
6. Snyder, N. W., Effect of Air Rate, Water Rate, Temperature, and Packing Density in a Cross-flow Cooling Tower, *Chemical Engineering Progress*, vol. 52, no. 18, 1955, p. 61.
7. Zivi, S. M. and B. B. Brand, An Analysis of the Cross-Flow Cooling Tower, *Refrigeration Engineering*, vol. 64, 1956, pp. 31–34.
8. Lowe, H. J. and D. G. Christie, *Heat Transfer and Pressure Drop in Cooling Tower Packing and Model Studies of the Resistance of Natural-Draft Towers to Airflow, paper at International Division of Heat Transfer*, Part V, p. 933, ASME, New York, 1961.
9. Hallett, G. F., Performance Curves for Mechanical Draft Cooling Towers, *Journal of Engineering for Power*, vol. 97, 1975, pp. 503–508.
10. Kelly, N. W., *Kelly's Handbook of Crossflow Cooling Tower Performance*, Neil W. Kelly and Associates, Kansas City, MI, 1976.
11. Penney, T. R. and D. B. Spalding, *Validation of Cooling Tower Analyzer (VERA)*, Vols. 1 and 2, EPRI Report FP-1279, Electric Power Research Institute, Palo Alto, CA, 1979.
12. Majumdar, A. K. and A. K. Singhal, *VERA2D-A Computer Program for Two-Dimensional Analysis of Flow, Heat and Mass Transfer in Evaporative Cooling Towers, Vol. II-User's Manual*, Electric Power Research Institute, Palo Alto, CA, 1981.
13. Bourillot, C., *On the Hypotheses of Calculating the Water Flowrate Evaporated in a Wet Cooling Tower*, EPRI CS-3144-SR, Electric Power Research Institute, Palo Alto, CA, 1983a.
14. Bell, D. M., B. M. Johnson, and E. V. Werry, *Cooling Tower Performance Prediction and Improvement*, EPRI GS-6370, Electric Power Research Institute, Palo Alto, CA, 1989.
15. Feltzin, A. E. and D. J. Benton, *A More Nearly Exact Representation of Cooling Tower Theory*, Cooling Tower Institute, Houston, TX, 1991.
16. Johnson, B. M., D. K. Kreid, and S. G. Hanson, A Method of Comparing Performance of Extended-Surface Heat Exchangers, *Heat Transfer Engineering*, vol. 4, no. 1, 1983, pp. 32–42.
17. Bourillot, C., *TEFERI, Numerical Model for Calculating the Performance of an Evaporative Cooling Tower*, EPRI CS-3212-SR, Electric Power Research Institute, Palo Alto, CA, 1983.

18. Benton, D. J., *A Numerical Simulation of Heat Transfer in Evaporative Cooling Towers*, Report WR28-1-900-110, Tennessee Valley Authority, Knoxville, TN, 1983.
19. Lefevre, M. R., "*Eliminating the Merkel Theory Approximations-Can It Replace the Empirical 'Temperature Correction Factor'?*", Cooling Tower Institute, Houston, TX, 1984.
20. Feltzin, A. E. and D. J. Benton, *A More Nearly Exact Representation of Cooling Tower Theory*, Cooling Tower Institute, Houston, TX, 1991.
21. Desjardins, R. J., *Using the EPRI Test Data to Verify a More Accurate Method of Predicting Cooling Tower Performance*, Cooling Tower Institute, Houston, TX, 1992.

11 Cooling Lakes

11.1 ADVANTAGES OF COOLING LAKES

A number of nuclear stations have been designed with a cooling lake to dissipate the waste heat in the CCW into the atmosphere. In areas where inexpensive land is abundant and where the topography is favorable, a dedicated cooling lake can be an attractive design feature, as it may provide an economic alternative to cooling towers. As discussed in Chapter 9, the advantages of a cooling lake include low maintenance and a low CCW pump head. Additionally, the inherent thermal inertia with a cooling lake, where the residence time is normally measured in days, prevents rapid swings in CCW inlet temperatures in response to changing meteorological conditions inherent to cooling towers. Cooling lakes also cause significantly less severe localized fogging and icing incidents than do MDCT. Some utilities have permitted private development around their cooling lakes at nuclear stations. However, in some cases, these lakes, especially those that were formed by impounding free-flowing streams, have been declared to be waters of the United States, resulting in states imposing thermal limitations on the CCW discharge to protect the aquatic life in the lake.

11.2 SIZING COOLING LAKES

As is the case with any heat dissipation device, determining the size of the cooling lake is a classical tradeoff of capital cost against future gains in electrical output as determined by economic policy and forecasting the value and future demand for electrical energy. The geographic location, which determines the net solar radiation and ambient wind speed (AWS) (two of the most important parameters in sizing a cooling lake), plays a significant role in determining the economic optimum cooling lake size. Another consideration is the shape of the proposed cooling lake. Normally, for a cooling lake that is arranged such that the CCW flow passes through the great majority of the lake, the effective surface area and volume would approach true area and volume. However, a cooling lake resulting from the impoundment of existing streams often has coves that are not in the direct flow path from the plant discharge to intake. As a result, some if not much of this area and volume of these coves must be discounted when estimating the heat rejection capability of the lake. The size of the existing cooling lake for a nuclear station is typically 1.0–1.5 acres per megawatt of electrical output.

11.3 DETERMINING THE AVERAGE COOLING LAKE SURFACE TEMPERATURE

The average temperature of a cooling lake required to dissipate the waste heat from a nuclear station is a function of the effective area, A, the equilibrium temperature, E, and the surface heat exchange coefficient, K. E is the water surface temperature at which no heat storage would occur if there were no heat added to the cooling lake by the nuclear station. Stated differently, E is the temperature that would be achieved if a pan of water were set out in the sun.

In 1965, Edinger and Geyer[1] showed that K may be defined as follows:

$$K = 15.7 + (\beta + 0.26) f(W) \tag{11.1}$$

where K is in units of Btu/ft²/day/°F.

The first term, 15.7, is the known background radiation. The β in the second term is the slope of the saturated vapor pressure curve shown in Figure 11.1 as defined by the following:

$$\beta = \frac{e_s - e_a}{AST - DPT} \tag{11.2}$$

where e_s is the saturated vapor pressure at the lake surface temperature, and e_a is the air vapor pressure in mmHg, and AST and DPT are the average surface temperature of the lake and the ambient dewpoint temperature, respectively in °F.

The 0.26 is the Bowen's conduction-evaporation coefficient in mmHg/°F, and the last term, $f(W)$, is the wind speed function.

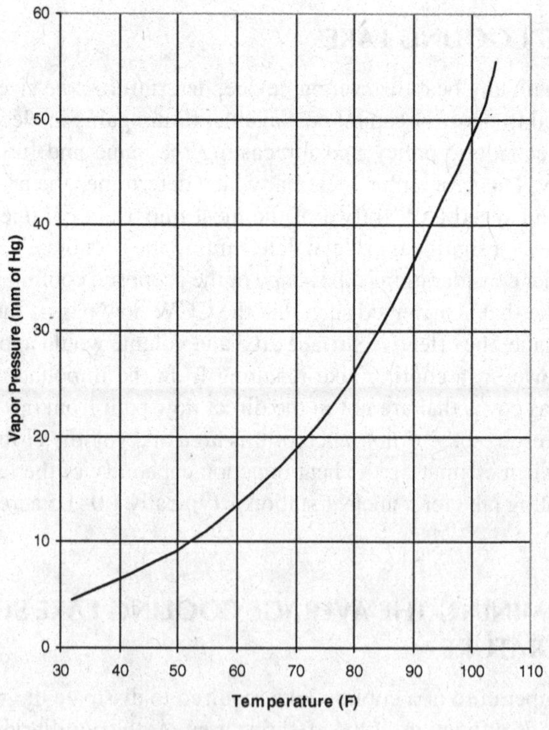

FIGURE 11.1 Vapor pressure changes with temperature.

Edinger and Geyer also concluded that the equilibrium temperature is simply the following:

$$E = DPT + \frac{H_{SN}}{K} \tag{11.3}$$

where H_{SN} is the net solar heating in Btu/ft^2/day and that

$$AST = E + \frac{H_{rj}}{K} \tag{11.4}$$

where H_{rj} is the plant heat rejection in Btu/ft^2/day.

Therefore, for a given set of ambient conditions (*DPT*, H_{SN}, and *AWS*), the *AST* required to dissipate the waste heat from a nuclear station is a function of β and $f(W)$.

In 1969, Brady et al.[2] after considering 10 possible sites, collected data at three cooling lakes in Texas. Site No. 3 at Coughlin Station is a shallow lake with many tree-shaded coves and negligible inflow under normal conditions. Site No. 7 is Smith Lake, a large shallow lake that serves the A. W. Parish Station, is roughly triangular with only a few small coves. Site No.11 at Wilkes Station is a deep lake with two main branches that are sheltered by a tree-covered hilly terrain and negligible inflow under normal conditions. The plant CCW is discharged near the end of one branch and withdrawn near the end of the other branch.

The following approximation of the slope of the vapor pressure curve is adopted:

$$\beta = 0.255 - 0.0085\, t_\beta + 0.000204\, t_\beta^2 \tag{11.5}$$

where

$$t_\beta = \frac{AST + DPT}{2} \tag{11.6}$$

A number of studies conducted by investigators to determine $f(W)$ assumed that the function was linear with *W*, with some assuming that the value was zero at zero wind speed. However, the 1969 data in Reference 2 showed that $f(W)$ was not zero at zero wind speed. Just as with a cooling tower, the humidity gained by the air over the water results in lighter air producing an updraft and wind even with no ambient wind. The function is best characterized by a polynomial regression of the form

$$f(W) = a_0 + a_1 + a_2^{\;2} + \cdots \tag{11.7}$$

After performing multiple regression analyses to address factors such as cloudiness, changes in heat storage in the lakes, the time periods of measurements, etc., Edinger and Geyer[1] determined by measuring the *AST* in the lakes that the appropriate expression for estimating the wind speed function is as follows:

$$(W) = 70 + 0.7\, W^2 \tag{11.8}$$

where *W* is in mi/hr and $f(W)$ is in units of Btu/ft^2/day/mmHg.

Therefore, one may estimate the average temperature of a cooling lake required to dissipate the waste heat from a nuclear station as follows:

$$\text{Assume } T_S \quad \text{then} \quad t_\beta = \frac{AST + DPT}{2} \tag{11.9}$$

$$\beta = 0.255 - 0.0085\, t_\beta + 0.000204\, t_\beta^2 \tag{11.10}$$

$$f(W) = 70 + 0.7\left(W^2\right) \tag{11.11}$$

$$K = 15.7 + (\beta + 0.26)\, f(W) \tag{11.12}$$

$$E = T_d + \frac{H_{SN}}{K} \tag{11.13}$$

$$T_S = E + \frac{H_{rj}}{K} \tag{11.14}$$

Of course, this process requires iteration because one must assume T_s in order to calculate T_s.

11.4 DETERMINING THE CCW INTAKE TEMPERATURE

Of course, the nuclear station operators are not interested in the *AST* but in the CCW intake temperature, which affects the main condenser backpressure and the electrical output of the plant. For specified meteorological conditions, the degree to which the lowest intake temperature is achieved is a function of the amount of longitudinal mixing (i.e., stratification) that occurs and the uniformity of flow distribution (i.e., no short-circuiting) between the plant discharge and intake. Both of the phenomena tend to convey warmer water from the discharge to the intake.

Figure 11.2 shows an illustration of a longitudinally mixed cooling lake. Brady et al.[2] showed that if the intake were located on the surface of the lake, the intake temperature could be computed as follows:

$$T_{CCW\text{-}in} = E + \theta \tag{11.15}$$

where

$$\theta = T_C \times NTU \tag{11.16}$$

and

$$T_C = T_{CCW\text{-}out} - T_{CCW\text{-}in} \tag{11.17}$$

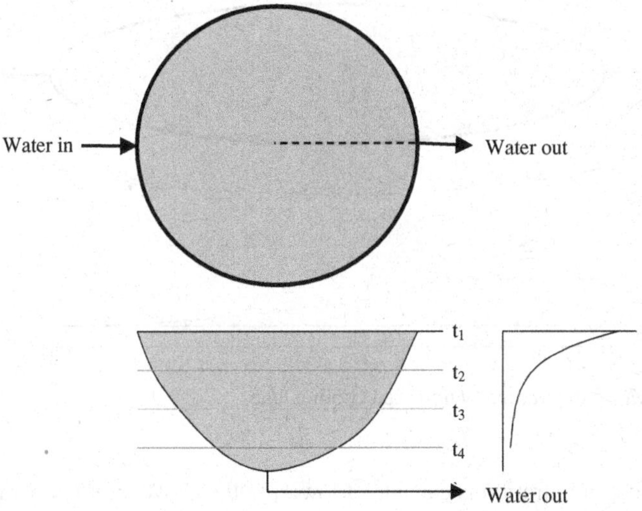

FIGURE 11.2 Longitudinally mixed cooling lake.

and

$$NTU = \frac{L\left(\text{gal}/\text{min}\right)\rho\left(\text{lbm}/\text{ft}^3\right)c_p\left(\text{Btu}/\text{lbm-°F}\right)1440\left(\text{min}/\text{day}\right)}{K\left(\text{Btu}/\text{ft}^2\text{-day-°F}\right)A\left(\text{ft}^2\right)7.48\left(\text{gal}/\text{ft}^3\right)} \tag{11.18}$$

Of course, if the intake were located on the bottom of the lake as shown, the CCW intake temperature would be reduced by the amount of the stratification in the lake.

Figure 11.3 shows an illustration of a longitudinally unmixed cooling lake. Longitudinally unmixed cooling lakes are normally relatively shallow, though deep enough that solar radiation does not reach the bottom of the lake but is reflected back as much as possible. These lakes are normally designed for uniformity of flow distribution where possible such that the "plug" flow of uniform transverse temperature progresses from one end to the other, decreasing as it progresses, as shown in Figure 11.3.

Brady et al.[2] showed that the intake temperature could be computed as follows:

$$T_{CCW\text{-}in} = E + \theta \tag{11.19}$$

where

$$\theta = T_C\left(\frac{X}{1-X}\right) \tag{11.20}$$

and

$$X = e^{-\frac{1}{NTU}} \tag{11.21}$$

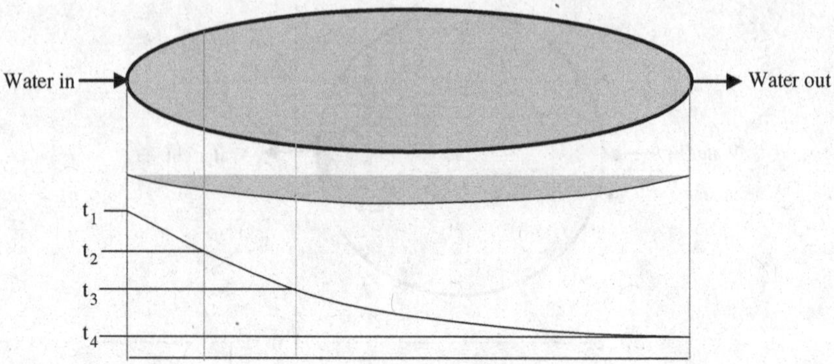

FIGURE 11.3 Longitudinally unmixed cooling lake.

Neither lake configuration is capable of lowering the CCW inlet temperature below the value of E. However, in the absence of significant stratification, the longitudinally unmixed configuration shown in Figure 11.3 would produce a much lower temperature because in the limit it would be capable of approaching the value of E. Without stratification, the longitudinally mixed configuration shown in Figure 11.2 would not be capable of approaching the value of E regardless of how large it might be.

11.5 DETERMINING THE EFFECTIVE SURFACE AREA

A critical parameter in performing the calculations described in Sections 11.3 and 11.4 is the effective surface area, A. With a well-configured cooling lake, the effective surface area would approach 90% of the actual surface area if dikes were employed to channel the CCW to avoid longitudinal mixing and short-circuiting between the plant discharge and intake. However, in practice, cooling lakes formed by impounding streams often have narrow coves where the heat transfer to the ambient is less than optimum. This is especially true where these coves are long, narrow, and shallow. The analysis of such configurations requires finite element techniques that are beyond the scope of this book but will be briefly discussed in the next section.

As a practical matter, if the shape of the cooling lake approximates one of the configurations described in Section 11.4, one may over time arrive at an approximate effective surface area of the cooling lake by recording the ambient conditions, the CCW intake temperature, and the amount of heat rejected by the nuclear station (see Section 6.3). By iteration, the effective area of the cooling lake that produces the measured CCW intake temperature may be determined.

11.6 NUMERICAL ANALYSIS OF COOLING LAKES

Quite obviously, the limitations imposed on the use of the relatively simple analytical tools described in Sections 11.3 and 11.4 are insufficient to address many cooling lake applications, and modern numerical analyses must be employed. The need for more rigorous analysis techniques was first recognized by the U.S. Atomic Energy

Commission (AEC) in conjunction with the proposed emergency cooling ponds to be used to safely shut down nuclear reactors. In 1972, Edinger et al.[3] employed the analysis methods described in Sections 11.3 and 11.4 along with two computer programs, Meteorological and Minimum Pond Computations and Hydrodynamic and Excess Temperature Analysis, in a report to the AEC entitled *Generic Emergency Cooling Pond Analysis*. For many years, these programs became the "gold standard" for licensing nuclear stations employing cooling ponds as their ultimate heat sink (UHS).

In the late 1970s and early 1980s, Buchak and Edinger[4,5] developed the Generalized Longitudinal, Lateral, and Vertical Hydrodynamic and Transport Model (GLLVHT) for J. E. Edinger Associates, Inc. (now a division of Environmental Resource Management, Inc.) GLLVHT is a time-varying three-dimensional (3D) numerical model in which the cooling lake is broken up into 3D grids, illustrated in Figure 11.4 for the Squaw Creek Lake at the Comanche Peak Nuclear Power Plant. The velocities and temperatures are computed at each grid point in the 3D representation of the cooling lake at approximately 1-min intervals in the simulation. Calibrating GLLVHT requires the review and/or collection of large amounts of data and the conduct of tests to establish E and K (both summer and winter values). These data including meteorological data (*DBT, DPT, AWS*, cloud cover, etc.), and CCW flow

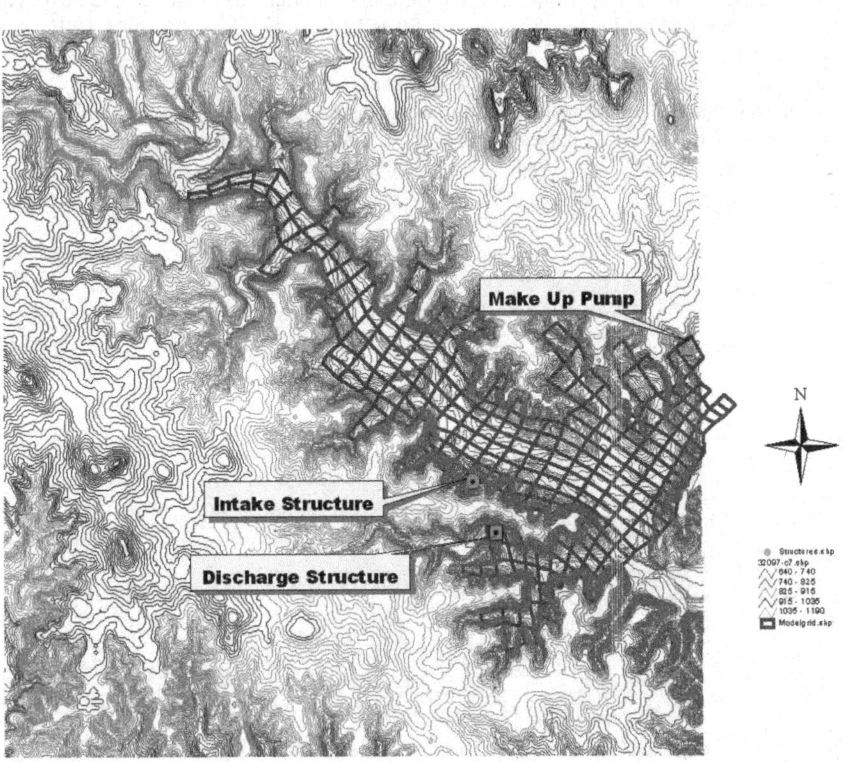

FIGURE 11.4 Grids for a numerical model of a cooling lake.

rates are used to compute a basic water/heat balance, taking into account inflows and outflows from the cooling lake. Calibrated battery-operated thermistors are used to measure the cooling lake temperatures at various points and various depths throughout the cooling lake. When this data is fed into GLLVHT, the program computes the temperature rise throughout the cooling lake as a result of the waste heat added by the nuclear station. The resulting model may be used to predict the CCW intake temperature based on projected meteorological data and to estimate the impact of modifications to the cooling lake such as the addition of dikes, changes in the surface area, and a submerged intake.

REFERENCES

1. Edinger, J. E. and J. C. Geyer, Heat Exchange in the Environment, Edison Electric Institute Report No. 65-902, June 1965.
2. Brady, D. K., W. L. Graves, Jr., and J. C. Geyer, Surface Heat Exchange at Power Plant Cooling Lakes, Cooling Water Discharge Project Report No. 5, Edison Electric Institute Publication No. RP-49, 1969.
3. Edinger, J. E., E. M. Buchak, E. Kaplan, and G. Socratos, Generic Emergency Cooling Pond Analysis – Emergency Cooling Pond Analysis and the Theoretical Basis of the GEPA Computational Program, Prepared for the U.S. Atomic Energy Commission, University of Pennsylvania, 1972.
4. Buchak, E. M. and J. E. Edinger. Hydrothermal Simulations of Comanche Peak Safe Shutdown Impoundment, Prepared for Texas Utilities Services, Inc. by J. E. Edinger Associates, Inc., 1980.
5. Buchak, E. M. and J. E. Edinger. Generalized, Longitudinal-Vertical Hydrodynamics and Transport: Development, Programming and Applications, Prepared for U.S. Army Corps of Engineers Waterways Experiment Station, Vicksburg, Mississippi by J. E. Edinger Associates, Inc., 1984.

12 Spray Pond Designs and Testing

12.1 ADVANTAGES OF SPRAY PONDS

A spray pond is a system of pipes and spray nozzles that spray water into the air to cool the water. They are similar in this way to cooling towers in that they dissipate waste heat into the atmosphere principally through evaporation. Spray ponds are employed at several nuclear stations as the UHS because they are inexpensive to maintain and the quantity of water required to ensure the safe shutdown of the station may be stored in the pond(s), which doubles as the site for the spray system. Much of the advancement in the understanding of the design of spray ponds has been gained because they play a critical role in the safe operation of these nuclear stations. Spray ponds require only about one-tenth of the area of a cooling lake. Unlike MDCT, spray ponds need no fan power or fill maintenance. However, predicting the thermal performance of a large spray pond has been an elusive goal. In some instances, fallacious assumptions have produced disastrous results.

12.2 CONVENTIONAL FLATBED SPRAY PONDS

Figure 12.1 shows the conventional flatbed spray pond (FBSP) that was the UHS for the now-defunct Rancho Seco nuclear station.[1] Typically, an FBSP consists of a series of trees mounted on straight header pipes in a rectangular pattern with each tree consisting of a riser pipe and four cross arms at 90° angles approximately 5 ft long with a spray nozzle at the end of each cross arm pointed vertically. The dimension of each of the two FBSP at Rancho Seco is 165 ft by 330 ft, with 304 nozzles, each delivering 53 gal/min at 7 lbf/in^2 nozzle pressure.

Figure 12.2 illustrates the flow pattern of two out of four spray nozzles on a conventional FBSP tree.

12.3 ORIENTED SPRAY COOLING SYSTEM

Figure 12.3 illustrates the flow pattern out of an oriented spray cooling system (OSCS). An OSCS consists of a series of trees mounted on a single circular header pipe with each tree comprising a riser pipe and several 4-ft-long arms spaced up the riser pipe at 90° angles to the riser pipe, with a spray nozzle at the end of each cross arm tilted toward the central axis of the circle. The OSCS employs the same type of spray nozzle as those used in an FBSP. However, the operating pressure and flow rate from each nozzle vary, depending on the elevation of the particular nozzle, but, in general, all of the nozzles operate at a higher pressure and deliver more flow than is the case with an FBSP.

FIGURE 12.1 Spray nozzles in a conventional flatbed spray pond.

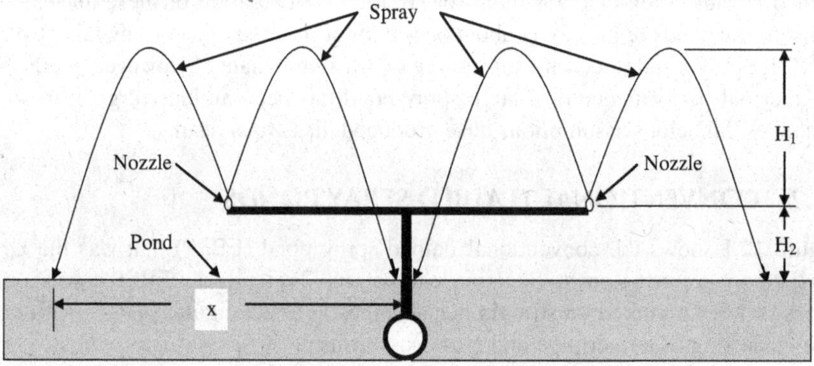

FIGURE 12.2 Spray from a conventional flatbed spray tree.

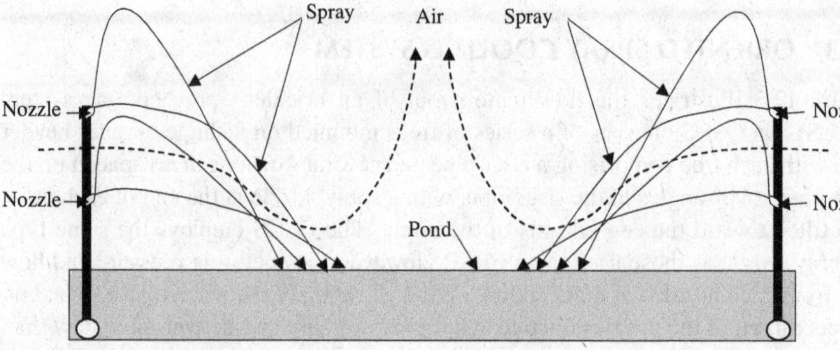

FIGURE 12.3 Spray from an oriented spray cooling system.

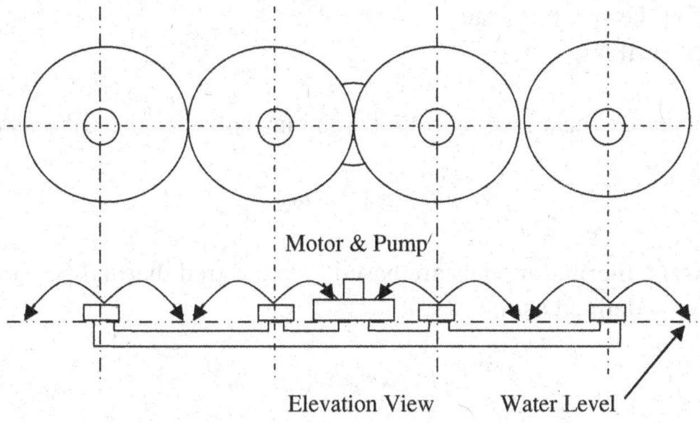

FIGURE 12.4 Power spray module.

12.4 POWER SPRAY MODULES

Figure 12.4 illustrates a power spray module (PSM). A PSM floats in a canal that transmits the CCW to the nuclear station intake. Each PSM is powered by a pump and motor located on the floating device. In the illustration shown in Figure 12.4, each pump and motor serve four spray modules connected by piping. With some PSM devices, each spray module has its own pump and motor and operates independently from the other modules. The PSMs shown in Figure 12.4 are spaced 40 ft apart. Each spray module delivers from 1,000 to 2,500 gal/min. In the case of the PSM shown in Figure 12.4, each pump delivers a total of 10,000 gal/min and requires a 75 hp motor. The sprayed drop size from a PSM is much larger than that from an FBSP or OSCS. Accordingly, the cooling achieved is much less, requiring the CCW to be pumped repeatedly through multiple PSMs on its way to the nuclear station intake, so many modules such as the ones shown in Figure 12.4 would be required.

12.5 PREDICTING THE THERMAL PERFORMANCE OF SPRAY PONDS

Much attention in both academic and industrial communities has been given to experimental and analytical investigations of spray cooling devices as a result of their widespread usage and the shortcomings of existing models.[2–4] The simplest measure of thermal performance is the efficiency, η, of the device as defined by the following standard formula:

$$\eta = \frac{t_1 - t_2}{t_1 - t_{WB}} \tag{12.1}$$

where:

η = spray efficiency of one spray nozzle
t_1 = HWT of the water entering the spray nozzle

t_2 = *CWT* of the sprayed water
t_{WB} = ambient *WBT*.

Assuming a constant heat load and *WBT*, one may solve for the *CWT* as follows:

$$t_2 = t_1 - \eta\left(t_1 - t_{WB\text{-}local}\right) \tag{12.2}$$

To a large extent, thermal models are based on measured thermal performance data such as the *NTU* defined as follows:[5,6]

$$NTU = \mu\, c_p dt/\left(h_s - h_a\right) \tag{12.3}$$

This is, of course, the Merkel equation described in detail in Chapter 10, where *NTU* is the equivalent of *KaV/L* in cooling towers.

Several investigators considered separately some real effects such as *WBT* degradation, internal resistance, buoyancy, and interference.[7–9] The first significant full-scale test of an FBSP was conducted by Schrock and Trezek[1] in 1973 at Rancho Seco. In 1976, Myers and Baird[10] conducted similar tests on the FBSP at the Okeelanta complex in Florida. Also, in 1976, Yang and Porter[11] conducted tests on a PSM system at the Dresden and Quad Cities nuclear stations. At Quad Cities, two different types of floating spray modules were tested, but both had similar spray characteristics. Each spray nozzle sprayed 2,500 GPM about 16 ft high with relatively large drops compared with the Rancho Seco spray pond.

12.6 SPRAY POND TEST RESULTS

Only a few quality full-scale spray pond efficiency tests are available in the open literature. In 1974, an analysis of the test data taken at Rancho Seco was performed to determine the local wet-bulb temperature (*LWBT*) as a function of the distance into the spray field.[12] In 1979, Conn[13] published the results of tests conducted on one of the two OSCS that serve as the UHS for the Columbia Generating Station (CGS). Table 12.1 shows the data taken during the Rancho Seco test.

TABLE 12.1
Rancho Seco Flatbed Spray Pond Test Data

Test No.	*AWS*, mi/hr	*WBT*, °F	*HWT*, °F	*CWT*, °F
1	1.0	54.1	101.5	84.7
2	1.0	48.6	77.4	69.1
3	6.7	69.6	81.1	77.2
4	6.9	72.3	80.1	77.0
5	8.3	66.6	80.8	74.3
6	12.5	61.5	80.1	71.2
7	13.0	61.0	79.9	72.0

TABLE 12.2
Okeelanta Flatbed Spray Pond Test Data

Test No.	*AWS*, mi/hr	*WBT*, °F	*HWT*, °F	*CWT*, °F
1	7	62	93	80
2	4	65	101	89
3	10	65	94	82
4	5	68	100	87

Table 12.2 shows the data taken during the Okeelanta test. Table 12.3 shows the data taken during the CGS test. Figure 12.5 shows the comparison among tests conducted on the conventional FBSPs at Rancho-Seco and Okeelanta Florida complex and the OSCS at the CGS. One may see from Figure 12.5 that the spray efficiency increases with an increase in *WBT*. This is the case for any type of evaporative cooling device such as a cooling tower.

Figure 12.6 shows the significant impact that *AWS* has on FBSP, as the spray efficiency increases with an increase in *AWS*. At zero wind speed, the OSCS is seen to have a very significant advantage in efficiency because the OSCS creates its own wind.

TABLE 12.3
Columbia Generating Station-Oriented Spray Cooling System Test Data

Test No.	*AWS*, mi/hr	*WBT*, °F	*HWT*, °F	*CWT*, °F
5	1.5	50.8	79.7	67.0
6	1.5	53.2	78.8	67.2
7	8.5	59.3	80.5	70.1
8	12	63.3	80.7	71.2
9	1	55.1	81.3	70.1
10	5.5	65.0	82.0	73.8
11	15	63.4	81.2	70.2
12	11	62.5	81.2	70.7
13	3	48.9	79.9	66.7
14	4	53.1	79.0	67.3
15	5	57.6	80.5	69.8
16	4	45.1	78.3	64.1
17	5	59	77.5	68.5
18	3.5	57.7	75.9	67.7
19	2	59.1	73.5	67
20	4	55	68.5	62.4

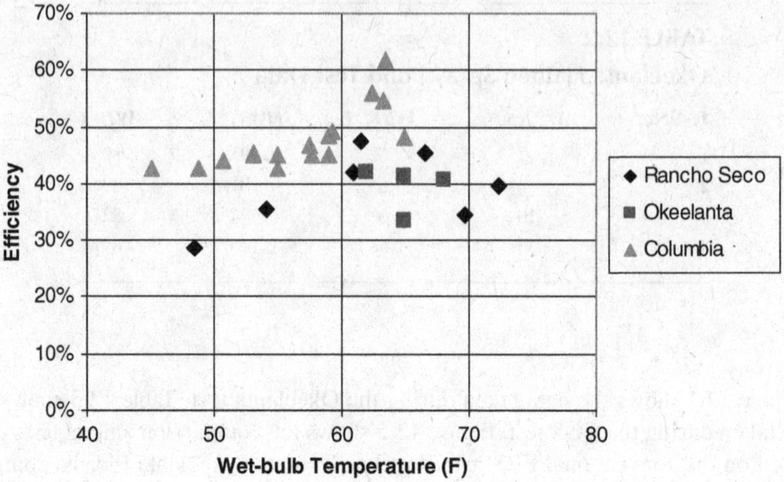

FIGURE 12.5 Flatbed spray ponds and oriented spray cooling system efficiencies vs. *WBT*.

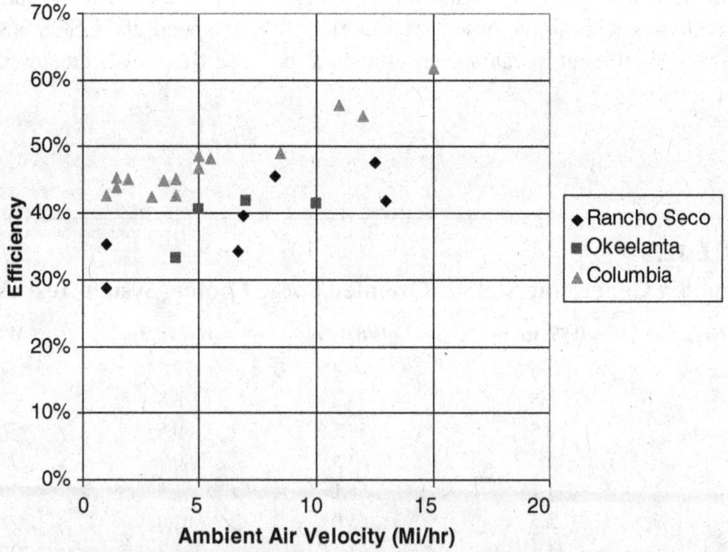

FIGURE 12.6 Flatbed spray ponds and oriented spray cooling system efficiencies vs. *AWS*.

12.7 SPRAY POND INTERFERENCE

A significant disadvantage of a conventional FBSP is interference. Spray nozzles in a conventional FBSP typically are arranged in a matrix about 5 ft above the pond surface with riser pipes at 13 ft 4 in on the center, making the nozzles approximately 7 ft apart. For a large collection of sprays, the fact that adjacent sprays interfere with each other by blocking ambient airflow to the adjacent sprays is well established.

The result is that the *LWBT* in the given spray field is elevated above the *WBT* and that the efficiency of the spray is reduced relative to the *WBT*. Therefore, in order to compensate for interference, the assumed *WBT* must be increased when calculating the expected cold-water temperature in a conventional FBSP. Such arrangements are highly sensitive to wind velocity and direction. For an OSCS, where nozzles are mounted on spray trees in a circle and tilted at an angle oriented toward the center of the circle, the jets of water droplets induce ambient air into the spray region and the warm air rises from the center of the circle. Therefore, interference is not a significant issue with an OSCS.

To evaluate the effect of the interference, Yang and Porter[11] defined an interference factor as follows:

$$f = \frac{t_{WB\text{-}local} - t_{WB\text{-}amb}}{t_1 - t_{WB\text{-}amb}} \tag{12.4}$$

where:

f = interference factor
$t_{WB\text{-}local}$ = *WBT* at the spray nozzle
$t_{WB\text{-}amb}$ = *WBT* outside the spray region
t_1 = temperature of the CCW entering the spray nozzle.

Figure 12.7 shows f as computed from the Rancho Seco tests. Figure 12.8 shows a plot of f for Quad Cities. Yang and Porter[11] took note of the difference in the shape of the curves for Tests 3(b) and 5(a)2 and commented that both results were within the uncertainty of the test and suggested that they are averaged.

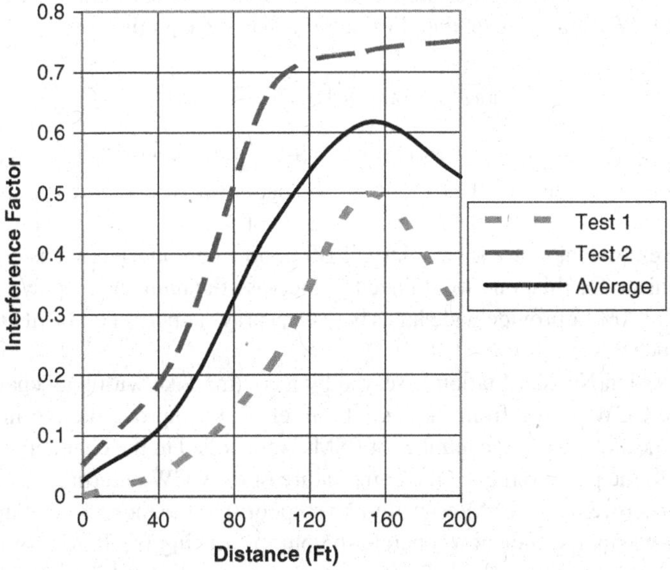

FIGURE 12.7 Interference factor computed for the Rancho Seco tests.

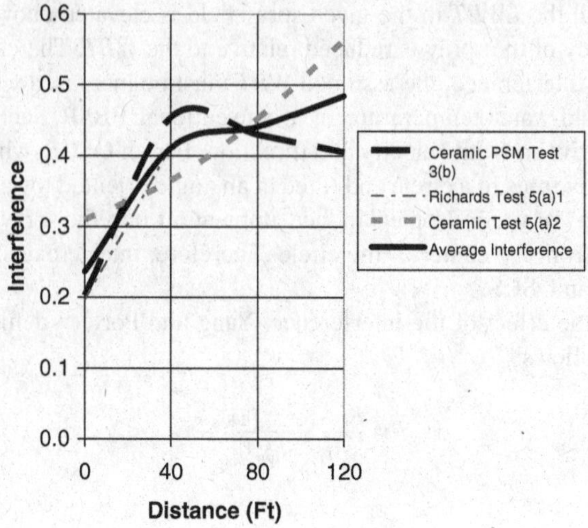

FIGURE 12.8 Interference factor computed for the Quad Cities tests.

Considering the fact that the spray patterns of the Quad Cities and Rancho Seco FBSP are quite different, the interference allowances are strikingly similar.

It should be noted that in the design of both the Quad Cities and Rancho Seco spray cooling systems, no provision was made for open lanes among the sprays for ambient air to reach the interior sprays. In that case, an appropriate interference allowance of perhaps 0.5 would not be unreasonable and some future similar spray systems might be sized accordingly.

Knowing the assumed interference allowance and the temperature entering the FBSP, the *LWBT* may be calculated from the following equation:

$$t_{WB\text{-}local} = t_{WB\text{-}amb} + f\left(t_1 - t_{WB\text{-}amb}\right) \tag{12.5}$$

For example, assuming an *HWT* of 100°F and a *WBT* of 80°F, the local *WBT* would be 90°F for an interference factor of 0.5. The net effect of this correction would be that a spray nozzle that might have an efficiency of 50% operating alone would have an effective efficiency of 25% in a large FBS pond when interference is considered. To compensate for this, one would need to increase the number of spray nozzles by as much as 100% or provide wide lanes between sprays to permit ambient air to reach the spray nozzles.

The Dresden Nuclear Station is served by a cooling lake with long channels that deliver the CCW to and from the lake. In an effort to improve the thermal performance of the station, a large number of PSMs were added to the channel that returns the CCW to the plant to reduce the temperature of the CCW entering the condenser. The performance of the PSM proved to be so poor that the increase in plant electrical output was not sufficient to operate the pumps serving the PSMs, so they were removed. Similarly, the Quad Cities Nuclear Station, located on the Mississippi

River, was designed with a channel surrounding the station that was filled with PSMs to dissipate the waste heat. These PSMs were also removed due to poor thermal performance. The failings at these and other locations where the utility relied on vendor performance estimates that failed to consider interference have given spray ponds of all types a bad reputation.

Any attempts at calculating the values of η, NTU, or f are applicable only to the particular FBSP that has been tested or to one of a similar design. Spray pond investigators have studiously avoided addressing the dynamics of and heat transfer from the individual water drops required to achieve a more universal solution applicable to some future spray pond design that has not previously been constructed and tested. The quest for such an approach is the subject of the next chapter.

REFERENCES

1. Schrock, V. E. and G. J. Trezek, National Science Foundation Waste Heat Management Report No. WHM-4, University of California, Berkeley, July 1, 1973.
2. Shrock, V. E., G. J. Trezek, and L. R. Keilman, Performance of a Spray Pond for Nuclear Power Plant Ultimate Heat Sink, ASME Paper 75-WAHT-41, 1975.
3. Chaturvedi, S. and R. W. Porter, Effect of Air-Vapor Dynamics on Interference for Spray Cooling Systems, Waste Energy Management Technical Report TR-77-1, March 1977.
4. Yadigaroglu, G., Heat and Mass Transfer between Droplets and the Atmosphere – State of the Art, Waste Heat Management Report No. WHM-21, July 1976.
5. Chen, K. H. and G. J. Trezek, Spray Energy Release (SER) Approach to Analyzing Spray System Performance, *Proceedings of the American Power Conference*, Volume 38, 1976, pp. 1434–1457.
6. Porter, R. W., U. Yang, and A. Yanik, Thermal Performance of Spray-Cooling Systems, *Proceedings of the American Power Conference*, Volume 38, 1976, pp. 1458–1472.
7. Elgawhary, A. M. and A. M. Rowe, Spray Pond Mathematical Model for Cooling Fresh Water and Brine, Environmental and Geophysical Heat Transfer, ASME HT-Volume 4, 1971.
8. Soo, S. L., Power Spray Cooling – Unit and System Performance, ASME Paper 75-WA/PWR-8, 1975.
9. Yao, S. and V. E. Schrock, Heat and Mass Transfer from Freely Falling Drops, *Journal of Heat Transfer*, vol. 98, Series C, no. 1, 1976, pp. 120–126.
10. Myers, D. M. and R. D. Baird, *Thermal Performance of Large UHS Spray Ponds*, Ford, Bacon & Davis, Utah Inc., 1977.
11. Yang, U. M. and R. W. Porter, *Thermal Performance of Spray Cooling Systems – Theoretical and Experimental Aspects*, National Science Foundation Waste Heat Management Report No. TR-76-1, Illinois Institute of Technology, Chicago, IL, 1976.
12. Schrock, V. E. and G. J. Trezek, National Science Foundation Waste Heat Management Report No. WHM-10, University of California, Berkeley, October 1974.
13. Conn, K. R., 1979 Ultimate Heat Sink Spray System Test Results, Washington Public Power Supply System Nuclear Project No. 2, WPPSS-EN-81-01, 1981.

13 Oriented Spray Cooling System

13.1 DESCRIPTION OF THE ORIENTED SPRAY COOLING SYSTEM

Spray ponds are simple and relatively inexpensive, essentially passive devices that do not require electrical power or excessive maintenance as is the case with MDCT. Most existing spray ponds employ the classical flatbed sprays oriented in the vertical direction, but the thermal performance of this design has proven to be very poor and the problems with this design have been the subject of extensive investigation.[1,2] In the flatbed design with all spray nozzles oriented in the vertical direction (discussed in Chapter 12), the bulk drag force of the water droplets (vertically downward) resists the natural buoyancy of the warm air (which is of the same order of magnitude and is directed vertically upward). Because the two forces acting on the air oppose each other, the result is a reduced airflow rate through the spray region and a large increase in the *LWBT*. Additionally, with large FBSP and spray canals, the falling water tends to block airflow into the central parts of the array, requiring an interference factor to be applied to determine the *LWBT* as a function of the *WBT*[3] (see Chapter 12).

The OSCS employs a radically different spray pond design that overcomes these problems. The OSCS was first proposed by Ecolaire Condenser Company (ECC) in the late 1970s.[4] Figure 13.1 shows the OSCS that was constructed, operated, and thoroughly tested by ECC at the Ingersoll-Rand pump and turbine factory in Phillipsburg, NJ. As one may readily see from Figure 13.1, the outstanding feature of the OSCS design is the circular arrangement of the spray nozzles on spray trees. The spray trees are spaced approximately 13 ft 9 in on the center with spray nozzles arranged in a helical pattern spaced approximately 4 ft from the mast of the tree and approximately 2 ft 8 in apart vertically and oriented at an angle from the vertical toward the center of the circle. In this design, both the bulk drag force of the water droplets on the air and the buoyant force promote ventilation of the spray region and ambient airflow to the spray nozzles is not obstructed. The result is a reduction in the *LWBT* in the spray region and improved cooling of the droplets as they fall through the spray region to the pond surface below.

FIGURE 13.1 Oriented spray cooling system. (Courtesy of ASME.)

13.2 ANALYSIS OF SPRAY NOZZLES

The testing conducted by ECC resulted in a simplified empirical model based on the Merkel equation (i.e., *NTU*). An analytical spray pond model (THERMAL) was developed in the late 1970s that does not rely on experimental thermal performance data other than correlations to predict enthalpy change in a drop during evaporative cooling and the experimentally derived drop spectrum for the particular spray nozzle.[5,6] The other elements of the model are analytical, based on classical heat and mass transfer and kinetic vector relationships for spherical water droplets rather than empirical models based on experimental thermal performance data from individual spray units or particular spray configurations. Therefore, the model is not limited in application with regard to spray pressure, nozzle spacing, or orientation and is not limited by droplet size considerations. The resulting spray pond model has since been validated by independent full-scale testing at the CGS near Richland, Washington, where two OSCS are employed as the safety-related UHS for the nuclear plant.

13.2.1 Nozzle Selection

In order to perform any detailed analysis of an OSCS, one must determine the relevant individual spray nozzle characteristic parameters. The trajectory of drops leaving a nozzle forms a pattern, the nature of which depends upon the type of nozzle. The selection of the nozzle and its operating pressure will determine the nozzle flow rate, the drop size distribution, and the velocity of the drop leaving the nozzle. This information, together with the nozzle height above the pond surface and the orientation, would determine the nozzle spray pattern for zero wind conditions. The testing conducted by ECC indicated the optimum orientation of an OSCS nozzle to be tilted 35° from the vertical in the direction of the center of the ring header. For a hollow-cone nozzle, the type normally used in spray pond applications and used exclusively herein, the spray pattern is easily determined by measuring the height and diameter of the spray for a given nozzle pressure. Both the Spray Engineering Co. SPRAYCO Type 1751 and the

Spraying Systems Co. Whirljet Type CX nozzles were investigated. Stainless steel nozzles are highly recommended due to the high exit velocities typically employed in the OSCS.

13.2.2 Drop Spectrum

Perhaps the most critical input data required to predict the performance of an OSCS is the drop spectrum. Because ECC selected the Whirljet Type $1^1/_2$ CX SS 27 nozzles for testing and marketing, that is the class of nozzles that will be exclusively analyzed herein. Considerable uncertainty exists in measuring the number and size of each drop from a spray nozzle. The methods include spatial sampling using flash photography and flux or temporal sampling in which one attempts to measure the drops passing through a specified area for a specified period of time. In general, the flux measurement technique yields larger drops. However, for the case under consideration, what matters is which method yields a more accurate prediction of the OSCS performance. Table 13.1 shows the drop spectrum for the Spraying Systems Whirljet Type CX nozzles. The $1^1/_2$ in and 2 in nozzles are rated to deliver 25 and 50 gal/min, respectively, at 7 lbf/in^{2}G.

Figure 13.2 shows a comparison between the drop spectrum for $1^1/_2$ in and 2 in nozzles at the same nozzle pressures. Figure 13.3 shows the drop spectrum as a function of the frequency of occurrence. Note that the smaller nozzle produces finer drops at the same pressure as would be expected. As an approximate rule of thumb, the drop size may be assumed to vary as the −0.3 power of pressure.

TABLE 13.1
Drop Spectrum for Spraying Systems Whirljet Nozzles (μm)

Accumulated	1-1/2 in	1-1/2 in	1-1/2 in	2 in	2 in	2 in
Volume (%)	10 psig	15 psig	20 psig	7 psig	10 psig	15 psig
1	860	770	725	1,060	1,020	950
2	1,020	915	875	1,250	1,200	1,100
5	1,288	1,170	1,120	1,580	1,470	1,320
10	1,590	1,460	1,390	1,910	1,780	1,670
20	2,000	1,780	1,680	2,520	2,250	2,100
30	2,360	2,220	2,100	3,010	2,710	2,500
40	2,670	2,530	2,400	3,480	3,100	2,850
50	3,040	2,860	2,700	4,000	3,580	3,210
60	3,400	3,200	3,000	4,520	4,080	3,680
70	3,800	3,530	3,350	5,180	4,680	4,200
80	4,250	3,980	3,730	5,950	5,410	4,850
90	4,830	4,560	4,300	7,050	6,450	5,650
95	5,300	5,000	4,750	7,900	7,200	6,150
98	5,700	5,380	5,150	8,650	8,000	6,700
99	5,900	5,600	5,350			

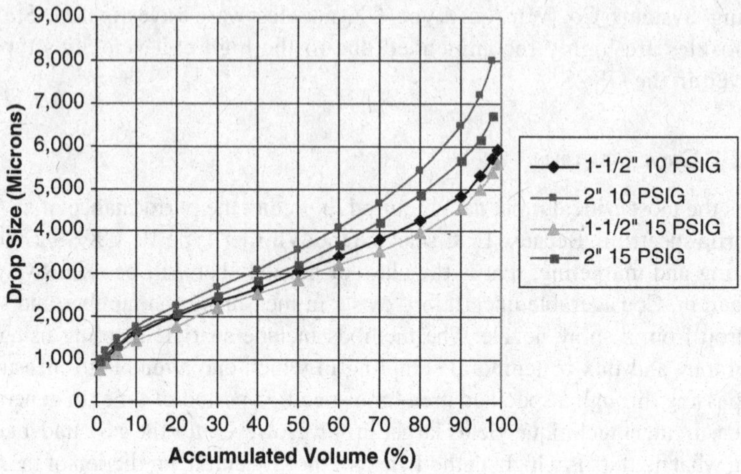

FIGURE 13.2 Percentage of volume smaller than the given drop size. (Courtesy of ASME.)

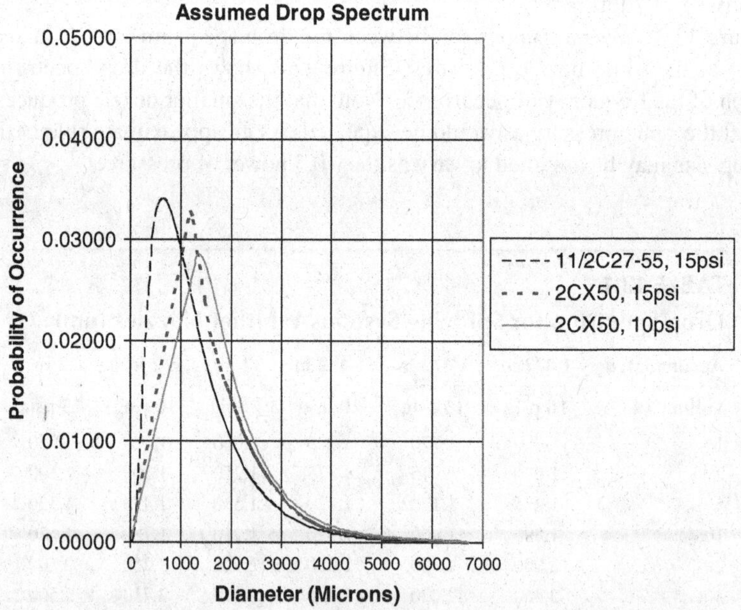

FIGURE 13.3 Drop spectrum as a frequency of occurrence. (Courtesy of ASME.)

13.2.3 Initial Drop Velocity

It is necessary to determine both the magnitude and the orientation of initial drop velocity vectors. These are determined for a given nozzle and nozzle pressure by measuring the height and diameter of the axisymmetric spray pattern produced by a vertically oriented hollow cone nozzle as well as the elevation of the nozzle above the drop impact surface. Referring to Figure 12.2 in Chapter 12, H_1 is the height of

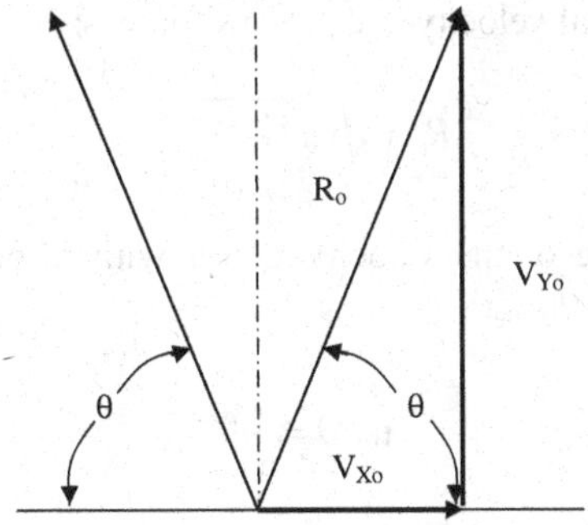

FIGURE 13.4 Initial velocity of drops for vertical spray.

TABLE 13.2
Spray Pattern Dimensions for Whirljet Nozzles in Feet

Pressure (psig)	1½ CX	25	2 CX	50
	Diameter (2*x*)	Height H_1	Diameter (2*x*)	Height H_1
7	27	7.5	30	8.0
10	30	9.0	35	9.5
15	35	10.5	39	11.0

the spray above the nozzle and H_2 is the height of the nozzle above the spray pond surface, Figure 13.4 illustrates the calculation of the initial velocity of the drops for a vertical spray.

Table 13.2 shows the results of tests that were conducted on the Spraying Systems Whirljet nozzles oriented in the vertical direction and located 1 ft above the water surface.

Based on these test results, the magnitude of the vertical and horizontal components of the initial drop velocity (v_{yo} and v_{xo}, respectively) may be calculated from data on the drop trajectory with the nozzle spraying upward, as in Table 13.2, as follows:

$$v_{yo} = \sqrt{2 g_c H_1} \tag{13.1}$$

$$t = \frac{v_{yo}}{g_c} + \sqrt{\left(\frac{v_{yo}}{g_c}\right)^2 + \frac{2\,H_2}{g_c}} \tag{13.2}$$

$$v_{xo} = x / t \tag{13.3}$$

where:

H_1 = peak height of the spray above the nozzle
H_2 = the height of the nozzle above the water level
x = the radial distance of the outer edge of the spray
t = the time of flight of the drop
g_c = proportionality constant.

The magnitude of the total velocity vector is as follows:

$$R_o = \sqrt{v_{xo}^2 + v_{yo}^2} \quad (13.4)$$

The tangent of the angle of the velocity vector with respect to the horizontal in Table 13.2 would be as follows:

$$\tan\theta = \frac{v_{yo}}{v_{xo}} \quad (13.5)$$

where θ is in radian.

Of course, these relationships neglect the drag force of the air on the drops, as do the test data in Table 13.2, because the measurements are made at the outer perimeter of the spray region, reflective of the larger drops where the effect of drag is negligible.

As previously stated, ECC determined experimentally that the optimum tilt angle (TA) for the OSCS is 35° (0.61 rad) from the vertical. As the nozzle is tilted toward the center of the OSCS (e.g. the right), the direction of the right side of the spray will shift downward to an angle α relative to horizontal and the left side will shift upward to an angle β above horizontal. Therefore, the minimum and maximum angles of inclination of droplets from the nozzle with reference to the horizontal would be as follows:

$$\alpha = \theta - TA \quad (13.6)$$

and

$$\beta = \theta + TA \quad (13.7)$$

and the minimum and maximum vertical velocities would be

$$v_{y,1} = R_0 \sin\alpha \quad (13.8)$$

and

$$v_{y,2} = R_0 \sin\beta \quad (13.9)$$

$$Y_{max1} = \frac{v_{y1}^2}{2\,g_c} + Y_0 \quad (13.10)$$

and

$$Y_{max2} = \frac{v_{y2}^2}{2\,g_c} + Y_0 \quad (13.11)$$

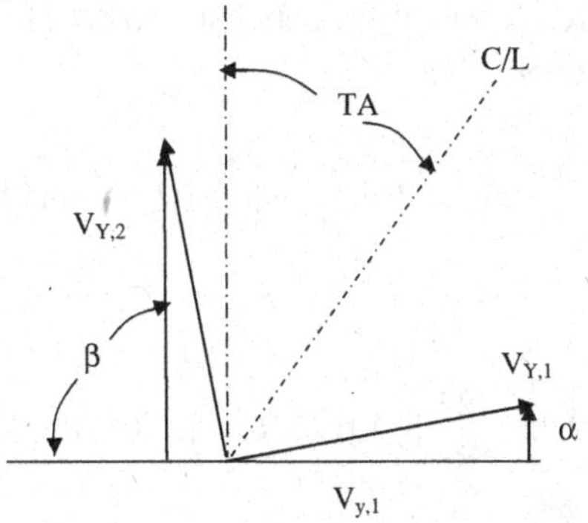

FIGURE 13.5 Initial velocity of drops for tilted spray.

where:

Y_0 = the elevation of the top nozzle on the OSCS tree

Y_{max1} = the maximum height of drops from the side of the nozzle closest to the central axis of the OSCS

Y_{max2} = the maximum height of drops from the side of the nozzle away from the central axis of the OSCS.

The top of the CV for a single spray tree is defined as follows:

$$Y_{max} = \frac{Y_{max1} + Y_{max2}}{2} \tag{13.12}$$

and the width of the CV is the distance between the spray trees. Figure 13.5 illustrates the calculation of the initial velocity of the drops for a tilted spray.

13.3 DESCRIPTION OF KXDRAG AND DRIFT MODELS

The trajectories of individual drops are calculated by numerical integration using their initial velocity vectors and the following assumptions:

- drops are spherical throughout their flight,
- collisions and interactions between drops are neglected,
- the nozzles, initial drop velocities, and drop size distribution are axisymmetric,
- heat transfer and evaporation are neglected,
- drop size distribution is known,
- air velocity and air properties are uniform across the entire pond,
- operation of all nozzles is uniform and steady state, and
- air buoyancy effects are neglected.

A drop in flight is subject to not only gravity but also drag forces. The magnitude of the drag force is as follows:[7]

$$d_f = \frac{3\,m}{4D}(RV)^2\,C_D\,\frac{\rho_a}{\rho_w} = \left(\frac{\pi}{8}\right)D^2\,\rho_a\,C_D\,(RV)^2 \tag{13.13}$$

and

$$C_D = 0.22 + \frac{24}{\text{Re}}\left(1 + 0.15\,\text{Re}^{0.6}\right),\quad 0 \le \text{Re} \le 3{,}000 \tag{13.14}$$

where:

d_f = magnitude of the drag force
m = mass of the drop
D = drop diameter
RV = velocity of the drop relative to the air
C_D = drag coefficient
ρ_a = density of air
ρ_w = density of water
Re = Reynolds number based on the drop diameter.

Trajectories are determined for a representative selection of classes of drops that have different diameters and that leave the nozzle in different directions. The result is a locus of impact points for each drop diameter, measured relative to the nozzle location. For a given nozzle location and drop diameter, the loci of impact points are compared with the distance to the pond perimeter to determine the percentage of drops that cross the perimeter and are lost as drift.

The information for various drops is then used for representative nozzle locations throughout the OSCS to determine the drift loss for these nozzles. The weighted sum of the drift loss rates for these nozzles gives the instantaneous fractional drift loss rate for the entire OSCS. When multiplied by the sprayed flow rate and integrated over a period of time, this gives the total drift loss over that period of time. The higher elevation of the top nozzle tends to give it a higher drift loss. However, the lower pressure in the top nozzle will result in larger drops that are less likely to be lost as drift.

13.4 DESCRIPTION OF THE THERMAL MODEL

13.4.1 Overview

In 2019, Bowman et al.[8] first described the THERMAL model in detail. The THERMAL model involves several steps. The thermal performance of the entire spray system is assumed to be represented by that of the control volume (CV) associated with a single spray tree. Further, the thermal performance of all nozzles in the CV is assumed to be represented by that of a single appropriately located nozzle. The *HWT* of the spray is determined based on pond temperature, heat load, and spray flow rate. CV airflow is determined by the total drag force on drops. The *LWBT* in the CV is determined by airflow, ambient air properties, and heat dissipated by the spray. Average *CWT* is found by integrating heat and mass transfer from the spectrum of sprayed drops. Drop heat and mass transfer coefficients are based on correlations developed by Ranz and Marshall (R–M) in 1952.[6] The sprayed flow is mixed with pond inventory to determine the new pond temperature. Iteration is required in some cases because of implicit relationships between parameters.

13.4.2 Control Volume and Wet-Bulb Degradation

A parcel of air undergoing displacement through a CV from ambient conditions, ∞, to exit conditions, *e*, may be described by a modified Bernoulli equation:

$$\frac{U_e^2 - U_\infty^2}{2} + g_c\left(\int_\infty^e \frac{dP}{\rho_a} - \int_\infty^e \frac{D_x}{\rho_a}dx - \int_\infty^e \frac{D_y}{\rho_a}dy - \int_\infty^e\left(\frac{\rho_{a\infty}}{\rho_a} - 1\right)dy\right) = 0 \quad (13.15)$$

where:

U = air velocity
g_c = proportionality constant
P = total static pressure
ρ_a = density of air–vapor mixture
D_x = Bulk drag force in the *x*-direction
D_y = Bulk drag force in the *y*-direction
subscript "*e*" denotes the CV exit
subscript "∞" denotes ambient conditions.

In 1978, Berger and Taylor[5] described the version of the THERMAL model to predict the performance of an FBSP. Figure 13.6 illustrates the CV for an FBSP.

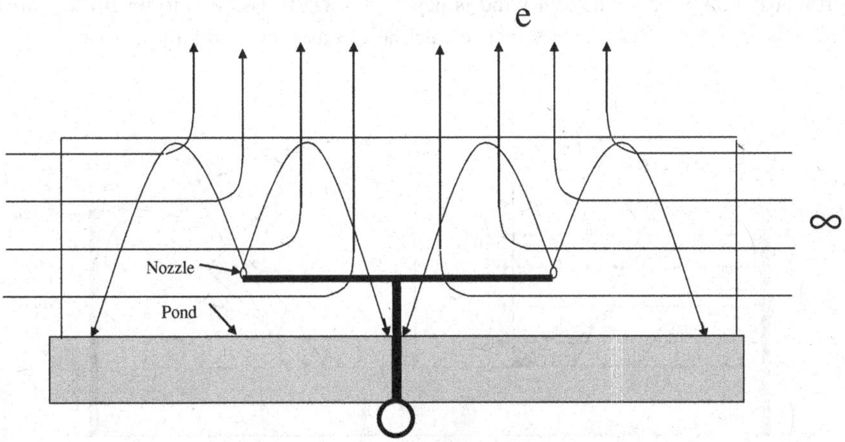

FIGURE 13.6 Control volume for a flatbed spray pond.

The CV height above the pond surface is the peak elevation of trajectories of drops leaving nozzle of the spray tree. The CV width is the distance between adjacent spray trees.

For the FBSP configuration, density variation within the CV represented by the first integral is neglected and exit pressure is assumed atmospheric. The second integral represents the effect of bulk horizontal drag, so the first two integrals in the above equation become zero. The two remaining integrals representing the vertical component of bulk drag force and buoyancy predominate and oppose each other because the net drag force of the water is down. Airflow is perpendicular to the inlet of the CV and exits the top of the CV. Air velocity is assumed to be uniform at the inlet and exit as well as within the CV. Inlet air is at known ambient psychometric conditions. Ambient wind is neglected. *RH* is assumed to be 100% within and at the exit of the CV. Psychometric calculations assume 1 atm air pressure.

The CV for the OSCS, shown in Figure 13.7, is a rectangular box with the *x*-axis toward the center of the spray pond. The CV height above the pond surface is the average peak elevation of trajectories of drops leaving the top and bottom edges of the highest nozzle of the spray tree. The CV width is the distance between adjacent spray trees. The CV length is just enough to include the full spray pattern, but its value is not used in calculations.

Airflow is entrained into the CV by the drag interaction between drops and air. Assuming that the bulk drag force is uniformly dispersed throughout a stream tube passing through the CV, the drag force acts upon the air as if it were a body force similar to gravity. Density variation within the CV is neglected and exit pressure is assumed atmospheric, so the first integral in the above equation becomes zero and the second integral represents the effect of bulk horizontal drag. The vertical components of bulk drag force and buoyancy are also neglected, so the last two integrals also become zero. Airflow is perpendicular to the inlet and exit of the CV, and no air exits the top of the CV. Air velocity is assumed to be uniform at the inlet and exit as well as within the CV. Air properties at the inlet, exit, and within the CV differ from each other but at each location are each uniform in the *y*- and *z*-directions. Inlet air is at known ambient psychrometric conditions. Ambient wind is neglected. *RH* is assumed to be 100% within and at the exit of the CV. Psychrometric calculations assume 1 atm air pressure.

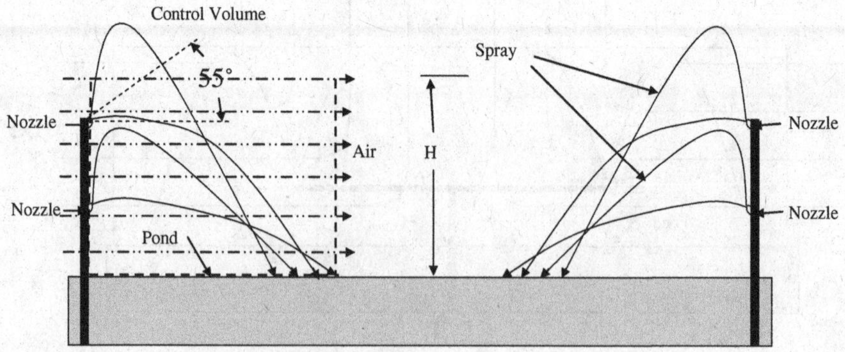

FIGURE 13.7 Control volume for an oriented spray cooling system.

The momentum equation as simplified by the above assumptions and applied to the CV results in the following:

$$U_e = \left(\frac{2\, g_c\, D_x}{\rho_a\, A} + U_\infty^2 \right)^{1/2} \quad \text{where} \quad U_\infty = 0 \tag{13.16}$$

where:
- A = cross-sectional area of the CV
- U_i and U_e = horizontal components of inlet and exit air velocity, respectively
- D_x = horizontal component of bulk drag force (positive radially inward).

The mass flow rate of air may then be calculated as

$$M = U_i\, A\, \rho_{a\infty} \tag{13.17}$$

where:
- M = airflow through the CV
- $\rho_{a\infty}$ = inlet air density.

Although nozzle pressure, flow rate, and drop size distribution vary with nozzle elevation in the spray tree, a single nozzle is analyzed herein. The nozzle is assumed to be placed at an elevation above the pond surface, resulting in it having the average flow rate of all nozzles of the spray tree and corresponding pressure.

The bulk horizontal drag force, D_x, exerted by the drops is equal and opposite to the total drag of air upon all drops sprayed into the CV throughout their trajectories. The bulk drag for flow from a single nozzle is calculated using the same approach as described earlier for drift calculations and scaled up based upon total spray flow in the CV.

Because the inlet and exit air velocities and drag force are interdependent, determining inlet air velocity requires an iterative process. A table of bulk drag force as a function of entrained air velocity is developed and used to determine the entrained airflow through the CV.

From the inlet air properties, the heat load, and the mass airflow, one may calculate the exit air enthalpy from the conservation of energy as

$$h_e = h_i + \frac{Q}{M} \tag{13.18}$$

where:
- h_i and h_e = inlet and exit specific enthalpies, respectively
- Q = rate of heat release to the air.

The exit *WBT* is calculated based on *RH* = 100%. The *LWBT* in the CV is assumed to be the average of inlet and exit values.

13.4.3 Heat and Mass Transfer of Drops

Cooling of drops with various sizes and initial directions is calculated by numerical integration along the trajectories of individual drops using the same assumptions and methods that were employed in the previously described trajectory calculations. Drops are assumed to have uniform internal temperature during flight, as high-speed photography has shown that drops actually oscillate during flight, producing mixing. Drop impact temperature and evaporation for each drop's flight are calculated by integrating the following equations for rates of convective and evaporative cooling:

$$q_c = \mathrm{Nu}\, k\pi D(T_d - LWBT) \tag{13.19}$$

$$m_e = \frac{\mathrm{Sh} D_V \rho_a \pi D\left[P_{sat\text{-}T_d} - RH\, P_{sat\text{-}LWBT}\right]}{P - P_a} \tag{13.20}$$

$$q_e = h_{fg} m_e \tag{13.21}$$

and

$$q_T = q_c + q_e \tag{13.22}$$

where:

q_c = convective heat transfer
m_e = mass evaporation
q_e = evaporative heat transfer
q_T = total heat transfer from the drop
T_d = drop temperature
k = thermal conductivity of air
Nu = Nusselt number $[h_c(k/D)]$, where h_c is the heat transfer coefficient in forced convection
Sh = Sherwood number $[h_D(D/D_V)]$, where h_D is the mass transfer coefficient in forced convection and D_V is the mass diffusivity of water vapor in air
h_{fg} = latent heat of vaporization
P = total static pressure
$P_{sat\text{-}Td}$ = saturation pressure of the water vapor at the indicated temperature
P_a = partial pressure of dry air.

Nu and Sh are determined at any point in the drop path by the R–M[6] correlations.

The *HWT* of the spray, impact temperatures, evaporation, frequency, and mass of all drops are used to determine the overall *CWT* of the spray and total evaporation rate. In a transient analysis, the spray flow rate, *CWT*, instantaneous pond mass, and solar radiation are used to determine the change in mixed pond temperature. Instantaneous values of ambient meteorology and known equipment heat loads are determined by linear interpolation in tables of these parameters. The mixed pond temperature is the inlet water temperature seen by the equipment served by the system. The pond temperature, spray flow rate, and heat load are again used to determine the *HWT* of sprayed water for the next time interval.

THERMAL may be used to calculate parametric performance curves for steady-state conditions or to calculate response to transient conditions. The *CWT* and heat load dissipated by the OSCS depend not only upon *HWT* but also upon *LWBT*, which itself depends upon dissipated heat load and ambient conditions. Thus, depending upon the case to be analyzed, iteration may be required to obtain the solution.

13.4.4 Drop Heat and Mass Transfer Coefficients

Behavior of a single drop in air has been extensively investigated, including heat and mass transfer coefficients, drop size, shape, vibration, spontaneous breakup of large drops,[9] and so on. Interactions between drops (collisions, wake effects, etc.) have been less extensively investigated. Experimental data for thermal performance of spray cooling systems having significant interaction between drops have usually been limited to overall thermal performance for a specific configuration without investigating drop interactions. The R–M[6] correlations for prediction of drop heat and mass transfer coefficients have been widely used and are adopted in THERMAL. They are as follows:

$$\mathrm{Nu} = 2 + 0.6\,\mathrm{Pr}^{1/3}\mathrm{Re}^{1/2} \tag{13.23}$$

$$\mathrm{Sh} = 2 + 0.6\,\mathrm{Sc}^{1/3}\mathrm{Re}^{1/2} \tag{13.24}$$

where:

Pr = Prandtl number [$\mathrm{Pr} = c_p\mu/k$], where c_p is the specific heat of air and μ is the dynamic viscosity

Sc = Schmidt number [$\mathrm{Sc} = \mu/\rho_a D_v$].

The R–M experiments used stationary spherical drops with diameters of ~1,000 μm and $0 \le \mathrm{Re} \le 200$. Accurate extrapolation was claimed up to Re = 1,000, and the correlation agrees well with the results of other experimenters for Re approaching 40,000 using spheres of various liquids and solids.

The R–M correlations are used in THERMAL because drops of less than 3,000 μm comprise more than 60% of the OSCS spray volume and drop trajectories reach *X*/*D* of 2,000–20,000. Small drops with large travel distances are likely to closely approach *LWBT* regardless of which correlation is used. THERMAL assumes the temperature of drops smaller than 500 μm to be equal to the *LWBT*. The incidents of collisions between drops are ignored as a practical necessity in THERMAL. Although such collisions are probable in the OSCS, when drops collide, renewed oscillations and internal mixing are likely to result, depending upon the relative size and velocities of the colliding drops.

13.5 EVAPORATION

Because the OSCS creates its own airflow and because at many such sites the ambient *RH* is quite low, discharging a significant quantity of saturated air can also remove a significant amount of moisture from the pond. Consider a site with an ambient dry-bulb temperature (*DBT*) and *WBT* with an associated *RH* and *P*,

$$P_{sat} = \int (DBT) \text{ from the ASME Steam Tables} \tag{13.25}$$

$$RH = \frac{P_w}{P_{sat}} \quad \Rightarrow P_w = RH\, P_{sat} \tag{13.26}$$

$$P_a = P - P_w \tag{13.27}$$

$$\omega = \left(\frac{MW_w}{MW_a}\right)\frac{P_w}{P_a} = 0.62\,\frac{P_w}{P_a} \tag{13.28}$$

where:

P_{sat} = saturation pressure of the water vapor at *DBT*
P_w = partial pressure of the water vapor
P_a = partial pressure of the dry air
MW_w = molecular weight for water
MW_a = molecular weight for air.

The mass of the air per unit of volume is as follows:

$$m_a = \frac{144\, P_a}{R_a\,(DBT + 460)} \tag{13.29}$$

where:

R_a = gas constant for air.

Then the mass of the water vapor entering the OSCS is as follows:

$$m_{w\text{-}in} = \omega m_a \tag{13.30}$$

The OSCS produces an air velocity, *U*, through the spray region, resulting in

$$V = 60AU \tag{13.31}$$

where:

V = volumetric flow rate of air through the OSCS control volume.

The total mass flow rate through the OSCS CV is as follows:

$$M = V\left(m_a + m_w\right) \tag{13.32}$$

The specific heat of the mixture of water vapor and dry air is

$$c_{p,mix} = \frac{m_a\, c_{p\text{-}a} + m_w\, c_{p\text{-}w}}{m_a + m_w} \tag{13.33}$$

where:

$c_{p\text{-}a}$ = specific heat of the air
$c_{p\text{-}w}$ = specific heat of the water vapor
$c_{p,mix}$ = specific heat of the mixture.

Then, the increase in the temperature of the air through the OSCS is as follows:

$$\Delta T_{air} = \frac{Q}{Mc_{p\text{-}mix}} \tag{13.34}$$

where:

Q = rate of heat release to the air in the CV.

Therefore, the dry-bulb temperature leaving the CV is as follows:

$$DBT_{out} = DBT + \Delta T_{air} \tag{13.35}$$

Assuming that the RH leaving the CV is 100%, the mass flow rate of the water vapor leaving the OSCS CV may be calculated similarly to that entering, and the evaporation rate in lbm/min may be calculated as

$$M_e = V\left(m_{w\text{-}out} - m_{w\text{-}in}\right) \tag{13.36}$$

and the fraction of sprayed water that is evaporated is

$$e = \frac{M_e}{M_s} \tag{13.37}$$

where:

M_s = mass flow rate of the water sprayed.

13.6 MODEL VALIDATION

13.6.1 Validation of THERMAL Analytical Model

Validation of THERMAL against the results of tests conducted at the Rancho Seco nuclear station for a classic FBSP is presented in Table 13.3 and Figure 13.8 from data from Reference 5 where the spray efficiency is as defined in Chapter 12.

Over the seven tests conducted, the average difference between the *CWT* measured and that predicted was 0.6°F, with the largest discrepancy in the spray efficiency occurring at higher *AWS*, as FBSP relies heavily on ambient wind, whereas the THERMAL neglects ambient wind.

TABLE 13.3
Comparison between THERMAL and Rancho Seco Test

				Measured	THERMAL	Measured	THERMAL
Test No.	*AWS* (mi/hr)	*WBT* (°F)	*HWT* (°F)	*CWT* (°F)	*CWT* (°F)	Efficiency	Efficiency
1	1.0	54.14	101.48	84.74	82.58	35.4%	39.9%
2	1.0	48.56	77.36	69.08	70.16	28.8%	25.0%
3	6.7	69.62	81.14	77.18	78.08	34.4%	26.6%
4	6.9	72.32	80.06	77.00	77.72	39.5%	30.2%
5	8.3	66.56	80.78	74.30	76.10	45.6%	32.9%
6	12.5	61.52	80.06	71.24	72.68	47.6%	39.8%
7	13.0	60.98	79.88	71.96	72.32	41.9%	40.0%

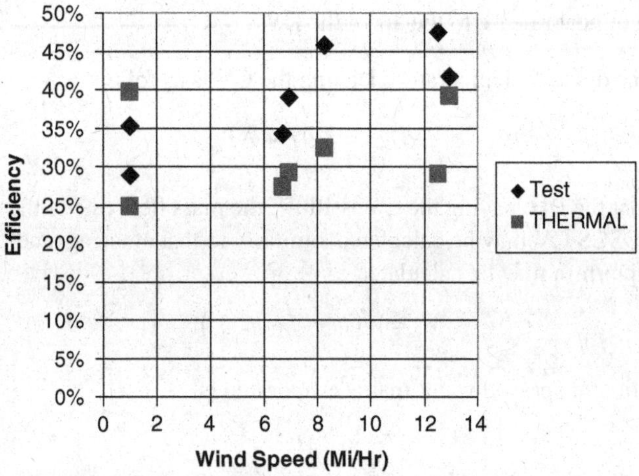

FIGURE 13.8 Comparison between THERMAL and Rancho Seco test efficiency vs. wind speed.

In 1981, Bowman et al.[10] presented a comparison between THERMAL and predictions based on the proprietary tests conducted by ECC at the same heat load and interface pressure, as shown in Figure 13.9.

This comparison indicated that the ECC predictions were overly optimistic, especially at lower ambient wet-bulb temperature (AWBT) conditions.

Figure 13.10 shows a satellite view of the CGS. The six circles at the bottom of the figure are round MDCTs that dissipate the nuclear station's waste heat from the condenser into the atmosphere. The two circles to the right of the figure are two OSCSs that are the UHS for the station. Each OSCS cools 10,400 gal/min.

Figure 13.11 shows the OSCS at CGS in operation. THERMAL was developed prior to the publicationof Reference 5 in 1978. In 1979, CGS issued their first report in Reference 11 of the tests conducted on their OSCS UHS. Verification of THERMAL

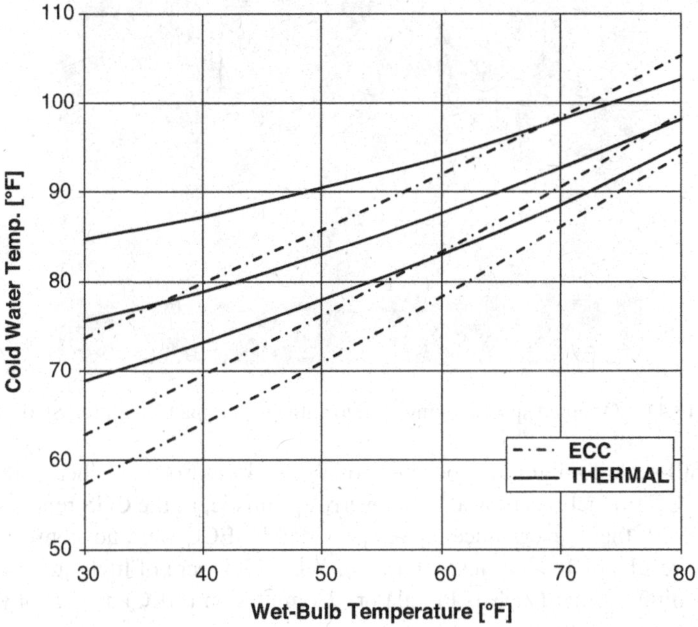

FIGURE 13.9 Comparison between THERMAL and Ecolaire Condenser Company tests.

FIGURE 13.10 Columbia Generating Station.

FIGURE 13.11 Oriented spray cooling system at the Columbia Generating Station.

for that full-scale configuration became possible and was first published in Reference 12. The ECC predictions proved to be overly optimistic, as the CGS report stated in Reference 10 that performance curves provided by ECC were nonconservative by 17%. The results of 16 CGS tests shown in Table 13.4 were of high quality, as they were submitted to the Nuclear Regulatory Commission (NRC) as part of the CGS licensing process.

Reference 11 reported that the overall uncertainty of the *HWT* was 0.1°F, the overall uncertainty of the *CWT* was 0.6°F, and the overall uncertainty of the *AWBT* was 1.2°F.

Table 13.4 also provides a comparison between the results of the tests conducted at CGS and the results predicted by THERMAL for the same operating

TABLE 13.4
Comparison between THERMAL and Columbia Generating Station Tests

Test No.	*AWS* (mi/hr)	*WBT* (°F)	*HWT* (°F)	*CWT* (°F)	THERMAL *CWT* (°F)
5	1.5	50.8	79.7	67.0	66.55
6	1.5	53.2	78.8	67.2	67.32
7	8.5	59.3	80.5	70.1	70.31
8	12.0	63.3	80.7	71.2	72.91
9	1.0	55.1	81.3	70.1	67.69
10	5.5	65.0	82.0	73.8	73.68
11	15.0	63.4	81.2	70.2	74.05
12	11.0	62.5	81.2	70.7	72.90
13	3.0	48.9	79.9	66.7	65.70
14	4.0	53.1	79.0	67.3	67.39
15	5.0	57.6	80.5	69.8	69.59
16	4.0	45.1	78.3	64.1	64.38
17	5.0	59.0	77.5	68.5	68.98
18	3.5	57.7	75.9	67.7	67.49
19	2.0	59.1	73.5	67.0	66.96
20	4.0	55.0	68.5	62.4	63.88

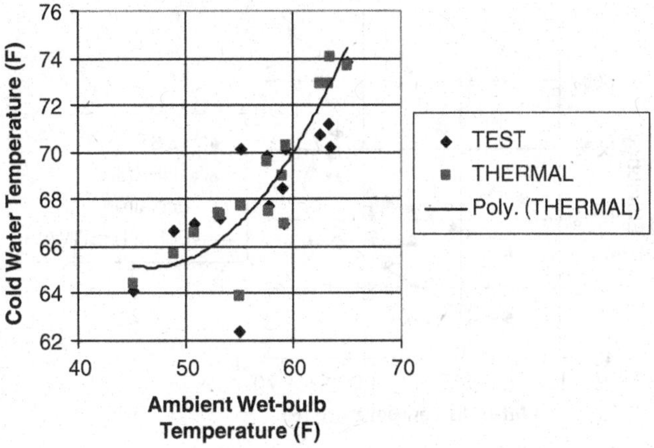

FIGURE 13.12 Comparison between THERMAL and Columbia Generating Station test (*CWT* vs. *AWBT*). (Courtesy of ASME.)

conditions with the *LWBT* set equal to the average of the *WBT* and *EWBT*. The average difference between the *CWT* measured at CGS and that calculated by THERMAL is −0.4°F.

Figure 13.12 shows the results of the comparison between the CGS test results and THERMAL predictions. The solid line is the polynomial curve fit of the THERMAL *CWT* data in Table 13.4. Figure 13.13 shows the results of the THERMAL model comparison with the *CWT* approach to the *AWBT* (*CWT–AWBT*) plotted as a function of the *HWT* minus the *CWT*, or the cooling range.

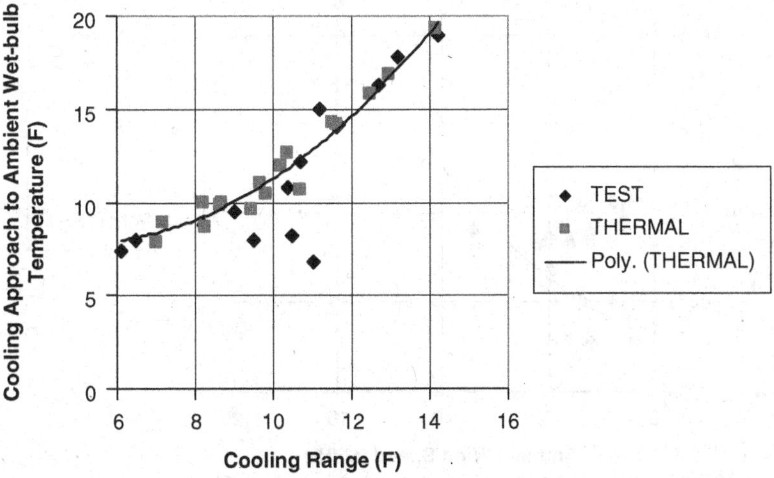

FIGURE 13.13 Comparison between THERMAL and Columbia Generating Station test (approach vs. cooling range). (Courtesy of ASME.)

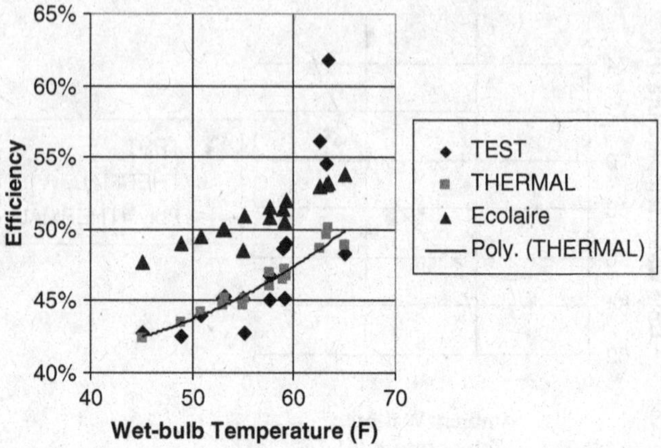

FIGURE 13.14 Comparison between THERMAL and Columbia Generating Station test (efficiency vs. *WBT*).

Figure 13.14 shows the comparison presented in terms of spray efficiency as a function of *WBT*. On Page 19 of Reference 11, the CGS report provides an equation for the cooling range (*HWT–CWT*) achievable based on the *CWT* predicted by the ECC empirical *NTU* model. Figure 13.13 shows the implied efficiency of the CGS OSCS. This confirms the CGS report statementin Reference 11 that performance curves provided by ECC were not conservative.

Figure 13.15 shows the comparison presented in terms of spray efficiency as a function of *AWS*. In this case, the solid line is the polynomial curve fit of the test data in Table 13.4.

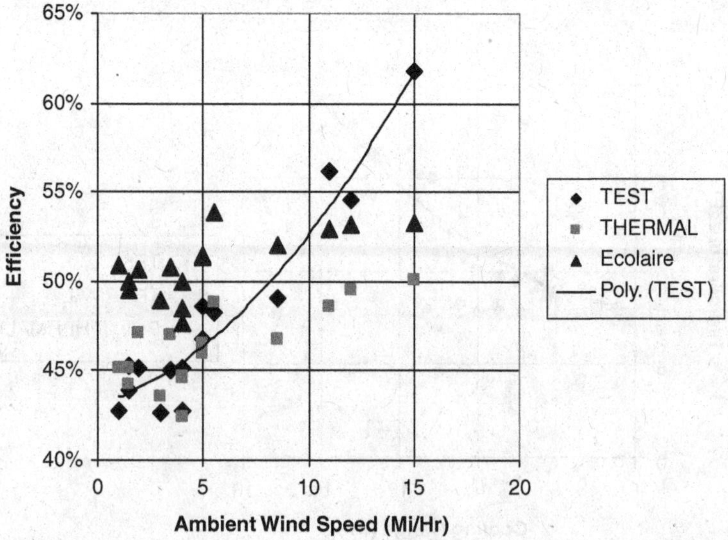

FIGURE 13.15 Comparison between THERMAL and Columbia Generating Station test (efficiency vs. *AWS*). (Courtesy of ASME.)

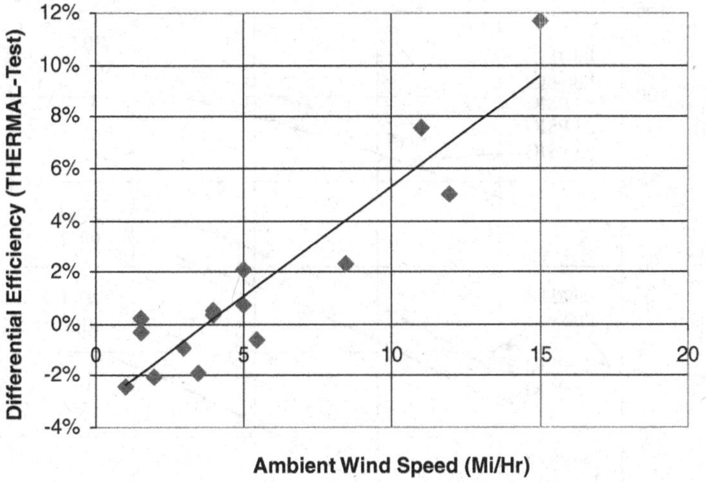

FIGURE 13.16 Difference in efficiency between THERMAL and CGS test results.

One may see from Figure 13.15 that THERMAL is increasingly conservative at higher *AWS*, as would be expected because THERMAL assumes zero *AWS*.

Figure 13.16 shows the difference in the efficiency as calculated by THERMAL minus that calculated from the CGS test results. The agreement between THERMAL predictions and the CGS test results for *AWS* of less than 10 mi/hr leaves little doubt as to the validity of the THERMAL analytical model at zero or very low *AWS*.

13.7 POTENTIAL APPLICATIONS

13.7.1 Power Plant Heat Rejection

One obvious potential application for the OSCS is to be used in lieu of MDCT where space is available due to their superior simplicity and operability, lower preferred power requirements, and lower capital and maintenance costs. The THERMAL results may be expressed as shown in Figure 13.17, from Reference 10 where the interface pressure is at the base of the riser pipe.

Reference 10 reported that the OSCS was tentatively selected to reject the main turbine cycle heat to the atmosphere for a proposed power plant that was later canceled. The paper suggested a cooling range of 29°F and an OSCS tree interface pressure of up to 33 lbf/in^2. Of course, the economics of that decision may have changed.

13.7.2 Nuclear Plant Ultimate Heat Sink

References 10–13 provide detailed analyses as to how the OSCS may be employed as the ultimate heat sink (UHS) for a nuclear power plant as is the case at CGS.

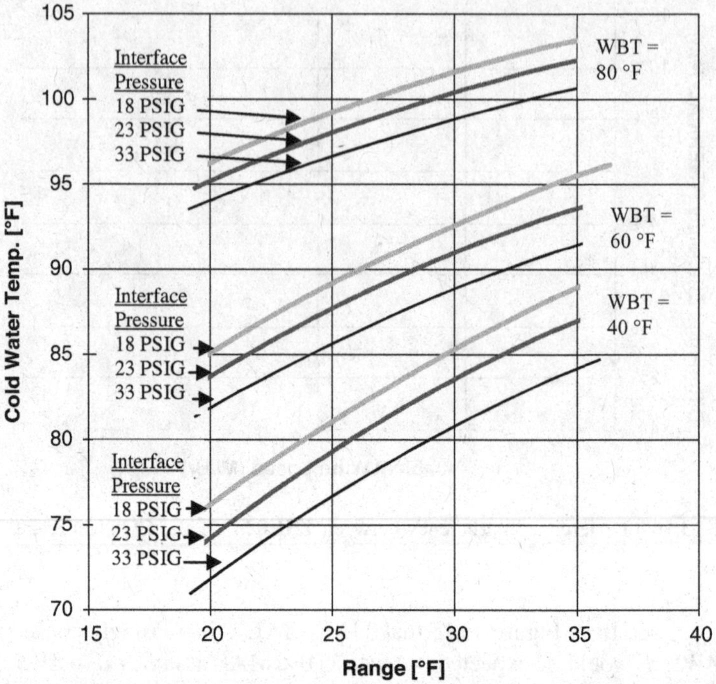

FIGURE 13.17 THERMAL predicted *CWT* for an array of cooling ranges and interface pressures.

Classical FBSP has been employed as the UHS for several nuclear plants in the United States. These ponds are normally sized to store sufficient water for continued plant operation for up to 30 days following an accident. Therefore, the source of makeup water is not required to be safety-related. However, because spray ponds cannot be protected from tornado missiles, a diverse source of cooling water is normally required to be tornado-missile-protected to serve as an alternative source of cooling water.

In 2008, Bowman[13] reported on the design of a UHS for the Advanced Boiling Water Reactor (ABWR) nuclear plant that was proposed to be constructed at the Bellefonte Nuclear Plant site in Scottsboro, Alabama, by Toshiba Nuclear. The UHS for the ABWR consists of three active safety-related Reactor Service Water System (RSWS) divisions, located in a single spray pond serving both nuclear units excavated from undisturbed earth and sized for a water volume adequate for 30 days of cooling under design basis conditions. Figure 13.18 shows the pond with three divisionalized ring headers.

Half of each ring header is dedicated to Unit 1 and the other half to Unit 2. Figure 13.19 illustrates the results of the transient spray pond analysis subsequent to a loss of cooling accident (LOCA).

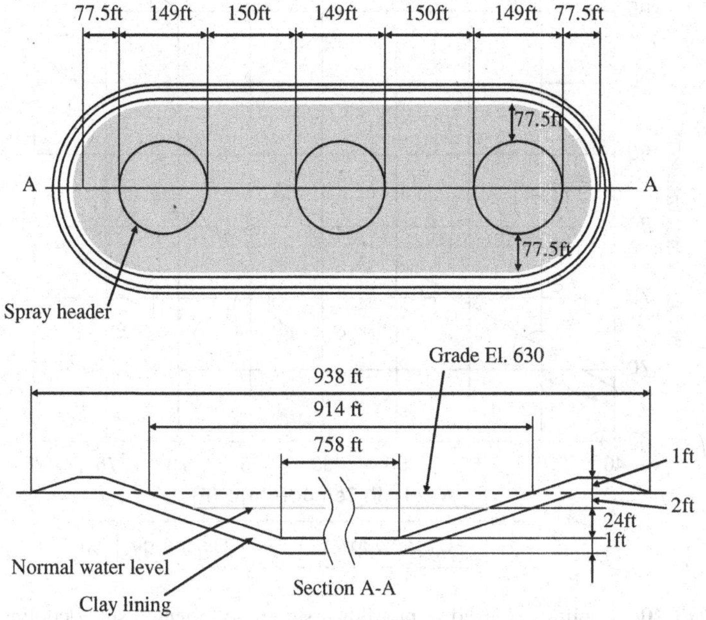

FIGURE 13.18 Ultimate heat sink oriented spray cooling system for an ABWR nuclear station.

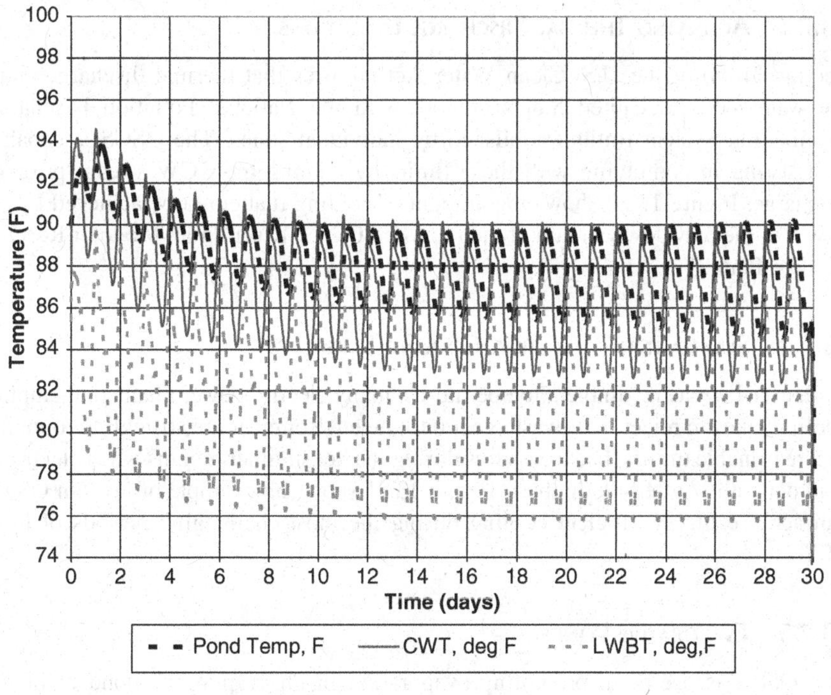

FIGURE 13.19 Transient spray pond analysis subsequent to a loss of cooling accident. (Courtesy of ASME).

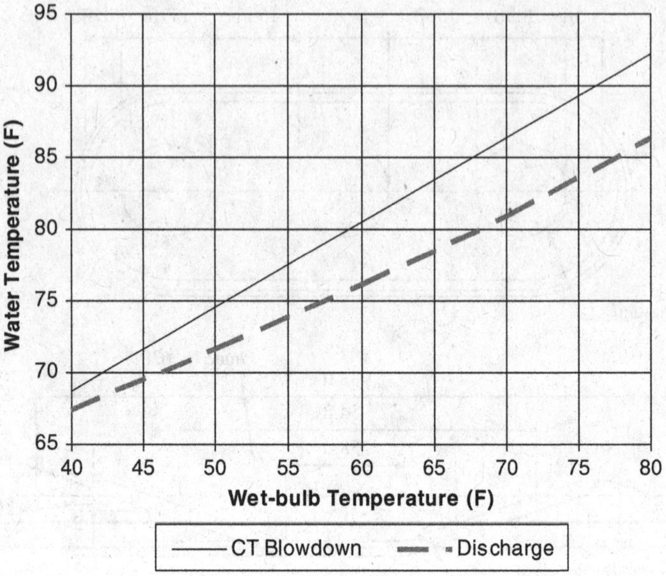

FIGURE 13.20 Cooling achieved in blowdown stream by oriented spray cooling system. (Courtesy of ASME.)

13.7.3 Achieving Thermal Discharge Limitations

Section 316(a) of the U.S. Clean Water Act requires that thermal discharges into the waters of the United States comply with the National Pollution Discharge Elimination System limits established by individual states. The OSCS is capable of assisting in complying with these limits by cooling the CCW stream prior to discharge. Figure 13.20 shows the amount of cooling that could be achieved by an OSCS in the blowdown stream of the proposed Clinch River Small Modular Reactor Nuclear Plant.[14]

13.7.4 Supplementing Existing Cooling Lakes

With ever-increasing ambient temperatures, many electric power plants that employ cooling lakes to reject their waste heat into the environment are struggling to maintain reasonable turbine backpressures during the hot summer months when electrical load demand is often the highest. Figure 13.21 shows an example of how an OSCS can achieve highly efficient cooling during the most challenging periods of high *AWBT*.[15]

13.7.5 Evaporative Ponds

The OSCS is useful in promoting evaporation in an evaporative pond at facilities requiring zero discharge in areas of low *RH* with or without heat, as seen in Figure 13.22.

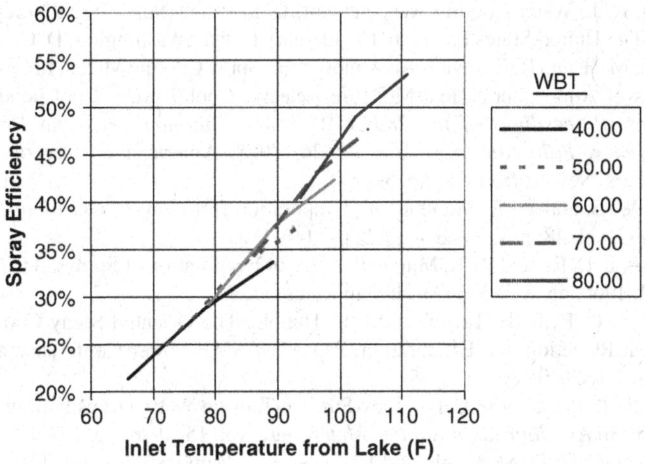

FIGURE 13.21 Oriented spray cooling system supplementing cooling lake. (Courtesy of ASME).

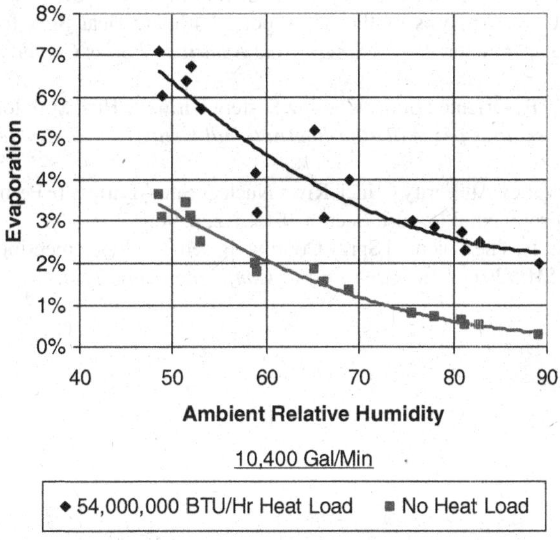

FIGURE 13.22 Oriented spray cooling system in an evaporative pond. (Courtesy of ASME).

REFERENCES

1. Shrock, V. E., G. J. Trezek, and L. R. Keilman, Performance of a Spray Pond for Nuclear Power Plant Ultimate Heat Sink, ASME Paper 75-WAHT-41, 1975.
2. Jain, M. L. and R. W. Porter, Heat, Mass, and Momentum Transfer from Sprays to Air in Cross Flow, IIT Waste Energy Management Technical Memorandum TM-79-2, Illinois Institute of Technology, July 1979.
3. Chaturvedi, S. and R. W. Porter, Effect of Air-Vapor Dynamics on Interference for Spray Cooling Systems, Waste Energy Management Technical Report TR-77-1, March 1977.

4. Stoker, R. J., Water Cooling Arrangement, U. S. Patent No. 3,983,192, September 28, 1976, The United States Patent and Trademark Office, Washington, D. C.
5. Berger, M. H. and R. E. Taylor, An Atmospheric Spray Cooling Model, In Environmental Effects of Atmospheric Heat/Moisture Release: Cool Towers, Cool ponds and Area Sources, *Proceedings of the 2nd AIAA/ASME Thermophysics and Heat Transfer Conference, Palo Alto, CA, May 24–26, 1978*, American Society of Mechanical Engineers, New York, 1978, pp. 59–64.
6. Ranz, W. E. and W. R. Marshall Jr., Evaporation from Drops, *Chemical Engineering Progress*, vol. 48, nos. 3 and 4, 1952, pp. 141–180.
7. Dickinson, D. R., and W. R. Marshall, Rates of Evaporation of Sprays, *AICHE Journal*, vol. 14, 1968, pp. 541–552 (As cited in Reference 2).
8. Bowman, C. F., R. E. Taylor, and J. D. Hubble, The Oriented Spray Cooling System for Heat Rejection and Evaporation, Paper at ASME Power Conference and Nuclear Forum, July 2019.
9. Ryan, R. T., Behavior of Large, Low-Surface-Tension Water Drops Falling at Terminal Velocity in Air, *Journal of Applied Meterology*, vol. 15, 1976, pp. 157–165.
10. Bowman, C. F., D. M. Smith, and J. S. Davidson, Application of the TVA Spray Pond Model to Steady-State and Transient Heat Dissipation Problems, *Proceedings of the American Power Conference*, Volumc 43, 1981.
11. Conn, K. R., 1979 Ultimate Heat Sink Spray System Test Results, Washington Public Power Supply System Nuclear Project No. 2, WPPSS-EN-81-01, 1981.
12. Bowman, C. F., Analysis of the Spray pond Ultimate Heat Sink for the Advanced Boiling Water Reactor, *Proceedings of the American Power Conference*, Volume 56, 1994.
13. Bowman, C. F., Oriented Spray Cooling System Ultimate Heat Sink for Future Nuclear Plants, *Proceedings of the at 16th International Conference on Nuclear Engineering*, May 2008.
14. Tennessee Valley Authority Clinch River Nuclear Site – Early Site Permit Application – Part 3, Environmental Report, Section 9.4.2.2.2, p. 96.
15. Bowman, C. F., The Oriented Spray Cooling System for Supplementing Cooling Lakes, *Paper at ASME 2017 Power and Energy Conference*, June 2017.

14 Oriented Spray-Assisted Cooling Tower

14.1 ORIGIN OF THE ORIENTED SPRAY-ASSISTED COOLING TOWER

As discussed in Section 10.7, most of the NDCTs that were designed and constructed based on the Merkel equation prior to the development of computerized CTSA were found to be deficient. This deficiency was particularly significant for nuclear stations, as it sometimes resulted in requiring the reactor power to be reduced to avoid exceeding LP turbine backpressure limitations (see Section 5.5). Increasing the capability of an existing NDCT is challenging. Simply increasing the amount of fill material increases the pressure drop and thus reduces the air flow through the NDCT, and increasing the height of the veil of an existing NDCT to compensate for increased pressure drop is not feasible. Although measures such as improving the distribution of CCW over the fill material and adding the fill material around the perimeter of counterflow NDCT can improve the performance somewhat, some nuclear stations with this problem have sought to increase the heat dissipation of their HRS by employing measures such as adding MDCT in parallel with the NDCT, etc.

In 1995, Bowman[1] patented the oriented spray-assisted cooling tower (OSACT), a new approach to NDCT design that may be applied to either a new or an existing NDCT without modifications to the existing structure. The OSACT design, shown in Figure 14.1, diverts a portion of the total amount of CCW flow from the NDCT through a ring header pipe to a series of spray trees. Each spray tree consists of a vertical riser pipe, spray arms, and spray nozzles that are evenly spaced externally to the NDCT to produce a uniform spray pattern oriented toward the central axis of the NDCT, which is the desired direction of airflow. The sprayed water then lands on an apron extending from the header pipe to the NDCT basin. The apron is sloped gently toward the NDCT basin so that the sprayed water drains into the basin. The water spray droplets apply a drag force to the air, increasing the air velocity and airflow into the NDCT over that achieved with conventional NDCT design. By diverting a portion of the water to be cooled to the spray trees external to the tower, the water loading in the cooling tower HX section is reduced and the resistance to airflow through the tower caused by the water falling through the HX section of the cooling tower is reduced. In spraying the water to be cooled in a region external to the tower in a manner such that the spray falls just short of the basin, the spray does not interfere with the operation of the NDCT proper, and the maximum increase in air velocity is achieved just above the basin where

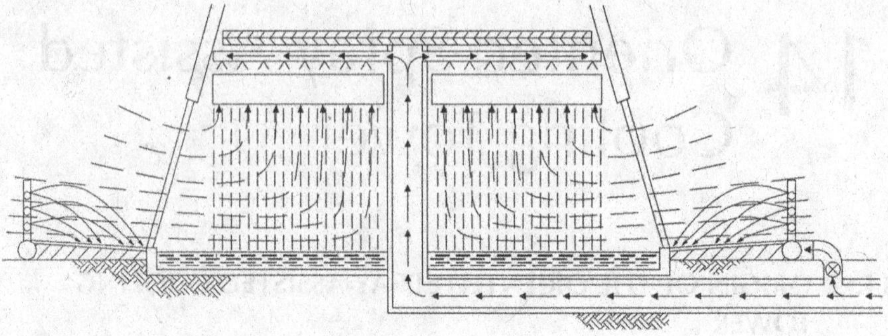

FIGURE 14.1 Oriented spray-assisted cooling tower.

it is the most effective by increasing the airflow into the center of the NDCT, where it is needed the most. Therefore, the effectiveness of evaporative cooling is improved. The resulting improvement in NDCT performance enhances the economic competitiveness of nuclear stations with counterflow NDCT by increasing the electrical output without an increase in the consumption of nuclear fuel or auxiliary power.

The technical basis for the improved capability of counterflow NDCT was documented by Bowman and Benton[2] in 1996. Although rigorous computer codes that have been validated by comparison with extensive test data are used to compute the improvement in heat transfer to increase the cooling tower capability over that of conventional counterflow NDCT design, the concept has not been embraced by the nuclear power industry.

14.2 ORIENTED SPRAY-ASSISTED COOLING TOWER DESIGN

The OSACT contains no moving parts other than a single motor-operated butterfly valve that permits the spray trees to be isolated. Therefore, because the spray trees rely entirely upon the residual pressure in the CCW system at the point where the flow is diverted to the spray header, the spray trees and header piping must be artfully designed hydraulically to produce the desired pressure at the spray tree nozzles.

As discussed in Section 13.7, in 2003 serious consideration was given to constructing an ABWR at the Bellefonte site by Toshiba that would utilize the existing NDCT. The design CCW flow rate for the proposed ABWR was 522,000 gal/min, whereas the design CCW flow rate for the existing NDCT at Bellefonte was only 435,000 gal/min, a difference of 87,000 gal/min. Serious consideration was given to converting the two counterflow NDCTs at the Bellefonte site to OSACTs. The diameter of each NDCT basin at Bellefonte is approximately 412 ft. Because the radius of the spray tree header should be approximately 20–30 ft longer than the basin radius and because the spray trees are spaced approximately 12 ft apart, there would be room for approximately 118 spray trees requiring a

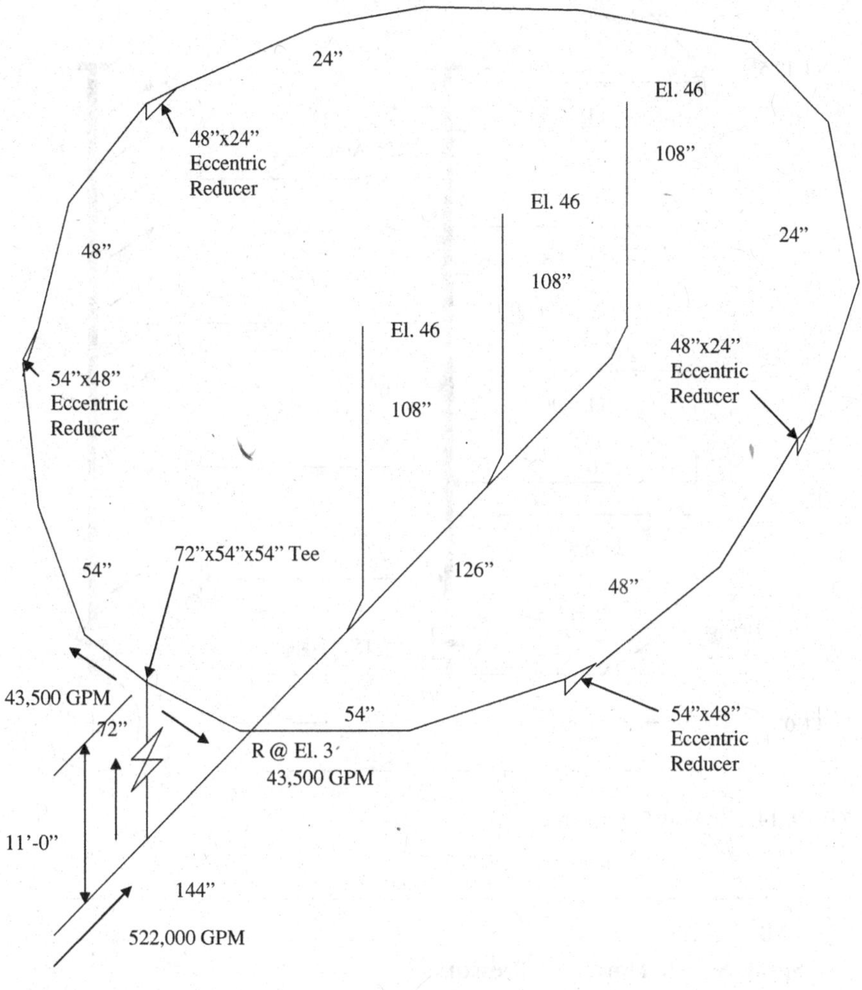

FIGURE 14.2 OSACT piping.

flow of 733 gal/min each. Figure 14.2 shows the selected piping, where pipe sizes were selected based on a hydraulic model analysis to maximize the available pressure at the spray trees.

The elevations shown in Figure 14.2 are with respect to the top of the NDCT basin wall. The results of the hydraulic model indicate a residual pressure of 14.1 lbf/in^{2}g at elevation 3.0 at the midpoint of the spray ring header. Figure 14.3 shows the proposed spray tree design, where the indicated elevations are relative to the top of the NDCT basin wall.

The design consists of (12) 2CX-SS50 Spray Engineering Co. spray nozzles rated at 50 gal/min at 7 lbf/in^2 supported by a 6-in riser pipe and 2-in spray arms. Table 14.1 shows the pressure and flow from each nozzle.

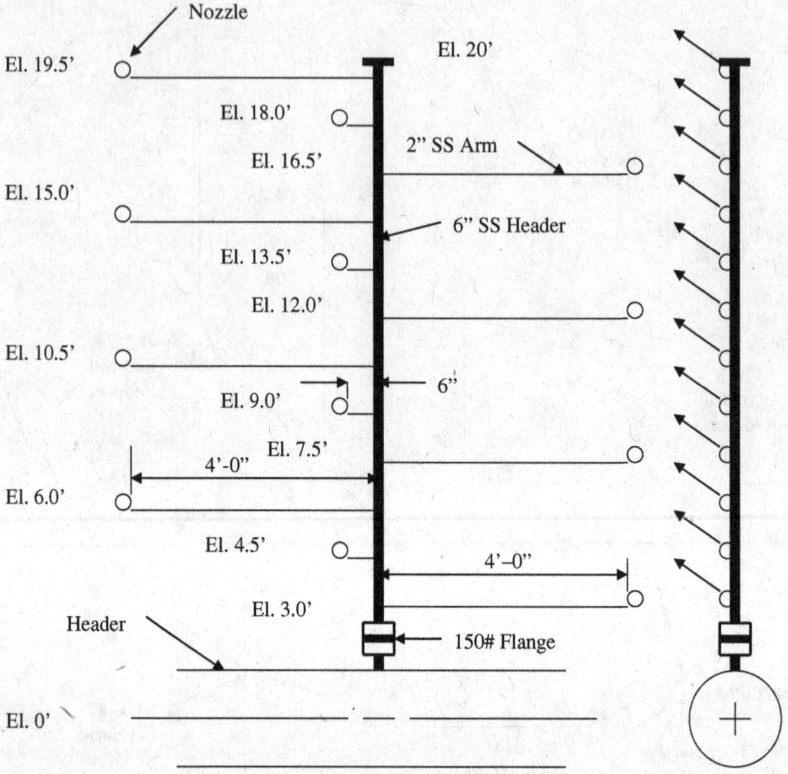

FIGURE 14.3 OSACT spray tree.

TABLE 14.1
Spray Nozzle Flows and Pressures

Nozzle Number	Elevation (ft)	Pressure (psig)	Flow (gal/min)
8L	19.50	7.0	50.00
8K	18.00	7.6	52.27
8J	16.50	8.3	54.44
8I	15.00	8.9	56.53
8H	13.50	9.6	58.55
8G	12.00	10.2	60.49
8F	10.50	10.9	62.38
8E	9.00	11.5	64.21
8D	7.50	12.2	65.99
8C	6.00	12.8	67.73
8B	4.50	13.5	69.42
8A	3.00	14.1	71.07
Total			733.09

14.3 TECHNICAL BASIS FOR THE ORIENTED SPRAY-ASSISTED COOLING TOWER

The technical basis for FACTS is documented in Section 10.8. and that of KXDRAG and THERMAL is documented in Sections 13.3 and 13.4, respectively. What remains is the marriage of these codes to be able to model the OSACT. KXDRAG and THERMAL must first be run to establish the exit condition of the OSCS that becomes the inlet conditions to the counterflow NDCT for the height of the inlet area equal to the control volume of the NDCT. For the Bellefonte spray tree described in Table 14.1, the average spray height is 25.4 ft above the basin wall. The air entering the NDCT above that elevation is modeled at ambient conditions.

Table 14.2 shows the calculated OSCS exit conditions that were calculated for the OSACT at Bellefonte.

Figure 14.4 shows the computed efficiency of the OSCS entering the NDCT. As one may see from Figure 14.4, the OSCS spray efficiency varies with the *WBT* as does that of an NDCT.

TABLE 14.2
Bellefonte Spray Tree Performance

Ambient *WBT* (°F)	Exit *WBT* (°F)	Exit *CWT* (°F)	Exit Wind Speed (ft/s)
40	72.86	88.26	6.46
55	76.57	91.78	6.47
80	93.02	104.45	6.56

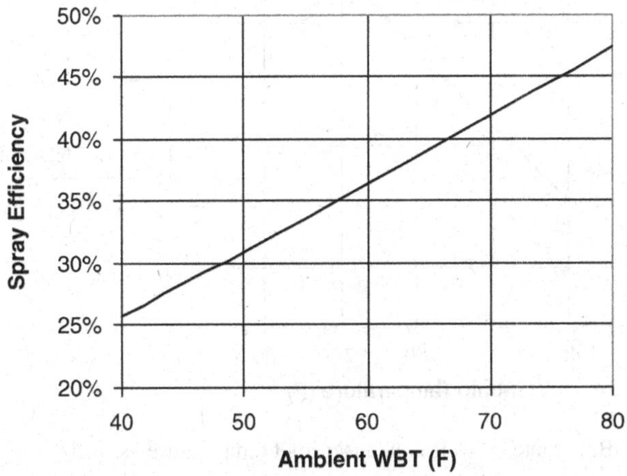

FIGURE 14.4 Efficiency of the proposed Bellefonte OSCS entering the existing NDCT.

Although the temperature of the water from the OSCS is typically several degrees higher than that coming off the NDCT fill material, the airflow through the NDCT is increased by approximately 12%, resulting in a reduction in the *L/G* ratio of approximately 20% and a reduction of the mixed temperature of the CCW coming from the NDCT.

14.4 RESULTS OF THE BELLEFONTE INVESTIGATION

The investigation into how to deal with the discrepancy between the effect of the CCW flow required by the Toshiba ABWR design and the design flow rate of the existing NDCT at the Bellefonte site involved a comparison among (1) increasing the water loading on the existing NDCT to 522,000 gal/min, (2) adding an MDCT, or (3) converting the existing NDCT to an OSACT to cool the 87,000 gal/min difference. Figure 14.5 shows the calculated condenser inlet temperature as a function of *WBT* for each alternative.

Figure 14.6 shows the calculated ABWR net station electrical output per nuclear unit as a function of *WBT*. Figure 14.7 shows the calculated average ABWR net station electrical output per nuclear unit as a function of the months of the year. The greatest benefit occurs during the summer.

Table 14.3 shows the monthly average ABWR net station electrical output per nuclear unit.

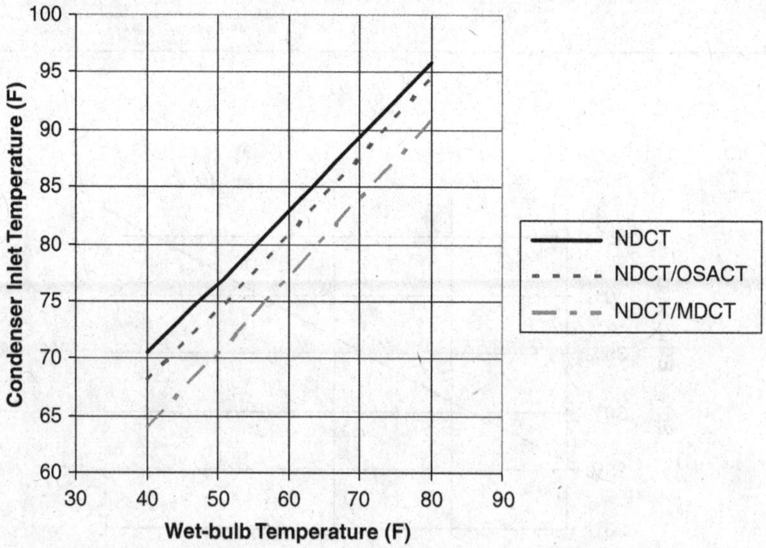

FIGURE 14.5 Bellefonte ABWR condenser inlet temperature vs. *WBT*.

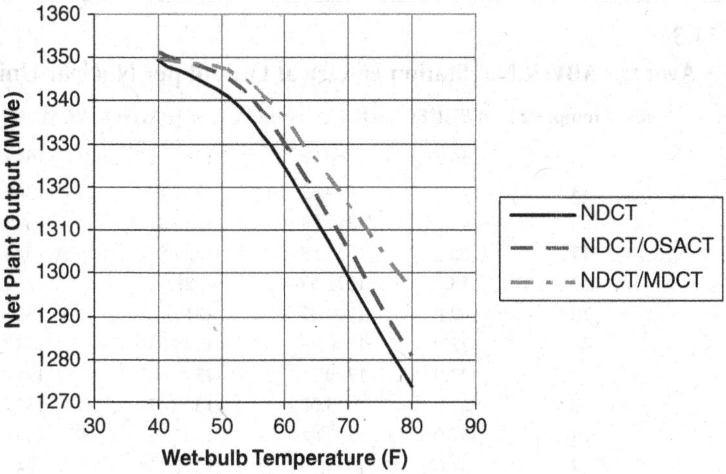

FIGURE 14.6 Bellefonte ABWR net station electrical output vs. *WBT.*

One may see that the increase in the net station electrical output with the OSACT modification is only about 60% of that with the addition of an MDCT. However, a cost estimate conducted in 2003 concluded that the OSACT modification to the existing NDCT costs only 25% of that of a new MDCT. Therefore, the benefit/cost ratio for the OSACT was 170% of that for the MDCT.

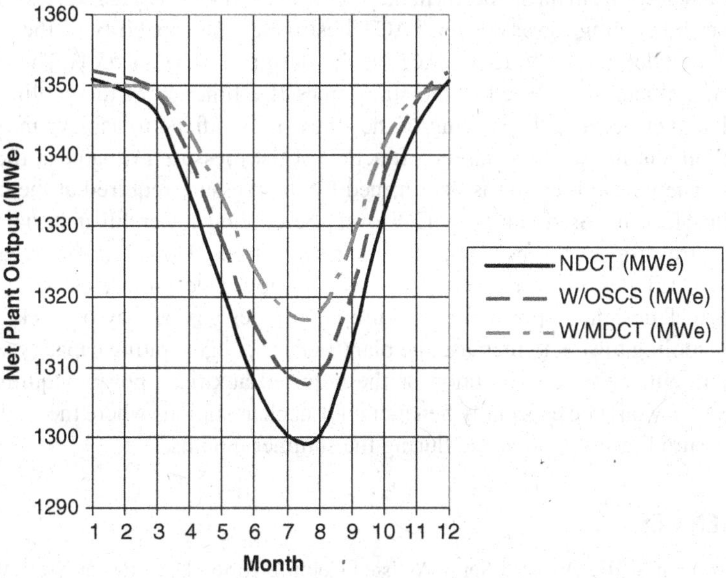

FIGURE 14.7 Bellefonte ABWR net station electrical output vs. *WBT.*

TABLE 14.3
Monthly Average ABWR Net Station Electrical Output per Nuclear Unit

Month	River Temp. (°F)	*WBT* (°F)	NDCT (MWe)	W/o SCS (MWe)	W/MDCT (MWe)
1		36.0	1350.88	1352.11	1349.86
2	33	40.0	1349.32	1351.19	1349.88
3	35	45.0	1345.94	1348.81	1349.01
4	45	55.0	1334.88	1339.86	1343.27
5	62	63.0	1321.57	1328.17	1334.17
6	70	69.0	1308.37	1316.14	1324.09
7	74	72.0	1300.26	1308.63	1317.61
8	77	72.0	1300.26	1308.63	1317.61
9	77	67.0	1313.08	1320.47	1327.77
10	73	57.0	1332.01	1337.39	1341.44
11	64	47.0	1344.19	1347.48	1348.31
12	49	37.0	1350.53	1351.92	1349.91
Average	34	55.0	1329.28	1334.23	1337.75
Difference			Base	4.96	8.47
Difference				3.51	Base

14.5 PERFORMANCE IMPROVEMENT WITH AN OSACT

The actual improvement in performance with the OSACT design varies depending upon the cooling tower design. FACTS estimates the capability of the existing NDCT at Bellefonte to be only 90.6% at the designed *WBT* of 55°F. The amount of CCW that can be diverted to the spray trees is a function of the cooling tower basin diameter because the spacing of the spray trees is fixed to achieve maximum air entrainment in the spray and is a function of the pressure available at the spray nozzles. The nozzle pressure is determined by the pressure required at the cooling tower interface to distribute the CCW over the cooling tower fill material minus the pressure drop in the header piping and spray trees. However, the advantages of the OSACT design over conventional cooling tower design are clearly evident. The OSACT design is applicable to future NDCT designs or may be backfitted to existing cooling towers to increase the plant's electrical generating capacity without increasing either fuel consumption or the required auxiliary power requirements. The OSACT would be especially beneficial for nuclear stations where the LP turbine backpressure limit is approached during the summer months.

REFERENCES

1. Bowman, C. F., Oriented Spray-Assisted Cooling Tower, U. S. Patent No. 5,407,606, April 18, 1995, The United States Patent and Trademark Office, Washington, DC.
2. Bowman, C. F. and D. J. Benton, Oriented Spray-Assisted Cooling Tower, Cooling Tower Institute Paper No. TP96-08, February, 1996.

15 Waste Heat Utilization

15.1 INTRODUCTION TO WASTE HEAT UTILIZATION

Waste heat utilization as described herein is not to be confused with waste heat boilers or cogeneration. A waste heat boiler is an HX that recovers energy from a high-temperature source such as a gas turbine exhaust or a steel mill and uses the energy to produce steam that is expanded through a turbine to create electricity. That would be an example of cogeneration. Another example of cogeneration might be to send a portion of the hot reheat steam, as described in Section 4.2, to an industrial application rather than expanding it all through the LP turbine. Waste heat as described herein is the energy from an electric power plant that is too low grade to be usable in generating electricity.

The second law of thermodynamics requires that any electric power plant operating on the closed Rankine cycle must reject to the main condenser approximately 60% to 70% of the heat that is added to the cycle. This waste heat must then be released into the ambient environment (see Section 6.1). Although the temperature of the waste heat exiting power plants is too low for electric power generation, it may be suitable for other purposes such as heating greenhouses or maintaining an optimum temperature in an aquaculture facility. This is particularly true for those power plants that reject this waste heat directly into the atmosphere via cooling towers.

15.2 WASTE HEAT UTILIZATION CHALLENGES

One might easily agree that rejecting 60%–70% of the heat added to the turbine cycle of a typical power plant is wasteful. The challenges to utilizing this energy are institutional and economic, not technical. Most electric utilities see little benefit to the company in promoting waste heat utilization. The cost of delivering hot water to the waste heat user can be significant compared to the benefit to the end user. In the case of electric power plants operating in an open HRS without cooling towers, the temperature of the waste heat exiting the plant can be as low as 50°F–60°F. A few such facilities where waste heat is utilized in greenhouses and aquaculture actually exist, but these are relatively small demonstration projects because the economics make little sense. Much more practical from an economic perspective are waste heat facilities associated with electric power plants that utilize cooling towers to reject the waste heat directly into the atmosphere. In those cases, the CCW leaves the plant at minimum temperatures ranging from 80°F to 95°F, depending on the location of the plant and the HRS design (see Section 6.2). In those cases, waste heat utilization, though still economically challenging, exhibits more promise.

The institutional problems associated with waste heat utilization are perhaps even more challenging than the economic issues. Unless the electric utility and the waste heat users see some significant economic benefit in cooperating in a waste heat energy park (WHEP) that offsets the potential inconveniences, the project is not likely to mature. In the past, some electric utilities such as Northern States Power, Pennsylvania Power and Light, and the TVA perceived some public relations benefit in cooperating with such projects, but with widespread deregulation of the electric utility industry, greater attention is paid to the bottom line.

Using funds appropriated by the United States Congress, TVA conducted extensive research into the uses of waste heat from electric power plants in the 1970s and 1980s. TVA operated research greenhouses at their Muscle Shoals facility in Alabama and at the Browns Ferry Nuclear Plant in Alabama and conducted experiments at aquaculture facilities at their Gallatin Fossil Plant in Tennessee and at Browns Ferry. TVA and others have clearly demonstrated the technical feasibility of using waste heat to heat greenhouses and in aquaculture. However, TVA no longer receives appropriated funds and must continue to be competitive with its neighbor utilities. Therefore, the future of waste heat utilization hinges more than ever on its economic viability. The economic viability of waste heat utilization is tied to the cost of natural gas, as it is the fuel that is most frequently replaced by waste heat. Low natural gas prices due to horizontal drilling and hydraulic fracturing techniques make waste heat utilization economically challenging.

15.3 PROPOSED WATTS BAR WASTE HEAT ENERGY PARK

In 1978, the TVA investigated the feasibility of utilizing the waste heat from their Watts Bar Nuclear Plant (WBNP), then under construction midway between Knoxville and Chattanooga, Tennessee.[1] At that time, the cost of natural gas was relatively high and was expected to continue to rise in the future. As a result of that investigation, TVA installed waste heat piping from the cooling towers to a point outside of the security fence. The piping is still in existence. A 400-acre area west of the plant was identified for the proposed WHEP. The piping system between the NDCT and the WHEP was designed to deliver 100,000 gal/min. In 1980, TVA issued a final environmental impact statement concluding that the successful demonstration of commercial waste heat utilization would benefit the Tennessee Valley region and the nation and that granting an easement to a park management organization for the development of a WHEP would be an environmentally sound action with fewer adverse impacts than would be expected from similar development elsewhere utilizing conventional heat sources. TVA proceeded to implement an aggressive marketing campaign for the WHEP, receiving letters of interest from nine greenhouse companies and five manufacturing companies that engage in ethanol production, leather tanning, and wood preserving. In March 1982 in a letter from the TVA Manager of Power to the Executive of Rhea County, Tennessee, TVA promised Rhea County "If your consultants are able to obtain commitments of sufficient infrastructure funding from grants or other sources and a substantial user of hot water makes a firm commitment to locate in the park, TVA would install the waste heat piping required to serve his needs."[2] However, TVA subsequently deferred the second unit at Watts Bar due to the high cost of nuclear construction and declining electric power demand. With only one nuclear unit, the WHEP occupants would not be assured of a reliable source of waste heat. A WHEP that relies on the availability of a single electric generating unit is not considered feasible. The second nuclear unit was completed in 2016. However, the corporate memory of the once proposed WHEP at Watts Bar no longer exists at TVA. The characterization of waste heat in the following section is based on the studies conducted on the WHEP at Watts Bar as reported in Reference 1.

15.4 WASTE HEAT CHARACTERIZATION

The quality of waste heat is a function of the HRS design, the ambient *WBT*, and, to a lesser extent, the *RH* (see Section 9.3). As discussed in Section 6.2, multi-pressure main condensers are common where cooling towers are employed, resulting in a reduced CCW flow rate and a corresponding increase in the temperature rise through the main condenser. This, along with the fact that the *CWT* coming from the cooling tower is relatively high, results in a much higher *HWT* coming from the main condenser than would otherwise be the case.

Figure 15.1 shows the minimum, average, and maximum daily average *HWT* coming from the main condenser. Of course, as with the *WBT*, there is wide variation in the *HWT* as shown for January in Figure 15.2.

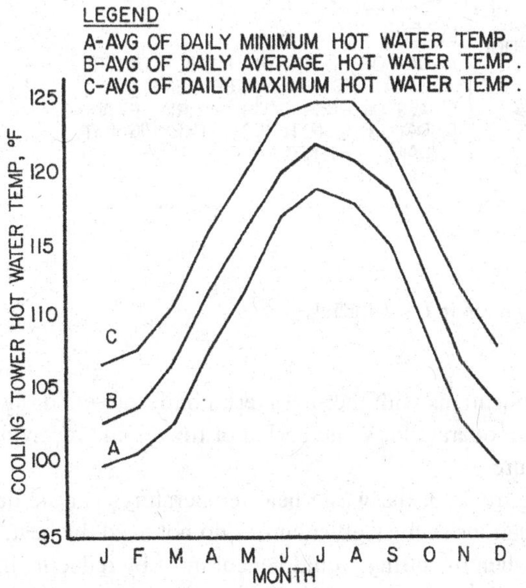

FIGURE 15.1 Average daily *HWT*.

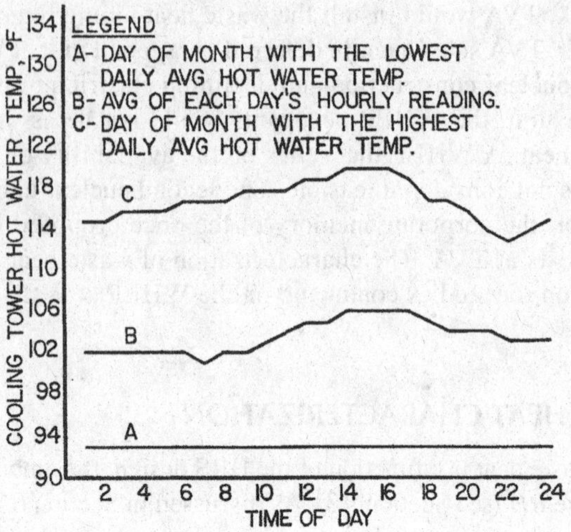

FIGURE 15.2 Variation in *HWT* for January.

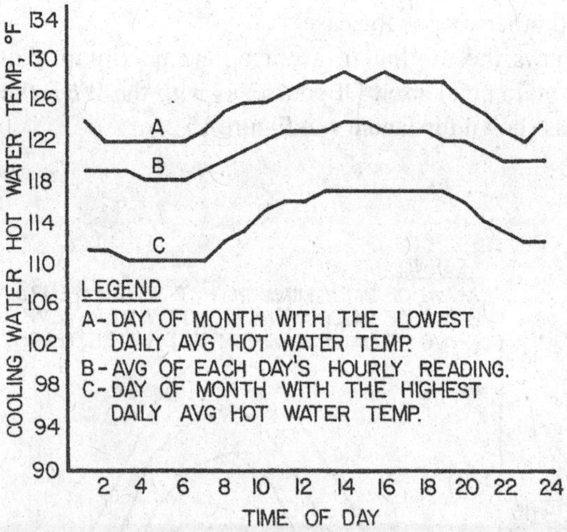

FIGURE 15.3 Variation in *HWT* for July.

Waste heat applications with thermal inertia built into the design of the application may be able to operate for some period of time while receiving the waste heat at lower temperatures.

As seen in Figure 15.3, the waste heat temperatures can be quite high in July when some applications such as greenhouses do not need the heat. However, this is the time of year when the utility would benefit most by reducing the heat that must be rejected through the cooling tower.

15.5 WASTE HEAT APPLICATIONS

Although some industrial applications such as ethanol production, leather tanning, and wood preserving might benefit from the use of waste heat, the vast majority of interest in using waste heat and the focus of much research has been on heating greenhouses and aquaculture.

15.5.1 Greenhouse Applications

Much research has been conducted and several demonstration projects have focused on using waste heat for heating greenhouses. Much of the research has focused on designing greenhouses to utilize very low-grade waste heat such as might be available from an electric generating station that has an open HRS (see Section 9.1).

In a conventional greenhouse, a common practice is to employ unit heaters to deliver forced hot air on the perimeter of the greenhouse and/or overhead heating pipes using steam or hot water as a heating source. TVA, Rutgers University, and others have conducted a considerable amount of research on waste heat greenhouses. One of the significant developments that is attractive for waste heat applications is to utilize floor heating using hot water heating tubing embedded in the floor. The floor heating system may include either a 3 in or 4 in slab of porous concrete poured over a bed of flooded stone aggregate containing the tubing or simply a 3 in or 4 in slab of porous concrete containing the tubing. This research has shown that the heat transfer coefficient is greater for a wet floor system compared to a dry floor system.[3] The recommended limit in the length of heating tubes in the floor is approximately 150 ft to maintain a uniform temperature in the greenhouse. Depending on the quality of the waste heat, additional overhead heating may be required during periods of colder weather.

For floor-heated greenhouses,

$$Q = U_F A\left(T_{wh-ave} - T_{GH}\right) = U_C A\left(T_{GH} - T_{amb}\right) \tag{15.1}$$

where:

Q = heat transfer rate, Btu/hr
U_F = floor heat transfer coefficient, Btu/hr-ft^2-°F
A = greenhouse floor area (neglecting sidewall area)
T_{wh-ave} = average waste heat temperature, °F
T_{GH} = greenhouse temperature, °F
U_C = covering heat transfer coefficient, Btu/hr-ft^2-°F
T_{amb} = ambient temperature, °F.

Published values for U_F from the heating pipes to the greenhouse interior vary from 0.73 to 1.23 Btu/hr-ft^2-°F, depending on the floor design and the configuration of the plants on the floor and/or on benches in the greenhouse.[4,5]

15.5.2 Aquaculture Applications

TVA and several universities have conducted a considerable amount of research on high-density raceway aquaculture (HDRA). Both catfish and tilapia are species that benefit from warm water. The vast majority of all of the catfish produced in the United States comes from farm ponds in the southern states.[6] Consumption of tilapia has increased dramatically as it has been transformed from a mostly ethnic cuisine to a staple on the menu of virtually every family restaurant chain in the United States. The vast majority of all tilapia that is consumed in the United States is imported from countries such as Taiwan, Costa Rica, and Ecuador.[7] Tilapia is the third largest aquaculture import, exceeded only by shrimp and Atlantic salmon.[8] Of the relatively small amount of domestically produced tilapia, 70% is produced in indoor water re-circulating aquaculture systems, perhaps because tilapia production in outside ponds is strictly regulated in the southern United States for fear that some fish may escape and encroach on the native sport fishing population.[8,9]

A feeding regimen optimizing the ratio of feed to fish weight is desired. The preferred water temperature range for optimum growth for catfish and tilapia is 82°F–86°F.[10] Figure 15.4 shows an estimated curve of feed conversion vs. water temperature. Factors other than water temperature also affect the feed conversion ratio.

Properly designed high-density raceways have a length-to-width-to-depth ratio of 30:3:1 with a flow rate of 6–12 gal/min per 100 pounds of fish, a water velocity of at least 6.5 ft/s, and 4–10 water changes per hour to support the oxygen requirements.[11,12] Table 15.1 shows typical HDRA design parameters.

Each raceway would be divided into eight sections. The first section at the CCW inlet end would contain fingerlings. As these grow to approximately 0.25 lb in weight, they would be moved to the next two sections where they would grow from 0.25 to 0.5 lb and finally to the next four sections where they would grow to approximately 1.0 lb before being harvested. The last section would be reserved for flushing out the waste. Any residual waste would be collected in a settling pond before the CCW is returned to the nuclear station.

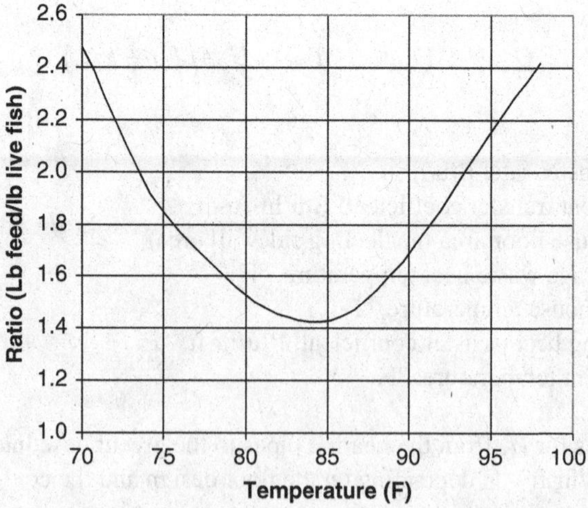

FIGURE 15.4 Warm water aquaculture feed conversion. (Courtesy of ASME, New York.)

TABLE 15.1
High-Density Raceway Design Parameters

Length, ft	90
Width, ft	9
Depth, ft	3
Water depth, ft	2.5
Number of raceways	54
Flow per raceway, gal/min	1,852

The entire growing process requires approximately 105 days.[13] Assuming a final stocking rate of approximately 9.0 lb/ft^3, each raceway would produce approximately 31,400 lbs/year for a total production of 1,700,000 lbs/year.

15.6 ECONOMICS OF WASTE HEAT UTILIZATION

In addition to the cost of any industrial park and that of the greenhouses and/or high-density raceways in a WHEP, the principal cost directly attributable to the waste heat is that of the waste heat piping to and from the cooling tower(s) and any additional cost to pump the CCW there and back to the nuclear station. Based on the proposed 400-acre Watts Bar WHEP with 100,000 gal/min delivered to the park, 200 acres of greenhouses is assumed. For greenhouses, a conservative value for U_F of 0.71 Btu/hr-ft^2-°F is assumed in the following analysis of waste heat cost. A wide variety of greenhouse designs is available. In the analysis contained herein, single glazed glass design with a U_G of 1.2 Btu/hr-ft^2-°F is assumed. Because the recommended limit in the length of heating tubes in the floor is approximately 150 ft to maintain a uniform temperature in the greenhouse, greenhouses would be arranged such that the CCW would cascade from the first greenhouse to the second with the operating temperatures and, perhaps, the crops being different in each. The temperatures shown in Figure 15.5 are without supplemental heat that would be provided as required for the particular crop being grown.

Because two greenhouses would share the same CCW flow, each acre of greenhouses would receive 1,000 gal/min flowing through 0.75-in I.D. pipes embedded in the floor spaced 10 in apart, each carrying a maximum flow of approximately 2.9 gal/min with a velocity of 2.1 ft/s. A bypass would be provided around each greenhouse so that the flow could be reduced as required to avoid overheating. Table 15.2 shows the greenhouse energy consumption.

Of course, during the warmer months greenhouses require cooling, not heating, and the analysis assumes that all heating would be terminated when the ambient temperature exceeds approximately 65°F. Each grower could tap into the pressurized waste heat piping to provide water for evaporative cooling pads as long as the remaining CCW is returned to the waste heat piping under pressure.

Based on the stated assumptions, the amount of heat that would be transferred from the CCW to each acre of greenhouse would be approximately 3.3 × 10^9 Btu/year, as shown in Table 15.2.

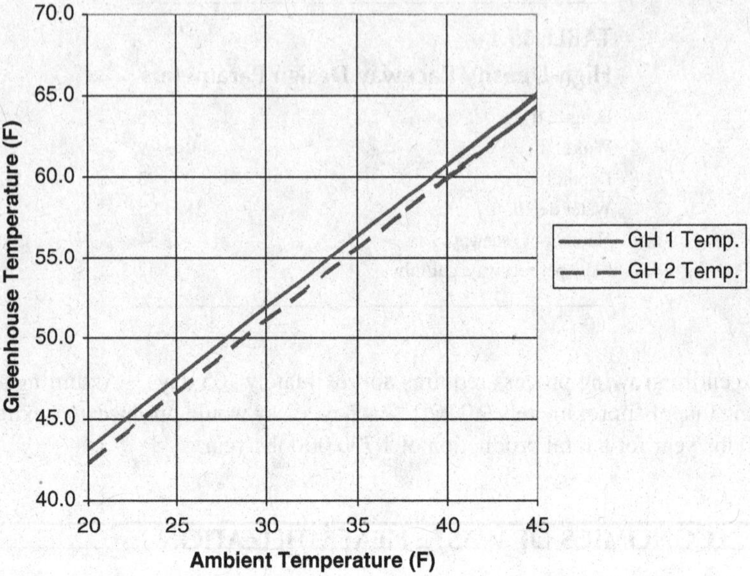

FIGURE 15.5 Greenhouse temperature without supplemental heat. (Courtesy of ASME, New York.)

TABLE 15.2
Greenhouse Heat Consumption per Year

Amb. Temp.	Hours per year	Deg. Hours	GH-1 Temp.	GH-1 Heat Loss (Btu)	GH-1 Temp.	GH-2 Heat Loss (Btu)
65	106	6,884	82.4	9.7E + 07	81.7	9.3E + 07
60	147	8,832	78.3	1.4E + 08	77.6	1.3E + 08
55	372	20,475	74.0	3.7E + 08	73.3	3.6E + 08
50	240	12,000	69.8	2.5E + 08	69.1	2.4E + 08
45	320	14,387	65.6	3.4E + 08	64.8	3.3E + 08
40	405	16,168	61.3	4.5E + 08	60.5	4.4E + 08
35	529	18,547	57.2	6.1E + 08	56.4	5.9E + 08
30	428	12,849	53.0	5.1E + 08	52.1	4.9E + 08
25	272	6,795	48.8	3.4E + 08	47.8	3.3E + 08
20	8	359	44.5	2.3E + 07	43.6	2.2E + 07
15	11	166	40.4	1.5E + 07	39.5	1.4E + 07
10	86	863	36.2	1.2E + 08	35.2	1.1E + 08
	37	185	31.9	5.2E + 07	30.9	5.0E + 07
0	10	0	27.7	1.4E + 07	26.7	1.4E + 07
−5	2	−10	23.5	3.0E + 06	22.4	2.9E + 06
				3.3E + 09		3.2E + 09

In 2011, the estimated capital cost of delivering and returning the 100,000 gal/min of CCW to and from the WHEP was approximately $28,000,000. This cost did not include any additional capital cost of the greenhouses to accommodate waste heat. The annual debt service to pay back the bonds at 5% interest required to construct just the piping and pumping station required to return the CCW to the cooling towers would be $2,200,000 or $11,000/acre plus an additional $4,500/acre/year for the electric power to return the CCW back to the cooling towers. Based on Table 15.2, the annual operating cost to supply the waste heat to the greenhouses would be $4.74/MBTU. The average cost of natural gas in 2011 was about $7.00/MBTU and dropping.

Figure 15.6 shows the average prices paid for catfish and tilapia as published by the United States Department of Agriculture (USDA). The USDA stopped publishing the price of imported tilapia when it discontinued its *Aquaculture Outlook* in 2006. However, the price for U.S.-farmed tilapia was approximately $2.50–$2.75 per pound at the end of 2011.[11] The sharp spike in the catfish price in 2011 because of the shortage of product due to high feed prices and the resulting loss of production capacity is expected to continue as producers continue to struggle to achieve profitability.[6] The decision as to whether the high-density raceway aquaculture facility would produce catfish or tilapia is beyond the scope of this investigation. However, for purposes of economic analysis, a farm price of $1.90/lb of live fish is assumed.

Fish feed cost averaged $353/ton in 2010 and was slightly higher in 2011.[6] For purposes of this investigation, a feed price of $400/ton is assumed. For both catfish and tilapia, the feed constitutes approximately 50% of the total cost of production.[6,7] Therefore, any feeding regimen that optimizes the ratio of feed to fish weight is to be desired. Although the preferred water temperature range for optimum growth is 82°F–86°F,[13] factors other than water temperature also affect the feed

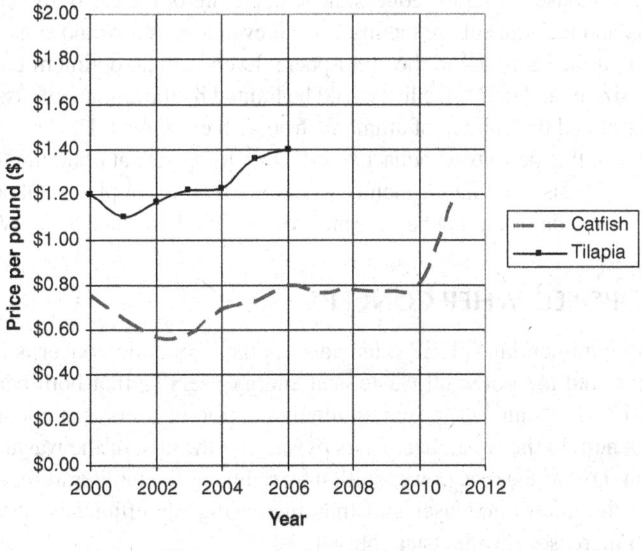

FIGURE 15.6 U.S. average warm water fish pond/import prices. (Courtesy of ASME, New York.)

TABLE 15.3
High-Density Raceway Economics

Live fish value, $/lb	$1.90
Production Costs per lb	
Total feed, $/lb	$0.36
Labor, $/lb	$0.16
Fingerlings, $/lb	$0.10
Utilities and fuel, $/lb	$0.03
Misc. fixed costs, $/lb	$0.14
Total production cost per pound, $/lb	$0.79
Net income, $/lb	$1.11
Net income per year	$1,900,000

conversion ratio. An annual average feed/live weight ratio of approximately 1.8:1 is assumed if the HDRA facility were to operate year-round.[13,14]

Table 15.3 shows the estimated net income from an HDRA facility in the proposed WHEP. As one may see, the net income from an HDRA would not even service the $28,000,000 debt for installing the WHEP piping.

One potential advantage of a WHEP to the utility might be to reduce the heat load on the cooling tower and thus the temperature of the CCW that is returned to the main condenser. A lower CCW temperature results in a lower main condenser pressure and more electrical output (see Section 6.1). However, neither the greenhouses nor aquaculture facilities cool the CCW to any significant extent. The temperature drop through each of the two greenhouses in series reduces the temperature of the CCW by only approximately 2.0°F, and the aquaculture facility cools it even less as it would consist of only 1 acre of water surface exposed to the atmosphere. Even with an optimum growing temperature, the size of an HDRA facility would be limited by the amount of dissolved oxygen in the water and the buildup of ammonia from fish excrement. However, in addition to the waste heat that permits an aquaculture facility to operate at optimum temperature on a year-round basis, the HDRA facility would not require supplemental aeration and a buildup of ammonia would not be a problem due to the abundance of CCW available.

15.7 PROPOSED WHEP CONCEPT

A successful commercial WHEP must address the economic concerns of both the electric utility and the potential waste heat energy users so that both benefit financially. In 2012, Bowman[15] proposed an amalgamation of users that could provide a significant benefit to the waste heat users by sharing the cost of the waste heat delivery system and to the electric power station by reducing the temperature of the CCW returning to the main condenser and thus increasing the efficiency of the turbine cycle, which increases the electrical output.

Figure 15.7 shows schematically the proposed WHEP. A portion of the CCW would be withdrawn from the CCW piping entering the cooling tower(s) and would be

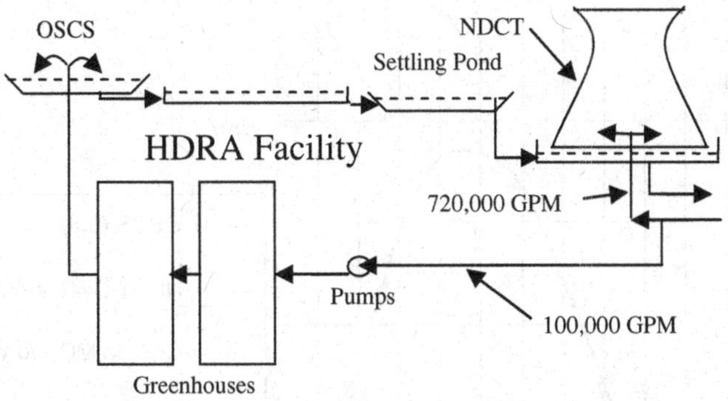

FIGURE 15.7 Proposed waste heat energy park. (Courtesy of ASME, New York.)

delivered to a pumping station by the residual pressure in the CCW piping at that point. From there, the CCW would be pumped through the waste heat piping system, through the two stages of greenhouses, and on to an OSCS (see Chapter 13) where it would be further cooled before entering the HDRA facility. From the HDRA facility, the CCW would drain by gravity through a settling pond and back to the cooling tower basin.

The NDCT shown in Figure 15.7 is illustrative of two NDCTs, each designed for a CCW flow of 410,000 gal/min and with 50,000 gal/min being drawn off from each to the WHEP.

Figure 15.8 shows the monthly average temperature of the CCW entering and leaving the greenhouses and the growing area temperatures in the second greenhouse growing area.

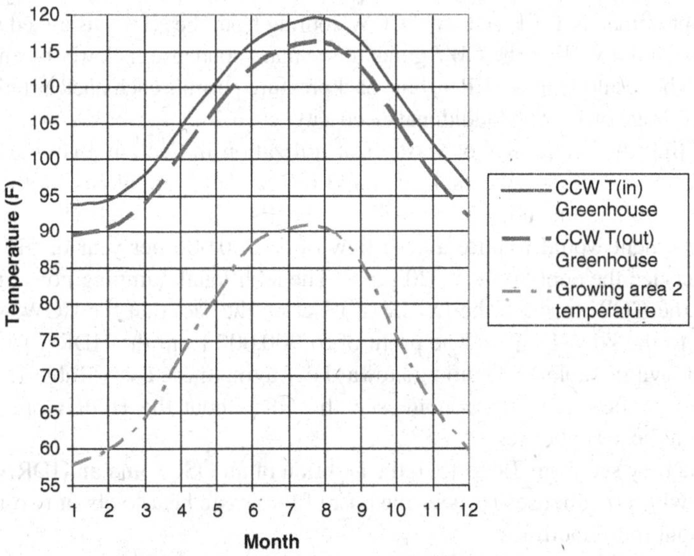

FIGURE 15.8 Waste heat greenhouse temperatures.

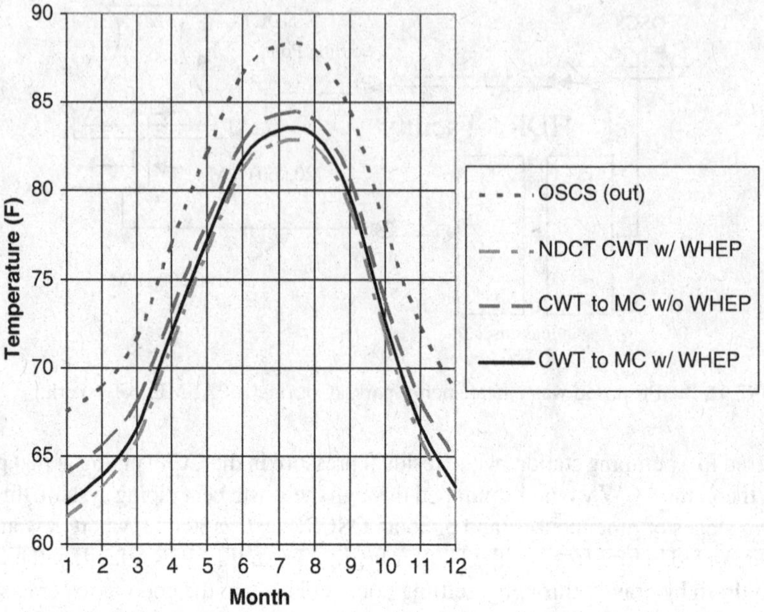

FIGURE 15.9 Benefit of the WHEP to the utility.

Figure 15.9 shows the benefit of the WHEP to the utility. Although the temperature leaving the OSCS/HDRA facility is higher than that coming from the NDCT either with or without the WHEP in service, the reduction of the water loading on the NDCT with the WHEP in service lowers the temperature coming from the NDCT by an average of approximately 1.9°F. After the CCW coming from the NDCT is mixed with that coming from the WHEP, the *CWT* going to the main condenser is lower by an average of 1.2°F. The resulting lower LP turbine backpressure permits each nuclear unit to generate an average of 1.5 MW additional electricity.

Revisiting the economics of waste heat utilization in Section 15.6, the addition of the OSCS and the HDRA facility would increase the capital cost of the WHEP to $38,000,000 to include the OSCS, the HDRA facility, and a settling pond. This investment would require a cash flow of $3,000,000 per year or $15,200 per acre to service the debt at 5% for 20 years. The additional pumping power required to spray the CCW through the OSCS is offset by the fact that the CCW would be returned to the NDCT basin. The profit of $1,900,000 from the HDRA facility per year as shown in Table 15.3 could go toward defraying these costs. Table 15.4 shows a comparison between the economics with and without the HDRA operating in series with the greenhouses.

As one may see from Table 15.4, the addition of an OSCS and an HDRA facility in series with greenhouses reduces the cost of the waste heat to about two-thirds of that without those facilities.

The economic analysis of the WHEP shown in Table 15.4 is limited to the direct benefits associated with the utilization of waste heat and the incremental costs associated

TABLE 15.4
Comparison of Cost of Energy with and without High-Density Raceway Aquaculture

	Without HDRA	With HDRA
Debt service/acre/year	$11,045	$15,200
Pumping cost/acre/year	$4,500	$4,500
HDRA profit/year		$9,650
Net cost/year	$15,545	$10,050
Energy consumed, Btu/year	3.28E + 03	3.28E + 03
Cost of energy, $/Btu	$4.74	$3.07

with providing this resource. For example, it does not include costs associated with any typical industrial park such as roads and utilities or the benefits to the community such as the additional jobs and the increase in the tax base. The projected net annual income does not include the projected benefit to the nuclear station. Nor does it include the income or expenses including debt service for the proposed 200 acres of floor-heated greenhouses, which may be more expensive than a conventional concrete floor. Nor does it include the profit to be realized from a successful greenhouse operation but only the fuel savings to the greenhouse operator due to the use of the waste heat for the base heating load. This approach is taken because many different types of possible greenhouse operations exist without the benefit of waste heat, each with its own economic parameters. Indeed, the consumptive use of waste heat as envisioned with greenhouses would be only one possible application. Others might include a pig farrowing or broiler house operation, lumber drying, tanning, and ethanol production. The cost and benefit of the HDRA facility are included because it is quite straightforward and because it would not be at all feasible without the waste heat being available.

The WHEP as envisioned in Figure 15.7 in which there exists an amalgamation of waste heat users enjoys a significant economic advantage over a conventional arrangement in which each waste heat user operates in parallel and consumes a portion of the available waste heat piping and pumping capacity, regardless of the nature of its need. In that case, the economic benefit would be significantly diminished. The new concept for utilizing waste heat presented here significantly reduces the cost of delivering waste heat by an amalgamation of users and provides a significant benefit to the utility by reducing the LP turbine backpressure, thus increasing the efficiency of the turbine cycle and the electrical output. Based on the assumed parameters, the technical and economic feasibility of the proposed WHEP concept is clearly demonstrated.

REFERENCES

1. Bowman, C. F. and R. E. Taylor, Design of the Proposed Watts Bar Waste Heat Park, *Proceedings of the Waste Heat Utilization and Management Conference*, Miami Beach, FL, Hemisphere Publishing Corp., 1983.
2. Dayton Herald, March 25, 1982.

3. Roberts, W. J. and D. R. Mears, Floor Heating of Greenhouses, ASAE Paper No. 80-4027, American Society of Agricultural Engineers, St. Joseph, Michigan, 1980.
4. Both, A. J., et al., Evaluating Energy Savings Strategies Using Heat Pumps and Energy Storage for Greenhouses, Paper No. 074011, American Society of Agricultural and Biological Engineers, St. Joseph, Michigan, 2007.
5. Manning, T. O., et al., Feasibility of Waste Heat Utilization in Greenhouses, ASAE Paper No. 80-401, American Society of Agricultural Engineers, St. Joseph, Michigan, 1980.
6. Hanson, T. and D. Sites, U.S. Farm-Raised Catfish Industry, 2009 Review and 2010 Outlook, Auburn University, 2009.
7. Fitzsimmons, K., Development of New Products and Markets for the Global Tilapia Trade, *Proceedings of the Sixth International Symposium on Tilapia in Aquaculture*, Manila, Philippines, September 12–16, 2004.
8. Kohler, C. C., A White Paper on the Status and Needs of Tilapia Aquaculture in the North Central Region, North Central Region Aquaculture Center, 2004.
9. Sell, R., Tilapia, North Dakota State University, 2005.
10. Rakocy, J. E., Tank Culture of Tilapia, Southern Regional Aquaculture Center Publication No. 282, Texas Agricultural Extension Service, The Texas A&M University, 1989.
11. Fitzsimmons, K., Personal communication, University of Arizona, December 9, 2011.
12. Soderberg, R. A., Tilapia Culture in Flowing Water, Mansfield University, 1990.
13. Goss, L. B., et al., Utilization of Waste Heat from Power Plants for Aquaculture, Gallatin Catfish Project, 1973 Annual Report, Power Research Staff, Tennessee Valley Authority, 1973.
14. Adams, C. and A. Lazur, A Preliminary Assessment of the Cost and Earnings of Commercial, Small-Scale, Outdoor Pond Culture of Tilapia in Florida, FE 210, Department of Food and Resource Economics, Florida Cooperative Extension Service, Institute of Food and Agricultural Sciences, University of Florida, October 2000.
15. Bowman, C. F., Electric Power Plant Waste Heat Utilization, *Proceedings of the ASME 2012 Summer Heat Transfer Conference*, 2012.

16 Heat Exchanger Analysis and Testing

16.1 TYPES OF HEAT EXCHANGERS IN A NUCLEAR STATION

An HX as described herein is a device used to transfer heat between two fluids that are normally separated by a solid wall to prevent the mixing of the fluids. Nuclear stations employ a wide variety of specialty HXs such as MSRs (Chapter 4), condensers (Chapter 6), and FWHs (Chapter 7). This chapter is devoted to the design and testing of the more common shell-and-tube HXs utilized in several applications in a nuclear station such as residual heat removal, component cooling, oil cooling, and others. Chapters 17 and 18 address the more exotic air-to-water and plate HXs, respectively.

16.2 FIRST LAW OF THERMODYNAMICS AS APPLIED TO TYPES OF HEAT EXCHANGERS

Figure 16.1 schematically illustrates the most basic form of an HX.

where:

$t_{c\text{-}i}$ = cold stream temperature in
$t_{c\text{-}o}$ = cold stream temperature out
$T_{h\text{-}i}$ = hot stream temperature in
$T_{h\text{-}o}$ = hot stream temperature out
m_c = mass flow rate of cold stream
m_h = mass flow rate of hot stream.

From the first law of thermodynamics,

$$m_h\, h_{h\text{-}i} + m_c\, h_{c\text{-}i} = m_h\, h_{h\text{-}o} + m_c\, h_{c\text{-}o} \tag{16.1}$$

$$Q = m_h\,(h_{h\text{-}i} - h_{h\text{-}o}) = m_c\,(h_{c\text{-}o} - h_{c\text{-}i}) \tag{16.2}$$

$$Q = m_h\, c_{p\text{-}h}(T_{h\text{-}i} - T_{h\text{-}o}) = m_c\, c_{p\text{-}c}\,(t_{c\text{-}o} - t_{c\text{-}i}) \tag{16.3}$$

where:

$c_{p\text{-}h}$ = hot-side specific heat
$c_{p\text{-}c}$ = cold-side specific heat
$h_{c\text{-}i}$ = cold stream enthalpy in
$h_{c\text{-}o}$ = cold stream enthalpy out

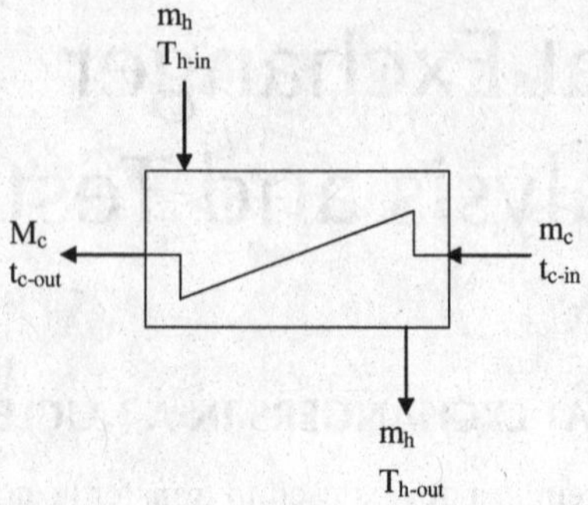

FIGURE 16.1 Heat exchanger.

$h_{h\text{-}i}$ = hot stream enthalpy in
$h_{h\text{-}o}$ = hot stream enthalpy out
Q = rate of heat transfer

and

$$\Delta h = c_p \, \Delta T \tag{16.4}$$

Any one of the six parameters may be calculated if the other five are known. For example,

$$m_h = \frac{m_c \, c_{p\text{-}c} \, (t_{c\text{-}o} - t_{c\text{-}i})}{c_{p\text{-}h} \, (T_{h\text{-}i} - T_{h\text{-}o})} \tag{16.5}$$

16.3 HEAT EXCHANGER ANALYSIS USING THE LOG MEAN TEMPERATURE DIFFERENCE METHOD

The oldest and most commonly used method of HX analysis to determine the heat transfer through an HX is the log mean temperature difference method defined as

$$Q = UA_h F \, LMTD \tag{16.6}$$

where:
U = overall heat transfer coefficient
A_h = effective hot-side HX surface area
F = $LMTD$ correction factor
$LMTD$ = log mean temperature difference.

The accepted practice is to take the hot-side (shell-side) effective surface area, A_h, as the HX reference surface area, where

$$A_h = N_{tubes} N_{pass} \pi d_o l_{eff} \tag{16.7}$$

where:

N_{tubes} = number of tubes per pass
N_{pass} = number of passes
d_o = outside tube diameter
l_{eff} = effective tube length.

The *LMTD* correction factor, *F*, is a function of the type of HX. The value of $F = 1$ for one-shell-pass/one-tube-pass counter-flow arrangements and other counter-flow HX arrangements (see Chapters 17 and 18). A formula for the value of *F* is available for a few other flow arrangements in the open literature. Appendix D of Reference 1 shows graphs of values for *F* for multiple tube pass HX where the *LMTD* correction factor is a function of the ratio of the hot-side to cold-side temperature change, *R*, and the effectiveness of the HX, *P*, as defined by the following:

$$R = \frac{T_{h,i} - T_{h,o}}{t_{c,o} - t_{c,i}} \tag{16.8}$$

$$P = \frac{t_{c,o} - t_{c,i}}{T_{h,i} - t_{c,i}} \tag{16.9}$$

The value of *LMTD* for a counter-flow HX is as follows:

$$LMTD = \frac{\Delta t_1 - \Delta t_2}{\ln\left(\dfrac{\Delta t_1}{\Delta t_2}\right)} \tag{16.10}$$

where

$$\Delta t_1 = T_{h,i} - t_{c,o} \tag{16.11}$$

$$\Delta t_2 = T_{h,o} - t_{c,i} \tag{16.12}$$

As one may see from the above equations, because both *Q* and the *LMTD* are functions of temperature, a calculation to design an HX using the log mean temperature difference method requires iteration.

16.4 HEAT EXCHANGER ANALYSIS USING THE EFFECTIVENESS AND EFFECTIVE MEAN TEMPERATURE DIFFERENCE METHODS

Alternative approaches to HX analysis have been developed that feature use of the relations among *P*, *R*, and *NTU* for specified geometries without dependence on *LMTD* or the correction factor *F*. Two of the methods that have proven to be of practical value in HX rating, testing, and uncertainty analysis for a wide range of basic and complex HX arrangements are the effectiveness and effective mean temperature difference (EMTD) methods. (See References 2 to 5.) Both of these methods are based on the defining relations for P, R, and NTU expressed in terms of cold stream reference as follows:

$$P = \frac{Q}{m_c c_{p\text{-}c}(T_{h\text{-}i} - t_{c\text{-}i})} \tag{16.13}$$

$$R = \frac{m_c c_{p\text{-}c}}{m_h c_{p\text{-}h}} \tag{16.14}$$

$$NTU = \frac{UA_h}{m_c c_{p\text{-}c}} \tag{16.15}$$

Explicit relations that express P in terms of R and NTU have been developed for many standard arrangements, with listing available in References 6 and 7. Equations 16.13 and 16.14 for P and R are related to Equations 16.8 and 16.9 by Equation 16.3 for Q.

The effectiveness method is based on the use of Equation 16.13 to express Q in terms of P by

$$Q = Pm_c c_{p\text{-}c}\left(T_{h\text{-}i} - t_{c\text{-}i}\right) \tag{16.16}$$

For rating applications, this equation normally can be coupled with inputs for R and NTU to calculate Q and two unspecified terminal temperatures without iteration, except for refinement of temperature-dependent fluid properties when required. The effectiveness method can also be adapted to testing applications that would be expected to involve iteration with a few exceptions.

The EMTD method is based on the use of Equations 16.3 and 16.15 to express NTU in terms of Q by

$$NTU = \frac{UA_h(t_{c\text{-}o} - t_{c\text{-}i})}{Q} \tag{16.17}$$

and the use of the defining relation for $EMTD$ (which corresponds to F $LMTD$) given by

$$Q = UA_h EMTD \tag{16.18}$$

to obtain

$$EMTD = \frac{t_{c\text{-}o} - t_{c\text{-}i}}{NTU} \tag{16.19}$$

It also follows that this relation can be combined with Equation 16.9 to express $EMTD$ in terms of P by

$$EMTD = \frac{P(T_{h\text{-}i} - t_{c\text{-}i})}{NTU} \tag{16.20}$$

Whereas Equation 16.19 can be used to evaluate U for basic testing applications without iteration, Equation 16.20 can be adapted for standard rating that normally involves iteration.

Because of the stability of the *P–R–NTU–Q* relations, the effectiveness and EMTD methods give rise to convergent calculations when iteration is required.

16.5 DETERMINATION OF THE OVERALL HEAT TRANSFER COEFFICIENT

The overall heat transfer coefficient, U, may be expressed as the inverse of the sum of the resistances to heat transfer through the following:

- Exterior convection layer, r_h
- Exterior fouling, $r_{f,h}$
- Tube wall, r_w
- Interior fouling, $r_{f,c}$
- Interior convection layer, r_c

$$\frac{1}{U} = r_h + r_{f,h} + \left(\frac{A_h}{A_w}\right) r_w + \left(\frac{A_h}{A_c}\right) r_{f,c} + \left(\frac{A_h}{A_c}\right) r_c \tag{16.21}$$

where

$$r_h = \frac{1}{h_h} \tag{16.22}$$

$$r_c = \frac{1}{h_c} \tag{16.23}$$

so that

$$U = \frac{1}{\dfrac{1}{h_h} + r_{f,h} + \left(\dfrac{A_h}{A_w}\right) r_w + \left(\dfrac{A_h}{A_c}\right) r_{f,c} + \left(\dfrac{A_h}{A_c}\right) \dfrac{1}{h_c}} \tag{16.24}$$

where:

h_h = hot-side convection coefficient
h_c = cold-side convection coefficient.

The value of the cold-side convection coefficient, h_c, is a function of the Nusselt number, Nu, as follows:

$$h_c = \text{Nu}\left(\frac{k_c}{d_i}\right) \tag{16.25}$$

where:

k_c = thermal conductivity of water

d_i = inside diameter of the tube.

Several algorithms are available for Nu with fully turbulent flow in Appendix F of Reference 1. The classic Colburn[8] equation as modified by Sieder and Tate[9] is as follows:

$$\mathrm{Nu} = 0.023\,\mathrm{Re}_t^{0.8}\,\mathrm{Pr}_t^{1/3}\left(\frac{\mu_t}{\mu_w}\right)^{0.14} \quad (16.26)$$

where:

Re_t = tube-side Reynolds number

$$\mathrm{Re}_t = \rho\, v_t d_i/\mu_t \quad (16.27)$$

and

Pr_t = tube-side Prandtl number

$$\mathrm{Pr}_t = c_{p,t}\,\mu_t/k_t \quad (16.28)$$

where:

ρ = density

μ_t = average tube viscosity

μ_w = viscosity at the wall surface temperature

v_t = tube velocity.

The last term is often neglected.

The Petukhov correlation is a more modern and more accurate equation for the Nusselt number as follows:

$$\mathrm{Nu} = \frac{\left(f/2\right)\mathrm{Re}_t\,\mathrm{Pr}_t}{1.07 + 12.7\left(f/2\right)^{1/2}\left(\mathrm{Pr}_t^{2/3} - 1\right)} \quad (16.29)$$

where

$$f = \frac{1}{\left(1.58\ln \mathrm{Re}_t - 3.28\right)^2} \quad (16.30)$$

The Colburn/Sieder Tate equation, although less accurate, is still commonly used.

No such similar direct method for calculating the hot-side convection coefficient, h_h, exists, as it is a complicated function of the HX shell geometry such as the type and spacing of the baffles, the leakage through the tube support plates, etc. Manufacturers generally hold this data to be proprietary. One method of estimating this value in the open literature is the Delaware Method, named for groundbreaking research on the subject performed at the University of Delaware by Dr. Kenneth J. Bell and others. The procedure for performing this calculation may be found in Appendix C of Reference 1.

If a manufacturer's HX datasheet is available, the most direct method for calculating h_h is as follows:

$$h_{h,\,design} = \frac{1}{\frac{1}{U} - r_{f,h} - \left(\frac{A_h}{A_w}\right) r_w - \left(\frac{A_h}{A_c}\right) r_{f,c} - \left(\frac{A_h}{A_c}\right) \frac{1}{h_c}} \quad (16.31)$$

where the value of U is available or may be calculated from the data-sheet if the value of F is known because

$$U = \frac{Q}{A_h \; F \; LMTD} \quad (16.32)$$

The resulting value of $h_{h,design}$ would vary depending on whether the manufacturer's value of U is based on the Colburn/Sieder Tate or Petukhov correlation.

16.6 DETERMINING THE VALUES OF A_w AND r_w

$$A_w = \frac{A_o - A_i}{\ln\left(\frac{A_o}{A_i}\right)} \quad (16.33)$$

$$r_w = \frac{d_o - d_i}{2\,k_t} \quad (16.34)$$

where d_o and d_i are the tube inside and outside diameters, respectively, and k_t is the tube material thermal conductivity.

16.7 DETERMINING APPARENT FOULING FROM THE TEST

When specifying a shell-and-tube HX, the preparer must assign values for the anticipated fouling on the inside and outside of the tube. In a nuclear station, the adequacy of an operating HX is often based on whether the apparent fouling as determined by the test is greater or less than the assumed design value. Because one cannot determine by the test whether the apparent fouling is located on the inside or the outside of the tube, one can only determine the total apparent fouling.

The total apparent fouling in an HX is the sum of the hot-side and tube-side fouling as follows:

$$r_f = r_{f,h} + \left(\frac{A_h}{A_c}\right) r_{f,c} \quad (16.35)$$

which may be determined as the result of HX tests.

The term "apparent fouling" is used in recognition of the fact that several factors such as internal damage to the HX, manufacturing errors, etc. would appear as fouling.

Total apparent fouling, r_f, is defined as

$$r_f = \frac{1}{U_{test}} - \frac{1}{U_{clean}} \tag{16.36}$$

where

$$U_{test} = \frac{Q_{test}}{A_h F\, LMTD_{test}} \tag{16.37}$$

and

$$U_{clean} = \frac{1}{h_{h,\,design}} + \left(\frac{A_h}{A_w}\right) r_w + \left(\frac{A_h}{A_c}\right)\frac{1}{h_{c\ design}} \tag{16.38}$$

As one can see, the uncertainty of r_f is a function of Q_{test}, which is a function of the inlet and outlet temperatures as well as the flow rates and of the analytical uncertainty of the hot- and cold-side convection coefficients. As a result, an uncertainty of fouling as measured by the test can be quite large.

16.8 DETERMINING HEAT TRANSFER RATE AT REFERENCE CONDITIONS

Because the objective of a heat transfer test in a nuclear station is to confirm that the HX will transfer the required heat under reference or design basis conditions, Section 5 of Reference 1 recommends that the heat transfer rate, Q^*, at reference conditions, not apparent fouling, be the test acceptance criterion. Appendix E of Reference 1 shows the derivation of the following equation for Q^* as follows:

$$Q^* = \frac{Q_{ave.\,test}\left((EMTD)^* \big/ (EMTD)\right)}{1 + U_{test}\left[\left(\frac{1}{h_h^*} - \frac{1}{h_h}\right) + \frac{A_h}{A_c}\left(\frac{1}{h_c^*} - \frac{1}{h_c}\right)\right] + \left(r_w^* - r_w\right)} \tag{16.39}$$

where

$$Q_{ave.test} = \left(\frac{u_{Qh}^2}{u_{Qc}^2 + u_{Qh}^2}\right) Q_{c\,test} + \left(\frac{u_{Qc}^2}{u_{Qc}^2 + u_{Qh}^2}\right) Q_{h\,test} \tag{16.40}$$

and

u_{Qc} = uncertainty of the cold-side rate of heat transfer
u_{Qh} = uncertainty of the hot-side rate of heat transfer.

$$Q_{c\,test} = m_c\,(h_{c\text{-}out} - h_{c\text{-}in}) \tag{16.41}$$

$$Q_{h\,test} = m_h\,(h_{h\text{-}in} - h_{h\text{-}out}) \tag{16.42}$$

As may be seen in the derivations above, the values necessary to calculate $h_{h,design}$ for the design conditions are normally available, but one must calculate a new value for h_h based on the test conditions on the hot side because Re and Pr for the hot side will be different from those in the case for the reference conditions. However, one is not required to calculate the absolute values of Re and Pr on the hot side. The value for h_h based on the test conditions on the hot side may be computed based on a correlation by the Delaware method discussed in Reference 10. If one assumes that the actual value for h_h is a function of some ideal value for the hot-side thermal conductivity for pure cross-flow such that

$$h_h = J_T\,h_{h,ideal} \tag{16.43}$$

where J_T is a correction factor that is a function of the various leakage and bypass paths, then J_T remains constant for a given HX. Therefore,

$$J_T = \left(\frac{h_h}{h_{h,ideal}}\right)_{design} = \left(\frac{h_h}{h_{h,ideal}}\right)_{test} \tag{16.44}$$

and

$$(h_h)_{test} = \left(\frac{h_{h,test}}{h_{h,design}}\right)_{ideal} h_{h,design} \tag{16.45}$$

From a simple method proposed by Taborek[10] for estimating the shell-side coefficient of heat transfer

$$h_h = C\,\text{Re}^m\,\text{Pr}^n\left(\frac{k}{D}\right) \tag{16.46}$$

For turbulent flow across a tube bank where Re_h is greater than 200, $m = 0.6$ and $n = 1/3$. The value of C and all geometric parameters cancel out because they do not change between test and reference conditions.

Therefore

$$h_{h,test} = \frac{\left[\left(\frac{m_h}{\mu_h}\right)^{0.6}\left(\frac{\mu_h c_{p,h}}{k_h}\right)^{0.333} k_h\right]_{test}}{\left[\left(\frac{m_h}{\mu_h}\right)^{0.6}\left(\frac{\mu_h c_{p,h}}{k_h}\right)^{0.333} k_h\right]_{design}} h_{h,design} \tag{16.47}$$

16.9 TEST UNCERTAINTY

Every measurement has error resulting in a difference between the measured value, X, and the true value. The difference between the measured value and the true value is the total uncertainty. Because the true value is unknown, the total uncertainty in a measurement cannot be known and therefore can only be estimated. The total measurement uncertainty consists of two components: precision (random) error and bias (systematic) error. Accurate measurement requires minimizing both precision and bias errors.

Precision error is the portion of the total measurement error that varies randomly in repeated measurements of the true value. The total precision error in a measurement is usually the sum of the contributions of several elemental precision error sources. Elemental precision error sources include those that are known and controlled, those that are negligible and are ignored, and those that are unknown and must be estimated. The sources of precision error are generally limited to process random variation that inhibits the prediction of the true average value of the parameter over the test duration based on finite sampling of the parameter. An example of process random variations would be surging service water (SW) flows. Precision error may be reduced by increasing the number of measurements taken. Therefore, with the advent of data logging techniques in which a large amount of data may be collected in a relatively short period, the precision error has become an almost insignificant aspect of measurement uncertainty.

Bias error is the portion of the total measurement error that remains constant in the same direction in repeated measurements of the true value. The total bias error in a measurement is usually the sum of the contributions of several elemental bias errors. Elemental bias errors include those that are known and can be calibrated out, those that are negligible and are ignored, and those that are unknown and must be estimated. Elemental bias errors may arise from imperfect calibration corrections, data acquisition systems, data reduction techniques, and measurement methods. The main categories of bias errors are instrument bias and spatial bias. Instrument bias errors are systemic errors in the instruments comprising a measurement system including errors due to calibration and installation effects such as locating a resistance temperature detector (RTD) on the outside of a pipe to measure the temperature in the pipe.

Spatial bias errors are systemic errors in the measurement of the bulk average value of a parameter due to the use of instruments that only measure a finite portion of the total parameter and are placed in a region that may not be representative of the true bulk average value. Spatial bias error may be reduced by increasing the number of test instruments used to measure the same parameter. An example would be to increase the number of RTDs located on the outside surface of a pipe to detect temperature stratification that may be occurring within the pipe.

The overall test uncertainty is as follows:

$$u_{overall} = \sqrt{b_{cal}^2 + b_{spat\,\mathrm{var}}^2 + u_{pv}^2} \tag{16.48}$$

where:

b_{cal} = instrument calibration

$b_{spat\,\mathrm{var}}$ = spatial variation

u_{pv} = process variation.

The standard deviation during the test is as follows:

$$S_X = \sqrt{\sum_{i=1}^{J} \frac{\left(X_i - X_{ave}\right)^2}{J-1}} \qquad (16.49)$$

where:

J = number of measurement points for each parameter measured
X_i = measured values
X_{ave} = average measured value at a particular instrument location during the test.

The bias due to special variation is as follows:

$$b_{spat\,var} = t\sqrt{\left(\frac{1}{J}\right)\sum_{i=1}^{J} \frac{(X_{ave} - X_{bulk\,ave})^2}{J-1}} \qquad (16.50)$$

where:

t = Student's t for the number of measurement points
$X_{bulk\,ave}$ = bulk average temperature at flow cross-section.

The uncertainty of the process variation during the test is as follows:

$$u_{pv} = 2\sqrt{\sum_{i=1}^{J} \frac{s_x^{\,2}}{J\ x\ N_{pv}}} \qquad (16.51)$$

where:

N_{pv} = number of measurements taken during the test.

As seen from the equation above for u_{pv}, one may minimize the importance of the contribution of process variation to the overall uncertainty by the use of a data logger for collecting data. This is the normal practice in nuclear power stations.

REFERENCES

1. ASME PTC 12.5-2000, Single Phase Heat Exchangers, September 2000.
2. Thomas, L. C., The EMTD Method: An Alternative Effective Mean Temperature Difference Approach to Heat Exchanger Analysis, *Heat Transfer Engineering*, vol. 31, 2010 pp. 193–200.
3. Thomas, L. C., *The P-NTUI Method: Classical Heat Exchanger Performance Analysis Methods*. EPRI, Palo Alto, CA, 2010, 1021065.
4. Thomas, L. C. and C. F. Bowman, Classical Heat Exchanger Analysis. EPRI, Palo Alto, CA, 2015, 3002005337.
5. Philpot, L. and S. Singletary, *Service Water, Heat Exchanger Testing Guidelines*, EPRI, Palo Alto, CA, 3002005340.
6. Kuppan, K., *Heat Exchanger Design Handbook*, Marcel Dekker, Inc., New York, 2013.

7. Shah, R. K. and D. P. Sekulic, Heat Exchangers (Chapter 17), in *Handbook of Heat Transfer*, 3rd ed., Edited by Rohsenow, W. M., Hartnet, J. P., and Cho, Y. I., McGraw-Hill Book Company, New York, 1998.
8. Colburn, A. P., A Method of Correlating Forced Convection Heat Transfer Data and a Comparison with Fluid Friction, *Transactions of AIChE*, vol. 29, 1933, pp. 174–219.
9. Sieder, E. and G. Tate, Heat Transfer and Pressure Drop of Liquids in Tubes, *Industrial Engineering and Chemistry*, vol. 28, no. 12, December 1936, pp. 1429–1435.
10. Thomas, L. C., *Heat Transfer Professional Version*, 2nd ed. Capstone Publishing Corporation, Tulsa, OK, 1999.

17 Air-to-Water Heat Exchangers

17.1 TYPES OF AIR-TO-WATER HEAT EXCHANGERS IN A NUCLEAR STATION

An air-to-water heat exchanger (AWHX) as described herein is a device that transfers heat between warm air and cooler water through an array of finned tubes. An AWHX is a box containing or attached to a fan that draws air through a filter medium and a series of serpentine coils and out of the box, frequently through a set of louvers. As such it is classified as a cross-flow HX. If the air passes through four or more sets of coils in series, the AWHX is generally treated as a counter-flow HX.

Nuclear stations employ a wide variety of specialty AWHXs that frequently perform important safety-related functions. These include containment air cooling (CAC) units (also known as containment fan cooling units and containment unit coolers), area and room coolers, diesel generator intercoolers, reactor building cooling units (RBCUs), and auxiliary building unit coolers. This chapter is devoted to the analysis of an RBCU that is normally associated with BWR nuclear units. The analysis of CACs that are normally associated with PWR is essentially the same. Due to the heat generated by the equipment inside the containment of a nuclear station, the AWHXs are normally called upon to operate during both normal operation and outages. At many nuclear stations, the AWHX is required to operate following an accident such as a leak or rupture of the primary piping inside the containment to limit the pressure in the containment due to the release of steam inside the containment. Some PWRs employ ice condensers that store sufficient ice inside the containment to limit the containment pressure by condensing the steam until other safety systems such as the containment spray system may be activated. During normal operation, an AWHX generally is called upon to transfer heat from relatively dry air, but when called upon to condense steam, the air is normally saturated with water vapor and the outside of the tubes can be flooded with water as it removes moisture from the air.

17.2 FINNED TUBES

The convection boundary resistance to heat transfer on the air side of an AWHX can be as much as 20–30 times that on the tube side where there is turbulent flow of water through the tubes. Therefore, to compensate for the poor heat transfer

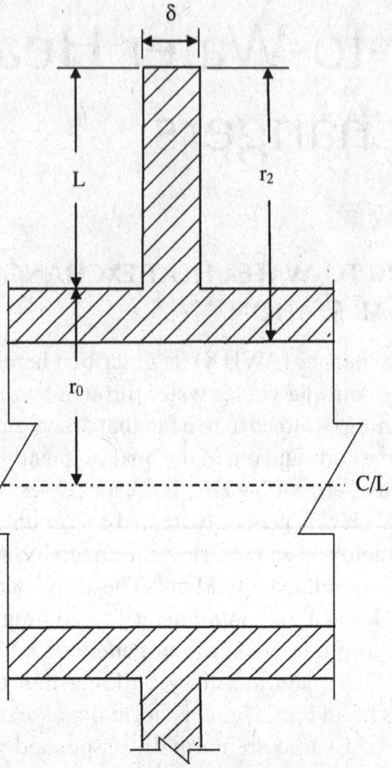

FIGURE 17.1 Finned tube.

on the air side, fins are provided to increase the surface area on the air side by as much as 20 or more times that of the water side. As a practical manufacturing consideration, these fins are normally rectangular, as shown in Figure 17.1. Accordingly, due to the resistance to conduction heat transfer down the fin to the surface of the water, increasing the length of the fin reaches a point of diminishing returns.[1] The measure of this reduction in heat transfer is referred to as the fin efficiency, η_{fin}. Some AWHXs employ plate fins that are analyzed herein as equivalent fins.

17.3 DEFINITION OF PHYSICAL DATA

Table 17.1 provides a list of the abbreviations and definitions for physical data for a typical AWHX.

Table 17.2 shows how to calculate some of the additional required parameters.

TABLE 17.1
AWHX Physical Data

N_r	Number of rows
S	Number of serpentines
L_{eff}	Effective tube length
N_t	Number of tubes per row
d_o	Outside tube diameter
t	Tube wall thickness
x_{air}	Tube spacing in line with air flow
δ	Fin thickness
N_f	Number of fins per unit of length
B	Face height

TABLE 17.2
Required AWHX Parameters

FA	Face area	$FA = L_{eff} B$	(17.1)
D	Bundle depth	$d = N_t x_{air}/12$	(17.2)
d_i	Inside tube diameter	$d_i = d_o - 2(t/12)$	(17.3)
A_t	Tube-side area	$A_t = \pi d_i L_{eff} N_r N_t$	(17.4)
$A_{c,f}$	Total cross-sectional area of plate fins	$A_{c,f} = BD$	(17.5)
$A_{c,t}$	Total cross-sectional area of tubes	$A_{c,t} = (\pi/4) d_o^2 N_t N_{rows}$	(17.6)
$A_{net,f}$	Net area of each plate	$A_{net,f} = 2(A_{c,f} - A_{c,t})$	(17.7)
NP	Number of plates	$NP = L_{eff} N_r$	(17.8)
A_f	Total area of plate fins	$A_f = A_{net,f} NP$	(17.9)
A_p	Net outside area of tubes (prime area)	$A_p = \pi d_o N_t N_r L_{eff}(1 - N_f \delta)$	(17.10)
A_h	Total effective area of finned surface	$A_h = A_f + A_p$	(17.11)
A_h/A_t	Ratio of finned side to inside area	A_h/A_t	(17.12)
a_o	Area of finned surface per fin	$a_o = A_f/N_r N_t NP$	(17.13)
d_{fin}	Effective fin diameter	$d_{fin} = \ (2/\pi)a_o + d_o{}^2)^{0.5}$	(17.14)
a_h	Area of finned side per fin	$a_h = A_h/N_r N_t N_f l_{eff}$	(17.15)
a_p	Area of the prime per fin	$a_p = \pi d_o (1/N_f - \delta)$	(17.16)
a_f	Area of the fin per fin	$a_f = a_h - a_p$	(17.17)
$A_{t,o}$	Outside tube area	$A_{t,o} = (d_o/d_i) A_t$	(17.18)
A_w	Tube wall area	$A_w = (A_{t,o} - A_t)/\ln(A_{t,o}/A_t)$	(17.19)
r_w	Tube wall resistance	$r_w = \left[d_o \ln(A_h)/A_t\right]/2k_t$	(17.20)

17.4 FIN EFFICIENCY

The analysis shown in Table 17.3 taken from Reference 1 calculates the fin efficiency of the AWHX.

TABLE 17.3
Calculation of Fin Efficiency

l	Characteristic fin length	$1 = \delta/2$	(17.21)
h_h	Assumed air-side film coefficient	h_h	(17.22)
Bi	Biot number	$Bi = h_h l/k_{fin}$	(17.23)
r_2	Fin radius	$r_2 = d_{fin}/2$	(17.24)
r_{2c}	Characteristic fin radius	$r_{2c} = r_2 + \delta/2$	(17.25)
L	Fin height	$L = r_2 - d_o/2$	(17.26)
L_c	Equivalent fin height	$L_c = L + \delta/2$	(17.27)
r_{2c}	Radius ratio	$r_{2c} = \left(d_o/2r_1\right)$	(17.28)
η_F	Consult Figure 17.2[2]	$\eta_F = \left[tanh\ (mL_c)\right]/\left(mL_c\right)$	(17.29)
	x-axis	$L_c^{1.5}[h_h/k_{fin}L_c\delta]^{0.5}$	(17.30)
η	Fin efficiency	$\eta_{fin,h}$	(17.31)
η_h	Surface efficiency	$\eta_h = (a_p + a_f\eta_{fin,h})/a_h$	(17.32)

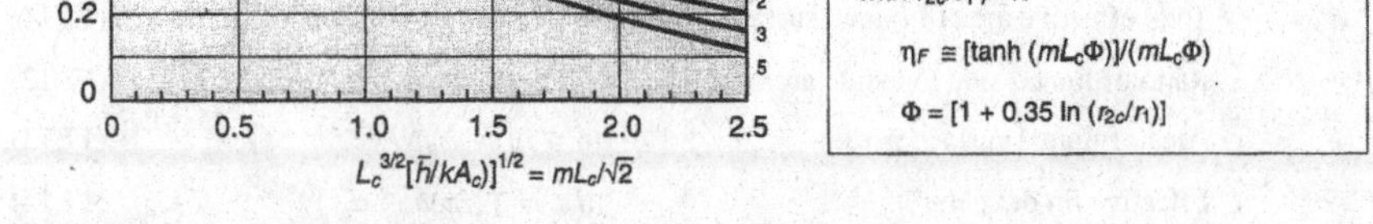

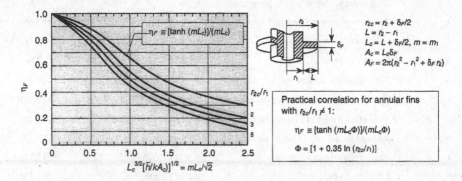

FIGURE 17.2 Fin efficiency. (Courtesy of Capstone Publishing.)

17.5 BACK CALCULATION OF AIR-SIDE COEFFICIENT FROM DESIGN CONDITIONS

As was the case with water-to-water heat exchangers discussed in Chapter 16, the air-side convection heat transfer coefficient for the AWHX may be found for the noncondensing normal mode of operation from vendor data to eliminate the need for hot-side heat transfer correlations specific to the geometry and configuration of the HX. Table 17.4

TABLE 17.4
Design Basis or Test Conditions for Normal Noncondensing Mode of Operation

G	Total service water flow rate
$g_{water\text{-}coil}$	Service water flow rate per coil
$t_{c,i}$	Entering water temperature
$t_{c,o}$	Exiting water temperature
t_c	Average water temperature
k_c	Tube-side thermal conductivity
μ_c	Tube-side absolute viscosity
$c_{p,c}$	Tube-side specific heat
ρ_c	Tube-side density
dt_c	Tube-side temperature rise
Q_c	Tube-side heat transfer rate
$T_{h,i}$	Entering air dry-bulb temperature
$T_{h,o}$	Exiting air dry-bulb temperature
T_h	Average air dry-bulb temperature
k_h	Hot-side thermal conductivity
μ_h	Hot-side absolute viscosity
$c_{p,h}$	Hot-side specific heat
F	*LMTD* correction factor
$m_{c\text{-}coil}$	Tube-side flow rate per coil
FV	Face velocity out of air side
RH	Relative humidity
P	Total atmospheric pressure

presents a list of the required design basis parameters for operating in the normal, noncondensing mode to be able to back calculate the air-side coefficient of heat transfer.

The design fouling resistance on the air side and the tube side are specified based on the tube and fin materials and the corresponding resistances to heat transfer.

The overall heat transfer coefficient at design conditions, U_{design}, is calculated for the normal noncondensing mode of operation in Table 17.5.

The tube-side convection coefficient may be computed from the Petukhov correlation as in Chapter 16 (Table 17.6).

TABLE 17.5
Verification of U_{design} for Normal Noncondensing Mode of Operation

ΔT_1	Greater terminal temperature difference	$\Delta T_1 = T_{h,i} - t_{c,o}$	(17.33)
ΔT_2	Lesser terminal temperature difference	$\Delta T_2 = T_{h,o} - t_{c,i}$	(17.34)
$LMTD$	Log mean temperature difference	$LMTD = (\Delta T_1 - \Delta T_2)/\ln(\Delta T_1/\Delta T_2)$	(17.35)
$EMTD$	Effective mean temperature difference	$EMTD = F(LMTD)$	(17.36)
U_{design}	Design overall heat transfer coefficient	$U_{design} = Q/(A_h EMTD)$	(17.37)

TABLE 17.6
Tube-Side Convection Coefficient

m_t	Mass flow rate/tube	$m_t = m_{c\text{-}coil}/N_t$	(17.38)
V_t	Volumetric flowrate/tube	$V_t = m_t/\rho$	(17.39)
a_t	Tube area	$a_t = \pi D_i^2$	(17.40)
v_t	Tube velocity	$v_t = V_t/a_t$	(17.41)
Pr_c	Prandtl number	$\mathrm{Pr}_c = c_{p,c}\,\mu_c/k_c$	(17.42)
Re_c	Reynolds number	$\mathrm{Re}_c = \rho v_t D_i/\mu_c$	(17.43)
F	Fanning friction factor	$f = (1.58 \ln \mathrm{Re}_c - 3.28)^{-2}$	(17.44)
Nu	Nusselt number	$\mathrm{Nu} = ((f/2)\mathrm{Re}_c\mathrm{Pr}_c)/(1.07 + 12.7(f/2)^{1/2}(\mathrm{Pr}_c^{2/3} - 1))$	(17.45)
h_c	Tube-side film coefficient	$h_c = \mathrm{Nu}(k_c/D_i)$	(17.46)

The hot-side convection coefficient may be computed as follows:

$$h_{h,\,design} = \frac{1/\eta_h}{\dfrac{1}{U_{design}} - r_{f,h} - r_w - \left(\dfrac{A_h}{A_c}\right) r_{f,c} - \left(\dfrac{A_h}{A_c}\right)\dfrac{1}{h_c}} \tag{17.47}$$

where

$$A_w = \frac{A_o - A_i}{\ln\left(\dfrac{A_o}{A_i}\right)} \tag{17.48}$$

and

$$r_w = \frac{d_o \ln\left(A_h/A_t\right)}{2\,k_w} \tag{17.49}$$

and k_t is the tube material thermal conductivity.

17.6 CALCULATION OF AIR-SIDE MASS FLOW RATE AND HEAT TRANSFER

The air-side volumetric flow is converted to mass flow using the ideal gas laws, and the air-side heat transfer rate is calculated as shown in Table 17.7.

TABLE 17.7
Air-Side Mass Flow Rate and Heat Transfer

V	Volumetric flow rate	$V = B\, l_{eff} FV$	(17.50)
P_{sat}	Saturation pressure at $T_{h,i}$	ASME Steam Tables	(17.51)
P_{wv}	Partial pressure of the water vapor in the air	$P_{wv} = RH\, P_{sat}$	(17.52)
P_a	Partial pressure of the air[3]	$P_a = P - P_{wv}$	(17.53)
w	Humidity ratio[2]	$w = (M_{wv}/M_a)\,(P_{wv}/P_a)$	(17.54)
m_a	Mass flow rate of air[4,5]	$m_a = \left[P_a 60V/R(T+460)\right]$	(17.55)
m_{wv}	Mass of the water vapor	$m_{wv} = wm_a$	(17.56)
m_h	Total mass flow rate on the air side	$m_h = m_a + m_{wv}$	(17.57)
$c_{p\text{-}MA}$	Specific heat of moist air[1]	$c_{p\text{-}MA} = \left(c_{p\text{-}DA} m_a + c_{p\text{-}WV} m_{wv}\right)/\,m_h$	(17.58)
ρ_a	Density of air	$\rho_a = m_h\,/\,V$	(17.59)
m_h	Hot-side flow measurement per coil	$m_h = V\,\rho_a$	(17.60)
Q_h	Air-side heat transfer rate	$Q_h = m_h c_p \left(T_{h,i} - T_{h,o}\right)$	(17.61)

Note: 1. $c_{p\text{-}DA}$ and $c_{p\text{-}WV}$ are the specific heats of dry air and water vapor, respectively.
2. m_{wv} and m_a are the molecular weights of water and air, respectively.
3. P_a is the pressure of the air.
4. V is the volumetric flow rate.
5. R is the gas constant, and $T + 460$ is the temperature of the air in °Rankine.

17.7 CALCULATION OF AIR-SIDE COEFFICIENT AT TEST CONDITIONS

The value for h_h on the air side based on the test conditions may be computed based on a correlation by Taborek.[1] If one assumes that the actual value for h_h is a function of some ideal value for the air-side thermal conductivity for pure cross-flow, then

$$h_h = J_T h_{h,ideal} \tag{17.62}$$

where J_T is a correction factor that is a function of the various leakage and bypass paths and, therefore, remains constant for a given HX.

Therefore,

$$J_T = \left(\frac{h_h}{h_{h,ideal}}\right)_{design} = \left(\frac{h_h}{h_{h,ideal}}\right) \tag{17.63}$$

and

$$h_{h,test} = \left(\frac{h_{h,test}}{h_{h,design}}\right)_{ideal} h_{h,design} \tag{17.64}$$

Reference 3 suggests the following correlation for AWHX:

$$h_h \approx \mathrm{Re}_h^{0.681}\,\mathrm{Pr}^{1/3}\left(\frac{k_h}{D_o}\right) \tag{17.65}$$

Substituting,

$$h_{h,test} = \frac{\left[\left(m_h/\mu_h\right)^{0.681}\left(\mu_h c_{p,h}/k_h\right)^{0.333} k_h\right]}{\left[\left(m_h/\mu_h\right)^{0.681}\left(\mu_h c_{p,h}/k_h\right)^{0.333} k_h\right]_{design}} h_{h,design} \tag{17.66}$$

17.8 COMPUTING "APPARENT" FOULING RESISTANCE

Therefore, the total "apparent" fouling resistance as determined by the test may be computed as follows:

$$r_{f,test} = r_{f,h} + \left(\frac{A_h}{A_c}\right) r_{f,c} = \frac{1}{\dfrac{1}{U_{test}} - \left(\dfrac{1}{\eta_h h_{h,test}}\right) - r_w - \left(\dfrac{A_h}{A_c}\right)\dfrac{1}{h_{c,test}}} \tag{17.67}$$

17.9 CALCULATING HEAT TRANSFER RATE AT REFERENCE DESIGN BASIS ACCIDENT CONDITIONS

The overall heat transfer coefficient at the design basis limiting conditions, U^*, may be expressed as follows:

$$\frac{1}{U^*} = \frac{1}{\eta_h h_h^*} + r_w + \left(\frac{A_h}{A_c}\right)\frac{1}{h_c^*} + r_{f,test} \tag{17.68}$$

where h_c^* is calculated as in the case of the design and test conditions (see Section 16.8).

Therefore,

$$Q^* = U^* A_h\, EMTD^* \tag{17.69}$$

It should be noted that the above analysis is applicable only if condensation is not occurring during the AWHX test. The calculation of U^* is not valid for accident conditions when water is being condensed from saturated air as a result of a pipe rupture. Such analyses are the purview of the nuclear steam supply system (NSSS) provider based on proprietary tests.

17.10 UNCERTAINTY

The method for calculating the uncertainties contained herein is the same as in Section 16.9. However, the uncertainty of the "apparent" fouling as it applies to AWHX deserves additional emphasis.

Fouling resistance measured during a test is defined as follows:

$$R_{f,test} = \frac{1}{U_{test}} - \frac{1}{U_{clean}} \quad (17.70)$$

As may be seen from the equations above, the uncertainty of the fouling resistance is a function of the uncertainties of U_{test}, h_h, and h_c. The sensitivity coefficients, θ, and uncertainty of the fouling are determined numerically by perturbing each parameter plus and minus by the amount of the uncertainty to determine the impact on the fouling

$$u_{f,c} = \left[\left(u_{U_{test}} \theta_{U_{test}} \right)^2 + \left(u_{h_c} \theta_{h_c} \right)^2 + \left(u_{h_h} \theta_{h_h} \right)^2 \right]^{1/2} \quad (17.71)$$

The sensitivity coefficients are

$$\theta_{U,test} = \frac{\Delta r_{f,test}}{\Delta U_{test}} \quad (17.72)$$

and

$$\theta_{h,c} = \frac{\Delta r_{f,test}}{\Delta h_c} \quad (17.73)$$

$$\theta_{h,h} = \frac{\Delta r_{f,test}}{\Delta h_h} \quad (17.74)$$

The overall coefficient of heat transfer indicated by the test conditions, U_{test}, may be calculated as follows:

$$U_{test} = \frac{Q_{ave}}{A \times EMTD} \quad (17.75)$$

The uncertainty of U_{test} may be determined by first calculating the sensitivity coefficients for the variables Q_{ave} and *EMTD*. The area, *A*, is assumed to be constant.

$$\theta_{Q_{ave}} = \frac{1}{A \times EMTD} \quad (17.76)$$

$$\theta_{EMTD} = \frac{Q_{ave}}{A \times EMTD} \quad (17.77)$$

The weighted average heat transfer rate and the uncertainty of the average heat transfer rate may be calculated as follows:

$$Q_{ave} = \left(\frac{u_{Q,h}^2}{u_{Q,c}^2 + u_{Q,h}^2} \right) Q_c + \left(\frac{u_c^2}{u_{Q,c}^2 + u_{Q,h}^2} \right) Q_h \quad (17.78)$$

$$u_{Q_{ave}} = \frac{\left(u_{Q,c}^4 \, u_{Q,h}^2 + u_{Q,h}^4 \, u_{Q,c}^2\right)^{1/2}}{u_{Q,c}^2 + u_{Q,h}^2} \tag{17.79}$$

One may calculate the uncertainty of the EMTD as a function of the uncertainties of the temperatures measured. In addition, Example K.1 in Reference 4 includes additional uncertainty to account for the uncertainty associated with possible variation in the HX fouling by assuming a Biot number of 0.5 for a water-to-water HX (see Appendix G of Reference 4).

Defining

$$\frac{b_{EMTD,u}}{EMTD} = 1 - \frac{2\left(\mathrm{Bi} - \Delta T_1/\Delta T_2\right)\ln\left(\Delta T_1/\Delta T_2\right)}{(1+\mathrm{Bi})\left(\left(\Delta T_1/\Delta T_2\right) - 1\right)\ln\left(\dfrac{\mathrm{Bi}}{\Delta T_1/\Delta T_2}\right)} \tag{17.80}$$

An additional uncertainty in the *EMTD* of 2% is assumed for incomplete mixing in the HX.

The sensitivity coefficients for *EMTD* to temperature as found in Appendix B of Reference 4 may be computed as follows:

$$\theta_{EMTD,T_i} = EMTD \frac{\left(1 - EMTD/\Delta T_1\right)}{\left(\Delta T_1 - \Delta T_2\right)} \tag{17.81}$$

$$\theta_{EMTD,T_o} = EMTD \frac{\left(1 - EMTD/\Delta T_2\right)}{\left(\Delta T_1 - \Delta T_2\right)} \tag{17.82}$$

$$\theta_{EMTD,t_i} = -EMTD \frac{\left(1 - EMTD/\Delta T_2\right)}{\left(\Delta T_1 - \Delta T_2\right)} \tag{17.83}$$

$$\theta_{EMTD,t_o} = -EMTD \frac{\left(1 - EMTD/\Delta T_1\right)}{\left(\Delta T_1 - \Delta T_2\right)} \tag{17.84}$$

The sensitivity coefficients for the variation in fouling and incomplete mixing are 1.0.

The hot-stream and cold-stream heat transfer rates may now be calculated as follows:

$$Q_h = m_h \, c_{p,h} \left(T_i - T_0\right) \tag{17.85}$$

$$Q_c = m_c \, c_{p,c} \left(t_i - t_0\right) \tag{17.86}$$

For the heat transfer rate, Q, the independent variables are the mass flow rate, the specific heat, and the inlet and outlet temperatures. The overall uncertainty of the heat transfer rate is the root mean square of the sum of the uncertainty contributions. For example, the uncertainty of Q_c would be as follows:

$$u_{Q,c} = \left[\left(u_{t_i}\theta_{t_i}\right)^2 + \left(u_{t_o}\theta_{t_o}\right)^2 + \left(u_{m_c}\theta_{m_c}\right)^2 + \left(u_{c_{p,c}}\theta_{c_{p,c}}\right)^2 \right]^{1/2} \quad (17.87)$$

Table B.2 of Reference 4 shows the resulting equations for the sensitivity coefficient for each independent variable:

$$\theta_{Q,T_i} = -m_h\, c_{p,h} \quad (17.88)$$

$$\theta_{Q,T_o} = m_h\, c_{p,h} \quad (17.89)$$

$$\theta_{Q,c_p h} = m_h\left(T_i - T_o\right) \quad (17.90)$$

$$\theta_{Q,t_i} = -m_c\, c_{p,c} \quad (17.91)$$

$$\theta_{Q,t_o} = m_c\, c_{p,c} \quad (17.92)$$

$$\theta_{Q,m_c} = c_{p,c}\left(t_o - t_i\right) \quad (17.93)$$

$$\theta_{Q,c_p c} = m_c\left(t_o - t_i\right) \quad (17.94)$$

REFERENCES

1. Thomas, L. C., *Heat Transfer –Professional Version*, Prentice Hall, Englewood Cliffs, NJ, 1993, p. 739.
2. Gardner, K. A., Fin Efficiency of Extended Surfaces, *Transactions of the ASME*, vol. 67, 1945, p. 621.
3. ASME PTC 30-1991, *Air Cooled Heat Exchangers*, American Society of Mechanical Engineers, 1991.
4. ASME PTC 12.5-2000, *Single Phase Heat Exchangers*, American Society of Mechanical Engineers, September, 2000.

18 Plate Heat Exchangers

18.1 DESCRIPTION OF PLATE HEAT EXCHANGERS

Figure 18.1 illustrates a typical plate heat exchanger (PHX) consisting of a frame and a pack of corrugated plates separated from each other with gaskets and clamped together between two end covers with bolts. The fluids to be heated and cooled flow through the channels created between the plates. The fluids enter and exit through ports located in the four corners of the plates, and the gaskets seal the plates at their outer edges and around the ports except as required to achieve the desired flow between the plates. A variety of flow patterns is made possible by the judicious design of the gaskets. There are many plate designs. A very common plate design is the "herringbone" in which the plates are stamped with a chevron pattern on either a 30° or 60° angle with the horizontal and alternate plates inverted so that the plates form crisscross passages with frequent points of contact between the plates.

PHXs are compact and are generally easily cleaned with a wire brush in a single day, and their surface area can be easily increased by simply adding more plates. When compared to shell-and-tube HXs, PHXs weigh less, are less expensive, occupy less floor space (no tube-pulling space required), and have higher coefficients of heat transfer. Therefore, they require less cooling water flow to achieve the same degree of cooling.

18.2 APPLICATIONS OF PLATE HEAT EXCHANGERS IN NUCLEAR POWER STATIONS

Although PHXs have found multiple applications in nuclear power stations (everything from component cooling to spent fuel cooling), the virtue of requiring less flow is particularly salient for these serving in systems cooled by SW systems. As documented by Bowman and Bain[1] and Bowman,[2] the reduction in the flow-passing capability of the carbon steel SW piping commonly utilized in SW systems is principally due to microbiologically induced corrosion that has been the bane of many nuclear power stations. Experience has shown that the surfaces of PHXs do not foul as readily as those of shell-and-tube HXs due to the high turbulence in the channels, but they are subject to both microfouling and macrofouling. Both proper straining and chemical treatment of SW at the source are essential for the satisfactory operation of PHXs in SW applications.

One distinct advantage of PHXs is that they can be assembled in place. PHXs have conveniently replaced very large shell-and-tube HXs in nuclear power stations, avoiding having to move the larger HXs in and out of buildings.

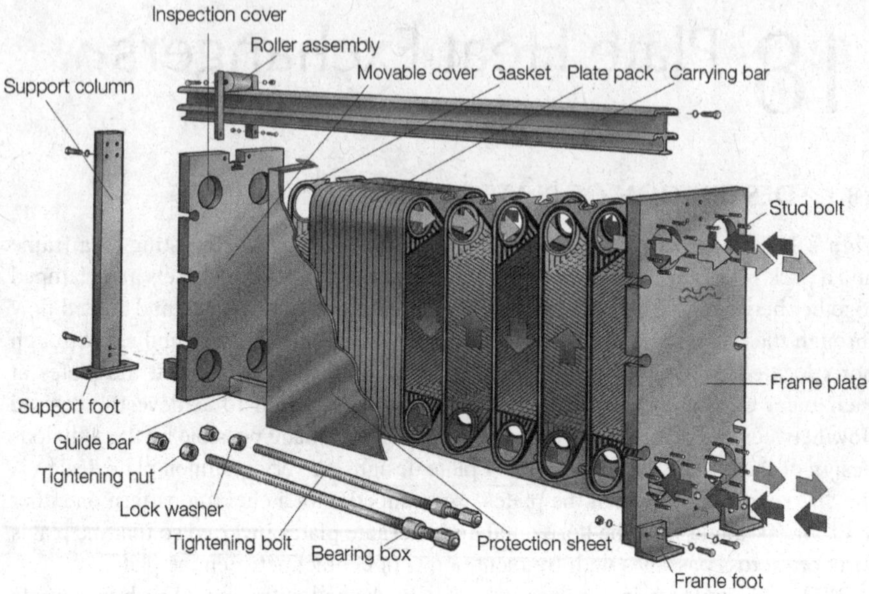

FIGURE 18.1 Plate heat exchanger. (Courtesy of Alfa Laval.)

PHXs are primarily suited for liquid-to-liquid heat transfer, although working fluids involving single-phase gases are possible. Alfa Laval, a PHX manufacturer, suggests the following performance limits in its literature:

Pressure, p (k Pa)	2,500
Temperature, t (°C)	150
Total effective area, A_{te} (m²)	2,200
Mass flow rate, w (kg/s)	1,000

Bond[3] stated that PHXs are capable of operating at temperatures up to 260°C.

18.3 TESTING PLATE HEAT EXCHANGERS

In 1995, Bowman and Craig[4] reported on extensive heat transfer testing of PHX at the TVA Sequoyah Nuclear Plant, conducted to satisfy NRC requirements. In the late 1980s, the TVA replaced three large CCS shell-and-tube HXs that are cooled with SW with six PHXs. The arrangement permits one of the PHXs to be removed from service for cleaning if the SW temperature is sufficiently low.

Alfa Laval, the PHX vendor, provided TVA with performance predictions for an array of plant operating conditions. Bowman and Craig[4] reported that when the data was plotted, the coefficient, C, required by the Colburn analogy to determine the coefficients of heat transfer on the hot and cold sides, could be determined. When the Nusselt number was plotted vs. the Reynolds number, it proved to be a linear function on a log–log plot so that a single value of C could be found to satisfy the following equation by trial and error:

$$U = \frac{1}{1/h_c + 1/h_h + r_w + r_f} = \frac{Q}{AF(LMTD)} \quad (18.1)$$

where:

U = overall heat transfer coefficient
h_c = cold-side convection coefficient
h_h = hot-side convection coefficient
r_w = wall resistance to heat transfer
r_f = fouling resistance to heat transfer
Q = heat transfer rate
A = surface area
F = *LMTF* correction factor
LMTD = log mean temperature difference.

Although the design fouling resistance for the PHX at Sequoyah was 0.0003 hr-ft^2-°F/Btu, the heat transfer tests indicated fouling resistances from 0.0002 to 0.00070 hr-ft^2-°F/Btu. When a fouling resistance was determined to be greater than the design value, an evaluation was performed considering the actual available SW flow and temperature to ensure that the plant could continue to operate safely and the PHX was scheduled to be cleaned in a timely manner. Each of the PHX has been disassembled and cleaned on several occasions, and the fouling has been observed on the SW side of the plates. This fouling is biological in nature including slime, algae, and Asiatic clams. A strong correlation was observed between PHX fouling and inadequate biocide treatment of the SW system. The PHXs that were tested at Sequoyah were not cleaned prior to testing because an adequate heat load was only available when one of the nuclear units was shut down for refueling.

18.4 ANALYSIS OF PLATE HEAT EXCHANGERS

In general, flow through adjacent passages in a PHX is countercurrent (i.e., without a cross-flow component as with a shell-and-tube HX). Therefore, $F = 1$ for most PHX applications. The PHX consists of a single type of plate in which the geometry of the plate is the same on both sides.

The value of the *LMTD* is as follows:

$$LMTD = \frac{\Delta t_1 - \Delta t_2}{\ln\left(\frac{\Delta t_1}{\Delta t_2}\right)} \quad (18.2)$$

where

$$\Delta t_1 = t_{h,i} - t_{c,o} \quad (18.3)$$

$$\Delta\tau_2 = t_{h,o} - t_{c,i} \tag{18.4}$$

The geometry of a PHX is defined by the following parameters:

NHX = number of PHX in service
L_H = height of plates
L_w = width of plates
L_{CP} = compressed height of plates
N_p = number of plates
ΔX = thickness of the plate
A = total effective area.

Because the plates are corrugated, the effective area of a plate is greater than the product of the width times the height by the surface enlargement factor, ϕ, which can be as large as 1.5, depending on the plate manufacture, and can easily be measured with a flexible tape measure. With this information, one may calculate the number of channel passes, the effective area, the spacing between plates, and the hydraulic diameter as follows:

$$N_{cp} = \left(N_p - 1\right)/2 \tag{18.5}$$

$$A_{eff} = \phi\, A \tag{18.6}$$

$$b = \left(L_c / N_p \right) - \Delta X \tag{18.7}$$

$$D_e = \frac{4 L_w b}{(2L_w + 2b)} \tag{18.8}$$

The coefficient of heat transfer, h, is

$$h = \mathrm{Nu}\left(k / D_e \right) \tag{18.9}$$

Raju and Chand[5] suggested using the Colburn analogy for determining the Nusselt number, Nu, for PHX:

$$\mathrm{Nu} = C\,\mathrm{Re}^m\,\mathrm{Pr}^n \left(\mu_b / \mu_w \right)^x \tag{18.10}$$

Arpaci[6] showed that for turbulent flow through flat plates, the Nusselt number is directly proportional to the Reynolds number, Re, to the 0.75 power and the Prandtl number, Pr, to the 0.333 power, where Pr > 1. The last term in the equation above may be neglected for turbulent flow (Re > 500). Therefore,

$$h_c = C\,\mathrm{Re}^{3/4}\,\mathrm{Pr}^{1/3} \left(k_c / D_e \right) \tag{18.11}$$

and

$$h_h = C\,\mathrm{Re}^{3/4}\,\mathrm{Pr}^{1/3}\left(k_h/D_e\right) \tag{18.12}$$

The Reynolds number is

$$\mathrm{Re} = D_e G/\mu \tag{18.13}$$

where μ is the viscosity and

$$G = \frac{m}{N_p\, b\, L_w} \tag{18.14}$$

and m is the mass flow rate.

The Prandtl number is

$$\mathrm{Pr} = \frac{\mu c_p}{k} \tag{18.15}$$

where c_p is the specific heat and k is the thermal conductivity of the fluid. These terms are equally applicable to both the hot and cold sides of the PHX.

In 1999, Bowman[7] pointed out that because the geometry on both sides of the PHX is the same, the expression for h for both the hot and cold sides is similar. Therefore, the value of C is the same for both sides, and one may solve for C for a clean PHX as follows:

$$U = 1/h_c + 1/h_h + r_w = \frac{A\,(LMTD)}{Q} \tag{18.16}$$

$$\frac{1}{C\,\mathrm{Re}_c^{3/4}\,\mathrm{Pr}_c^{1/3}\left(k_c/D_e\right)} + \frac{1}{C\,\mathrm{Re}_h^{3/4}\,\mathrm{Pr}_h^{1/3}\left(k_h/D_e\right)} = \frac{A\,(LMTD)}{Q} - r_w \tag{18.17}$$

$$C\frac{\mathrm{Re}_c^{3/4}\,\mathrm{Pr}_c^{1/3}\,k_c}{D_e} + C\frac{\mathrm{Re}_h^{3/4}\,\mathrm{Pr}_h^{1/3}\,k_h}{D_e} = \frac{1}{\dfrac{A\,(LMTD)}{Q} - r_w} \tag{18.18}$$

$$C = \frac{\dfrac{D_e}{\mathrm{Re}_c^{3/4}\,\mathrm{Pr}_c^{1/3}\,k_c} + \dfrac{D_e}{\mathrm{Re}_h^{3/4}\,\mathrm{Pr}_h^{1/3}\,k_h}}{\dfrac{A\,(LMTD)}{Q} - r_w} \tag{18.19}$$

This equation makes it possible to estimate the performance of a PHX at other than design conditions by simply knowing the physical parameters and the predicted heat transfer rate for a single set of design operating conditions (i.e., hot-side and cold-side mass flows and inlet and outlet temperatures). If the design data are not available, the value of *C* could be determined by conducting a series of heat transfer tests on a clean PHX over a range of operating conditions.

REFERENCES

1. Bowman, C. F. and W. S. Bain, *A New look at Design of Raw Water Piping, Power Engineering*, PennWell Publishing Co., Tulsa, OK, 1980.
2. Bowman, C. F., *Solving Raw Water Piping Corrosion Problems, Power Engineering*, PennWell Publishing Co., Tulsa, OK, 1994.
3. Bond, M. P., Plate Heat Exchanger for Effective Heat Transfer, *The Chemical Engineer*, April 1981, pp. 163–167.
4. Bowman, C. F. and E. F. Craig, Plate Heat Exchanger Performance in a Nuclear Safety-Related Service Water Application, *Proceedings of the International Joint Power Conference*, 1995.
5. Rauj, K. S. N. and J. Chand, *Consider the Plate Heat Exchanger, Chemical Engineering*, 1980, McGraw-Hill Inc., New York, 1980.
6. Arpaci, V. S., *Microscales of Turbulent Heat and Mass Transfer, Advances in Heat Transfer*, pp. 1–91, Academic Press, Cambridge, MA, 1997.
7. Bowman, C. F., Plate Heat Exchangers, *Proceedings of the Electric Power Research Institute Service Water System Reliability Improvement Seminar*, 1999.

19 Testing of Containment and Reactor Building Air Coolers

19.1 FUNCTION OF CONTAINMENT AND REACTOR BUILDING AIR COOLERS

As explained in Chapter 17, nuclear stations employ AWHXs that perform important safety-related functions following an accident—such as a leak or rupture of the primary piping inside containment or reactor building to limit the pressure due to the release of steam. These AWHXs sometimes also perform cooling functions during normal operation.

19.2 BASIS FOR TESTING

In 1989, in response to problems reported, the NRC issued Generic Letter 89-13, *Service Water System Problems Affecting Safety-Related Equipment*, which calls for all nuclear stations to conduct a test program to verify the heat transfer capability of all safety-related HXs cooled by SW. GL 89-13 listed the following specific actions for SW systems:

1. Implement and maintain an ongoing program of surveillance and control techniques to significantly reduce the incidence of flow blockage problems as a result of biofouling.
2. Conduct a test program to verify the heat transfer capability of all safety-related heat exchangers cooled by service water.
3. Ensure by establishing a routine inspection and maintenance program for open-cycle service water systems piping and components that corrosion, erosion, protective coating failure, silting, and biofouling cannot degrade the performance of the safety-related systems supplied by service water.

After three tests have been conducted, the nuclear station is to determine the best frequency for re-testing to provide assurance that the HX will perform its intended safety function. The test frequency is not to exceed once every five years. GL 89-13 also states that an equally effective program to ensure satisfaction of the heat removal requirements would be acceptable. An example of an alternative action acceptable to the NRC is frequent regular maintenance of the HX in lieu of testing them. Enclosure 2 of GL 89-13 states that if it is not possible to test AWHX, a program where the licensee measures air and water flow rates and trends, those results would be acceptable. The licensee would also be required to perform visual inspections, where possible, of both the air and water sides of the HX to ensure cleanliness. In GL 89-13, Supplement 1, the NRC was asked whether a heat transfer test is required if the pressure drop (*dP*) across the HX at design flow is less than the manufacturer's specification, provided the baffles have been inspected to ensure that the flow is not bypassing the coils. The NRC responded that the goal of the GL is to ensure that the heat removal requirements are satisfied, and if that can be achieved by showing the design flow to be necessary and sufficient, then heat transfer testing would be superfluous.

In 1991, in response to GL 89-13, EPRI produced EPRI Report NP-7552[1] to provide a menu of thermal performance monitoring methods to comply with the requirements of GL 89-13. In 1994, the American Society of Mechanical Engineers (ASME) produced ASME OM-S/G 1994 Standard Part 21[2] to establish the requirements for pre-service and in-service testing to assess the operational readiness of HX required to shut down and maintain a reactor in a safe condition following an accident. Among the methods suggested are the following:

- Parameter trending
- Visual inspection
- Periodic maintenance
- Pressure loss monitoring
- Temperature difference monitoring
- Temperature effectiveness test
- Heat transfer test
- Functional test.

Both documents provided guidance as to which method would be appropriate.

A number of nuclear stations have made licensing commitments to perform periodic heat transfer tests, extensive and frequent maintenance, or a combination thereof on an AWHX as a portion of their response to GL 89-13. Conducting a meaningful heat transfer test on an AWHX, such as the ones described in Chapter 16 for shell-and-tube HX, is very difficult if not impractical. Some of the factors that may make heat transfer testing of AWHX impractical include inadequate heat load, requirements to test at different flow conditions, high test uncertainty, and non-steady-state conditions.

In 1998, Bowman[3] investigated the influence of tube-side fouling and the heat transfer rate in an AWHX. Bowman concluded that only a weak correlation exists between tube-side fouling and the heat transfer rate in an AWHX. Calculations presented showed that the tube-side fouling could be acceptable by several times the value determined by test with only a modest reduction in the rate of heat transfer. Based on tests conducted at the Calvert Cliffs Power Plant as reported in Reference,[4] the combined effects of factors such as tube material, tube velocity, and tube wall temperature could be taken into consideration to determine asymptotic fouling (AF) beyond which biological microfouling does not continue to increase. Bowman noted that Hosterman[5] had suggested that typical heat load calculations used to size an AWHX for nuclear stations contain excess conservatism that might be removed to provide additional operating margin. Therefore, Bowman proposed that AWHX could be reanalyzed to employ fouling assumptions based on AF to predict AWHX performance under limiting conditions and that such an analysis in conjunction with periodic *dP* testing and regular inspection and cleaning (if required) should satisfy the requirements of GL 89-13.

The Bowman paper stimulated interest in the nuclear power industry to the extent that EPRI commissioned Bowman to prepare EPRI Report No. 1007248, *Alternative to Thermal Performance Testing and/or Tube-side Inspections of Air-to-Water Heat Exchangers.*[6] The report describes a pragmatic rationale by which utilities may justify a revision to their GL 89-13 program with respect to AWHX that may provide for a technically superior and more cost-effective alternative to existing efforts to conduct heat transfer tests on AWHX. As a result of this EPRI initiative, a number of nuclear stations have successfully revised their GL 89-13 licensing commitments to exclude performing periodic heat transfer testing of AWHX such as room coolers. The report addresses the CAC, stating that although tube-side micro-fouling resistance may not be a small part of the total resistance to heat transfer for a CAC operating in condensing mode, the program recommended in the report may be more technically sound than heat transfer testing. A number of utilities with PWR plants do not perform heat transfer tests on HX that limit containment pressure by condensing steam following an accident. Some of these plants perform frequent regular maintenance of the AWHX in lieu of testing, while others conduct periodic flow and differential *dP* tests that are trended to detect SW flow blockage and inspect the air side. In the latter case, the utility often relies on References 1 and 3.

19.3 IMPACT OF TUBE-SIDE FOULING ON AWHX HEAT TRANSFER RATE

For AWHX that are required to condense steam following an accident, the rate of heat transfer on the hot side in the condensing mode is quite high. Therefore, tube-side fouling represents a significant portion of the resistance to heat transfer during the accident. However, the AWHXs are not tested on the condensing mode but in the normal operating mode with dry air on the hot side. Because for air the rate of heat transfer is quite low, fouling represents a much smaller percentage of the total resistance to heat transfer in the test mode and any small error in the test results in a much larger error in tube-side fouling. Normally, the heat load available to test an AWHX that is designed for accident conditions is relatively small and changes in temperature in the hot and/or cold side are quite small. Further, AWHX tests are often performed under non-steady-state conditions such as during reactor shutdown. Therefore, the test uncertainty can be quite high. For AWHX required to condense steam following an accident, the mechanism for heat transfer on the hot side is completely different during the accident than it is during the test and entirely different algorithms are required to model the resistance to heat transfer on the hot side. Therefore, the analytical uncertainty of the apparent fouling computed as a result of the test and applied to the accident conditions is a function of the uncertainty of both of these algorithms. If the test flow rate is higher than the normal flow rate on

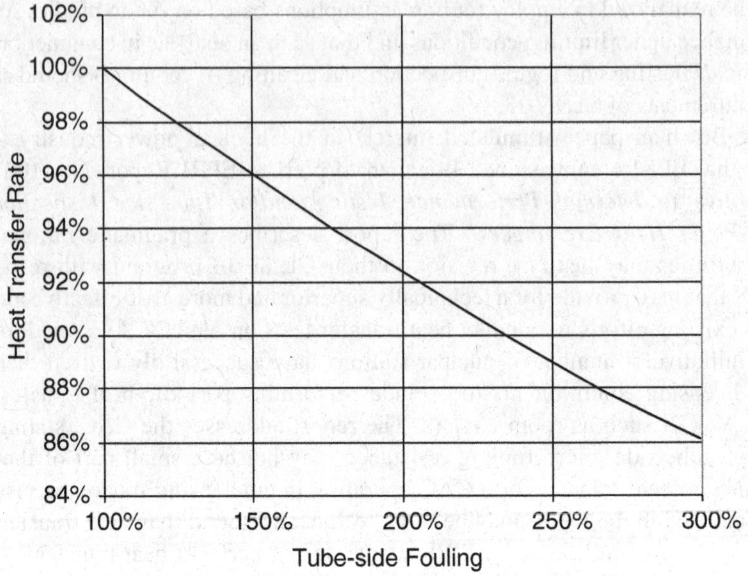

FIGURE 19.1 Impact of fouling resistance on AWHX heat transfer rate.

the tube side, the test may result in preconditioning the HX and may not be truly representative of the accident conditions.

Figure 19.1 illustrates the impact of the tube-side fouling resistance on the heat transfer rate for an AWHX. This figure was developed using the procedure described in Chapter 17 but by varying the tube-side outlet temperature and determining the impact that the change makes on the heat transfer rate and implied fouling resistance. The area ratio of the hot side to the tube side is 23:1.

As illustrated in Figure 19.1, tripling the tube-side fouling resistance only reduces the AWHX heat transfer rate by 14%. The weak correlation between tube-side fouling and the heat transfer rate is because the total resistance is dominated by the air-side convection boundary,

19.4 CHARACTERISTICS OF FOULING

The Tubular Exchange Manufacturer's Association[7] (TEMA) recommends for river water minimum design fouling values of 0.002–0.003 hr-ft^2-°F/Btu for tube-side velocities below 3.0 ft/s and 0.001–0.002 hr-ft^2-°F/Btu for velocities above 3.0 ft/s. However, Taborek[8] noted that the fouling rate is also a function of tube material and SW quality. Figure 19.2 shows the range of fouling rates recommended by Taborek, overlaid onto those recommended by TEMA for river water as well as data reported.[4,9] Taborek[8] stated that $r_f = f(V^{-1.75})$.

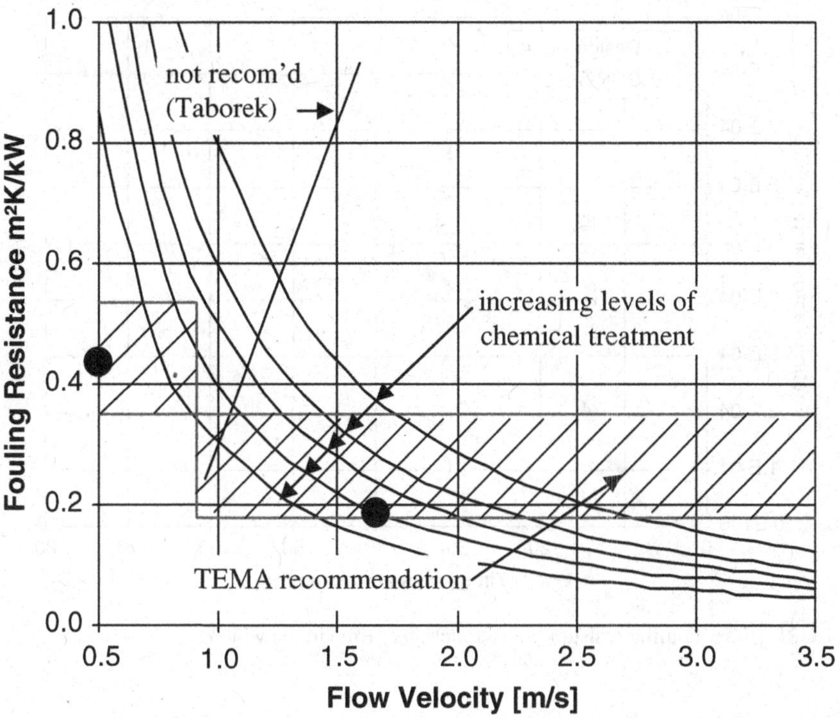

FIGURE 19.2 Fouling resistance.

19.5 ASYMPTOTIC FOULING

Somerscales[10] described AF as fouling where resistance to heat transfer initially increases rapidly when the tube is first exposed to SW but then decreases steadily until the fouling resistance is constant. AF is normally present when the fouling is due to biological microfouling but not due to sediment or calcium scale buildup, which is hard and resistant to sloughing off.

In Figures 19.3 and 19.4, Nolan and Scott[4] show the results of tests on a side stream HX with 90-10 copper–nickel tubes with a biofilm that was a combination of organic activity and inorganic deposits. The test data showed an inverse relationship with the fluid shear force created by the SW flow. AF was achieved after between 30 and 60 days depending on the temperature of the SW. The buildup of fouling was observed to be repeatable. Nolan and Scott reported that the test demonstrated that for 90-10 copper–nickel tubes the fouling layer achieved a maximum equilibrium value that is relatively constant and repeatable for a given tube velocity and temperature range.

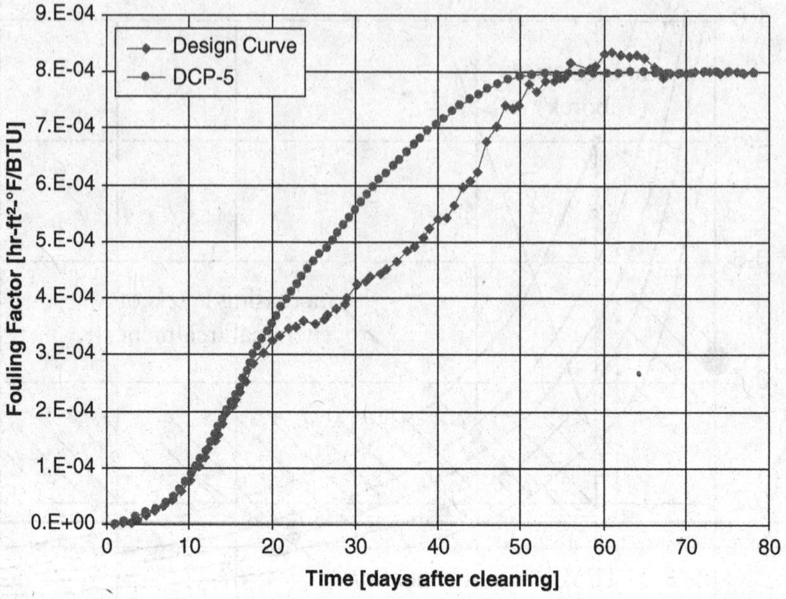

FIGURE 19.3 Fouling resistance as a function of time in the winter.

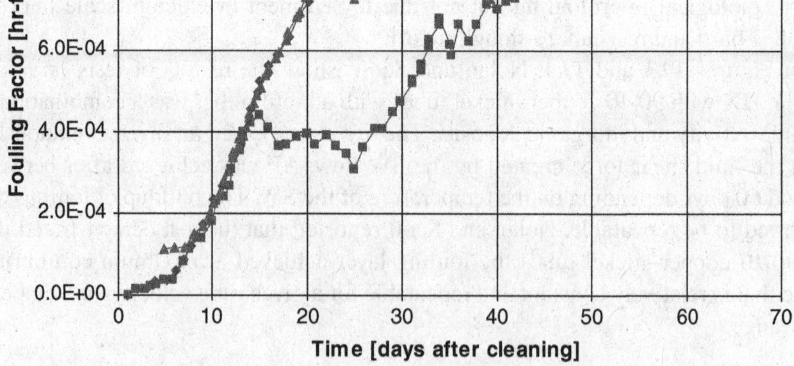

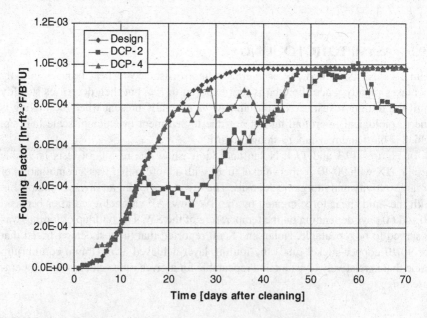

FIGURE 19.4 Fouling resistance as a function of time in the summer.

19.6 FLOW AND PRESSURE DROP TESTING

As indicated in the previous section, biological microfouling approaches asymptotically a maximum value that may not be exceeded beyond a certain point in time even if the HX is not cleaned due to the sloughing off of the material caused by the SW velocity in the tube. The same cannot be said for macrofouling (see Chapter 6). Indeed, a significant portion of the loss of heat transfer capability in AWHX may be due to macrofouling from foreign objects that are commonly found in SW. Additionally, due to the serpentine nature of AWHX and the inaccessibility of the SW water boxes, inspection and cleaning on the tube side is very difficult. Therefore, the alternative procedure to heat transfer testing proposed in Reference 3 and recommended in Reference 6 requires the nuclear station to conduct periodic flow and *dP* testing and regular inspection and cleaning on the air side to satisfy the requirements of GL 89-13.

A proper flow and *dP* test program described in Reference 6 requires that the value of AF first be determined by conducting precise tests on a clean HX that is subject to the same source of SW, operates in the same temperature and tube velocity range, and has the same tube material as the AWHX in question until AF is achieved. One should conduct the flow and *dP* test after the tube side has been inspected and cleaned to baseline the flow and *dP* under clean conditions. The AWHX must then be re-analyzed using that value of AF to demonstrate that the AWHX can perform its safety function with that level of fouling.

An alternative procedure used by some nuclear stations is to conduct three high-quality heat transfer tests on the actual AWHX in question to demonstrate that the AWHX has achieved AF and can still perform its safety function with the indicated flow and *dP* measured during those tests. Subsequent tests monitor only flow and *dP* and inspect the air side thereafter. A calculation should be performed to determine the *dP* from the inlet to outlet pressure measurement points based on the AWHX vendor-specified tube-side *dP* and the associated piping pressure drop. If prior heat transfer tests are the basis for determining the AF, one should conduct the initial flow and *dP* test during or shortly after the last heat transfer test to baseline the "as found" SW flow and tube-side *dP* measurements as corrected to the design basis flow rate. As in the case of the initial heat transfer tests, subsequent flow and *dP* testing must consider the uncertainty of the measured flow and *dP* measured during the test.

Flow and *dP* tests should be conducted simultaneously so that the results may be corrected to the SW flow and *dP* reference point for trending. All SW *dP* measurements across the AWHX should be corrected to the design basis flow by the square of the flow ratios so that all *dP* measurements are referenced to the design basis flow rate through the HX. The test flowrate should be as close as possible to the required design basis flow rate to minimize the error in this correction.

The critical performance parameters for the flow and *dP* test are the flow rate through the AWHX and the *dP* across the AWHX corrected to the design basis SW flow rate. A successful test may be achieved while tolerating some amount of tube plugging, even when including the uncertainty of the test. One should compare the corrected *dP* measured with the AWHX design *dP* or that measured when the HX has been cleaned. The corrected *dP* across the tube side of the AWHX plus the test uncertainty should be expressed as a function of the equivalent number of plugged tubes and should not be permitted to exceed the tube plugging criteria in the air-to-water HX analysis. One should trend the corrected tube-side *dP* to ensure that the progress of any fouling detected does not result in *dP* exceeding the acceptance criteria before the next test.

The test results including uncertainty analysis, calculations, trending procedure and results, and conclusions from the SW flow and tube-side *dP* testing should be made clear so that an independent party would be able to review the results and they should be documented in the permanent plant records along with the most recent trending results against the established acceptance criteria. Based on these trending results, one may establish a firm technical basis for the frequency of SW flow and tube-side *dP* testing.

If it may be demonstrated that the SW flow rate is equal to or greater than the design basis flow rate through the AWHX and that there is acceptable tube blockage due to macrofouling when tube-side microfouling at the AF level is considered, the AWHX will be able to transfer the design basis heat while maintaining the air outlet temperature at or below the design basis value with the SW at the design basis temperature.

19.7 FLOW AND PRESSURE DROP TEST UNCERTAINTY

Refer to Chapter 16 for details on performing the uncertainty analysis except that temperatures and air-side flow rates are not applicable performance parameters. The performance parameters of interest are the measured flow and the measured *dP* adjusted to the design basis SW flow rate. The uncertainties in the measured parameters, flow and *dP*, are projected to the uncertainty of the percent tubes unplugged through the following equation:

$$u_{\%\,test} = [(\theta_{dP} u_{dP})^2 + (\theta_G u_G)^2]^{1/2} \tag{19.1}$$

where:

θ_{dP} = sensitivity of % plugged to *dP*
θ_G = sensitivity of % plugged to flow
u_{dP} = uncertainty in measured *dP*
u_G = uncertainty in measured flow.

Sensitivity coefficients are normally determined by perturbing the given parameter up and down by a small amount to determine the effect of the change on the performance parameter of interest. An analytical uncertainty is combined with the test uncertainty to yield a total uncertainty in the number of tubes unplugged. The sources of analytical uncertainty include the Darcy equation, the tube-side convection coefficient (the shell-side coefficient remains constant because the shell flow is unchanged by plugging), and the uncertainty of the calculated *dP*:

$$u_{\%\,Plugged} = [(\theta_{test} u_{test})^2 + (\theta_{an} u_{an})^2]^{1/2} \tag{19.2}$$

19.8 EFFICIENCY TEST

GL 89-13 states that in addition to the requirements as stated in Section 19.2 above, the following is to be required for AWHX:

> Perform efficiency testing (for example, in conjunction with surveillance testing) with the heat exchanger operating under the maximum heat load that can be obtained practically. Test results should be corrected for the off-design conditions. Design heat removal capacity should be verified. Results should be trended, as explained above, to identify any degraded equipment.

The staff at the V. C. Summer Nuclear Generating Station has conducted four heat transfer tests on each of their four RBCUs during the period from 2002 to 2014. These tests established the AF value.

Although the results of the heat transfer tests conducted in 2014 indicated that the RBCU could transfer from between 40% and 70% more than the required heat under accident conditions, the results of the flow and *dP* tests were marginal due to the high uncertainty of the test. Therefore, a future procedure that relied solely on the results of the flow and *dP* test would indicate that some of the RBCUs might fail the test, whereas the heat transfer tests indicated that all of the RBCUs passed the test with margin to spare. Accordingly, a better, simplified test is required to avoid conducting the laborious and expensive heat transfer tests in the future. The primary difficulty in conducting a heat transfer test is associated with accurately measuring the air-side temperatures and air flow. Accordingly, the efficiency test relies on only the tube-side flows and temperature measurements while taking advantage of the demonstrated large heat transfer margin available as a result of the heat transfer tests.

During their last test conducted on May 22, 2014, the staff at V. C. Summer also conducted flow and *dP* tests. The results of these tests afforded the opportunity to conduct only efficiency tests based on the *NTU* in the future that fully meet the requirements of GL 89-13. The *NTU* is a dimensionless HX thermal performance parameter as defined by the following equation when referenced to the cold stream:

$$NTU = \frac{UA_{h\text{-}eff}}{m_c c_{p_c}} \tag{19.3}$$

where:

$A_{h\text{-}eff}$ = shell-side effective surface area.

$$U = \frac{Q_c}{A_h\ EMTD} \tag{19.4}$$

where:

A_h = shell-side surface area

and

$$Q_c = m_c\ c_{p_c}\left(t_0 - t_i\right) \tag{19.5}$$

$$EMTD = \frac{\Delta t_1 - \Delta t_2}{\ln\left(\frac{\Delta t_1}{\Delta t_2}\right)} \tag{19.6}$$

where

$$\Delta t_1 = t_{s,i} - t_{t,o} \tag{19.7}$$

$$\Delta t_2 = t_{s,o} - t_{t,i} \tag{19.8}$$

Figure 19.5 shows a plot of the *EMTD* as calculated from the RBCU tests. As Figures 19.5 shows, previous heat transfer tests have demonstrated that the *EMTD* is a linear function of the cold stream heat transfer rate. This is as expected because

$$Q = UA(EMTD) \tag{19.9}$$

The equation below yields the best agreement for all of the RBCUs:

$$EMTD = 0.000035\,Q_c \tag{19.10}$$

Therefore, one may determine the *EMTD* for a given RBCU in the future as a function of the cold stream heat transfer rate.

When compared to the results of the last set of RBCU tests when both the heat transfer test and the flow and *dP* tests were conducted, the average difference in the cold stream heat transfer rate and the weighted average heat transfer rate is less than 1%. Therefore, historically, the cold stream heat transfer rate is an accurate measure of the actual heat transfer rate.

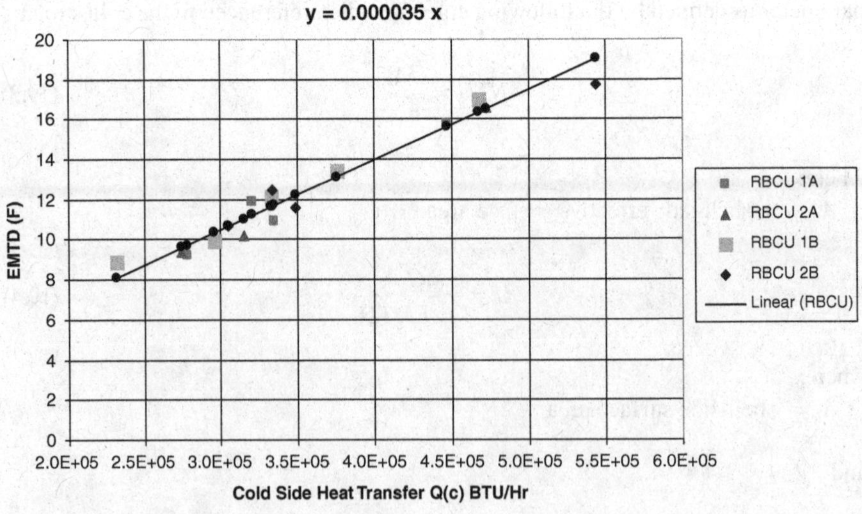

FIGURE 19.5 RBCU *EMTD* vs. heat transfer.

The only remaining parameter to be determined in the equation for *NTU* is A_{eff} One may correct the pressure drop from the flow and *dP* test to that corresponding to the reference flow rate using the Darcy equation as follows:

$$\Delta P_{corrected} = \left(\frac{G_{reference}}{G_{test}}\right)^2 \Delta P_{test} \tag{19.11}$$

Figure 19.6 shows the relationship between the corrected pressure drop across an RBCU and the percentage of tubes unplugged. In calculating the *NTU*, the effective hot-side area is calculated as follows:

$$A_{h\text{-}effective} = \%\ unplugged \times A_h \tag{19.12}$$

Figure 19.7 shows the LOCA heat transfer rate for the RBCU as a function of *NTU* when calculated as shown above. One may see that the heat transfer rate is essentially proportional to *NTU*. That is not to say that *NTU* under LOCA conditions is equal to that during a test, as the air-side conditions are quite different, but only that performance is proportional to *NTU*. The heat transfer rate for some future test may be estimated as follows:

$$(Q^* - u^*)_{future} = \left(\frac{NTU_{future}}{NTU_{test}}\right)(Q^* - u^*_{test}) \tag{19.13}$$

where:

Q^* = heat transfer rate at reference accident conditions
u^* = uncertainty of heat transfer rate at reference accident conditions.

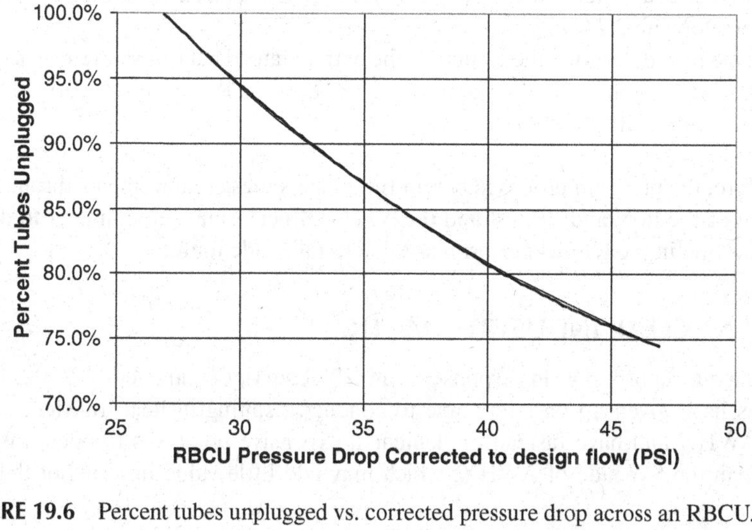

FIGURE 19.6 Percent tubes unplugged vs. corrected pressure drop across an RBCU.

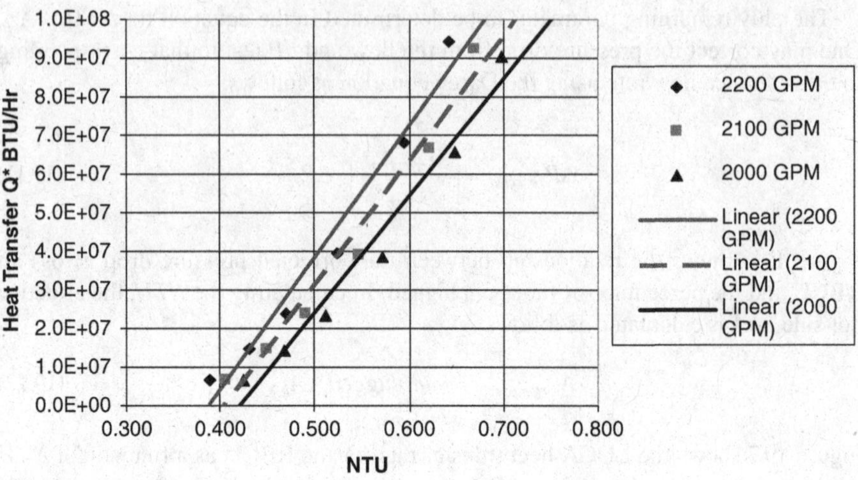

FIGURE 19.7 LOCA heat transfer rate RBCU vs. *NTU*.

Enclosure 2 of GL 89-13 defines a program that is acceptable to the NRC for HX testing. Enclosure 2 states that for all HXs the cooling water flow and inlet and outlet temperatures are to be monitored and recorded. Enclosure 2 further states that for AWHX efficiency tests are to be performed and corrected for off-design conditions and the design heat removal capacity is to be verified. One may conclude that by employing the procedure described herein, a flow and *dP* test utilizing the *NTU* parameter constitutes an efficiency test for the following reasons:

- It has been demonstrated herein that the *NTU* of an RBCU may be calculated from the results of a flow and *dP* test alone.
- The heat transfer rate extrapolated to reference accident conditions is proportional to *NTU*.
- One may determine the change in the extrapolated heat transfer rate at accident conditions from the change in the *NTU* of the RBCU as determined by subsequent flow and *dP* tests.

Therefore, the program proposed herein fully meets these requirements through utilization of the flow and *dP* test and the *NTU* HX performance parameter to detect deterioration in the RBCU performance due to tube-side fouling.

19.9 NUCLEAR INDUSTRY PRACTICE

Reference 6 reports that in response to an EPRI survey conducted in 2000, many utilities have given up on being able to conduct meaningful heat transfer tests of their AWHX and have instead implemented expensive programs to open, inspect, and clean the SW side of AWHX, which may add little value in terms of the safe and reliable operation of the plant. As a result of this EPRI initiative,[6] a number of nuclear stations have been successful in revising their GL 89-13 licensing commitments to exclude performing periodic heat transfer testing of some AWHXs.

REFERENCES

1. Stambaugh, N., W. Jr. Closser, and F. J. Mollerus, Heat Exchanger Performance Monitoring Guidelines, EPRI Report NP-7552, Electric Power Research Institute, Palo Alto, CA, 1991.
2. Stambaugh, N., and W. Closser, ASME OM–S/G 1994, Standard Part 21, ASME, 19994.
3. Bowman, C. F., Influence of Water-Side Fouling on Air-to-Water Heat Exchanger Performance, *Proceedings of the American Power Conference*, Vol. 60, 1998.
4. Nolan, C. M. and B. H. Scott, On Line Monitoring of Heat Exchangers Microfouling: An Alternative to Thermal Performance Testing, *Proceedings of the EPRI SW Reliability Improvement Seminar*, Electric Power Research Institute, Palo Alto, CA, 1995.
5. Hosterman, E. W., Reclaiming Heat Exchanger Design Margin: An Analytical Approach, *Proceedings of the EPRI SW Reliability Improvement Seminar*, Electric Power Research Institute, Palo Alto, CA, 1994.
6. Bowman, C. F., EPRI Report No. 1007248, Alternative to Thermal Performance Testing and/or Tube-Side Inspections of Air-to-Water Heat Exchangers, Electric Power Research Institute, Palo Alto, CA, 2002.
7. Bell, K. J., Standards of the Tubular Exchange Manufacturers Association (TEMA), 7th ed., New York, 1988.
8. Taborek, J., Assessment of Fouling Research on the Design of Heat Exchangers, *Proceedings of the Fouling Mitigation of Industrial Heat Exchangers Conference*, Shell Beach, CA, 1995.
9. Zelver, N., W. G. Characklis, J. A. Robinson, F. L. Roe, Z. Dicic, K. Chapple, and A. Ribaudo, Tube Material, Fluid Velocity, Surface Temperature and Fouling: A Field Study, CTI Paper TP-84-16, Cooling Tower Institute, Houston Texas, 1984.
10. Somerscales, E. F. C., Fouling of Heat Transfer Surfaces: A Historical Review, *Heat Transfer Engineering*, vol. 11, no. 1, 1990, pp. 19–36.

20 Nuclear Power Uprate

20.1 NUCLEAR POWER UPRATE OVERVIEW

A number of utilities have successfully increased their power level, resulting in an increase in the T-G electrical output of the station without a significant increase in the operating cost other than the nuclear fuel, with only modest changes in the plant equipment. These power uprates fall into the following two categories:

- Minor uprates of less than 2% authorized under the NRC's 10CFR50, Appendix K, when improved techniques are utilized to measure the FFW flow rate to encroach on the 2% reactor power margin required by NRC regulations. This power uprate does not require an increase in the licensed reactor power.
- Major uprates of more than 2% that result in an increase in the reactor power by taking advantage of improved accident analysis computer codes for the NSSS to reduce the margin that was built into the original design.

While major power uprates have been successful at both BWR and PWR nuclear power stations, the BWR stations have enjoyed larger increases in power.

20.2 DESCRIPTION OF THE WATTS BAR NUCLEAR PLANT

The WBNP is a PWR station originally licensed for operation at 3,475 MW_{th} SG power. (SG power is the reactor power plus the heat added by the reactor recirculating pumps.) The turbine cycle consists of a six-flow tandem-compound turbine with a single-pass, multi-pressure, multi-shell main condenser (similar to Figure 6.5 but with three zones) and seven stages of FWH plus a separate condenser for the main feed pump turbine located in the condensate cycle just upstream of the first stage FWH (see Section 7.1). The heater drains for the first and fifth stages of feedwater heating (designated as Heaters No. 7 and No. 3, respectively) are pumped forward. (Note that unlike most nuclear stations, TVA numbers their FWH from the highest pressure FWH down so that the top FWH is always designated as FWH No. 1.) Condensate is pumped from the main condenser to the two turbine-driven main feed pump suctions in three stages. An additional motor-driven feedwater pump is used on startup. The HRS consists of an NDCT supplemented by CCW taken from above the Watts Bar Dam.

20.3 POWER UPRATE STUDY METHODOLOGY

In 2002, Bowman et al.[1] reported on a power uprate study conducted at the WBNP to evaluate the ability of the major turbine cycle equipment to meet system requirements and determine the impacts on the balance-of-plant system requirements at up to 3,816 MW_{th} SG power during wintertime and summertime operation. The scope of the study included not only the non-safety-related turbine cycle systems but also the safety-related balance of plant mechanical systems including the SW system. However, this chapter addresses only the turbine cycle systems.

In conducting power uprate studies, common practice is to develop a reference Performance Evaluation of Power System Efficiencies (PEPSE) computer model (from Scientech, a division of Curtis-Wright Nuclear) based on the original SG power and to evaluate the turbine cycle system parameters based on the differences between the original design and those at the uprate conditions. This approach neglects actual plant conditions that should be considered, including equipment operating bias. At WBNP, the authors developed a Turbine Cycle Equipment Evaluation (TCEE) Excel workbook[2] to facilitate a comparison between the state point values predicted by PEPSE and those measured in the plant and to calculate additional parameters from simple mass and energy balances around the main condenser, FWH, and the MSR from data measured in the plant. The TCEE workbook takes enthalpies, moisture removal and leakage flows, and percent moisture from the PEPSE HB calculations.

Surveillance tests were conducted at extreme summer and winter conditions. Actual plant data were downloaded for the plant computer, averaged and checked for consistency, and loaded into the TCEE. A PEPSE model was constructed for operation at the boundary conditions existing during the surveillance test including the reactor power, SG pressure, HP turbine throttle pressure, main condenser pressure, and SG blowdown. When the data in the TCEE and the PEPSE models were compared, the percentage variance between design values and actual values was identified. The same percent variance was applied to these PEPSE HB models at the power uprate conditions to correct the PEPSE HB state point values to anticipated actual operating conditions. The variances between the plant data and the PEPSE model include the following:

- The main steam line pressure drop is less than the design, making the estimated main steam pressure at the stop valves greater than the PEPSE value.
- The first-stage pressure in the HP turbine is significantly greater than the PEPSE value.
- The measured HP exhaust pressure is less than that predicted by PEPSE.
- The pressure drop through the reheat is less than that predicted by PEPSE, resulting in a higher LP turbine inlet pressure than predicted by PEPSE.
- The estimated temperature going into the LP turbine is below that assumed by PEPSE due to poor reheater performance.
- PEPSE underestimates the hotwell temperature by several degrees due to the unique configuration of the WBNP condenser hotwell.
- PEPSE underestimates condensate temperatures up to the third stage of feedwater heating, which are influenced by the main condenser hotwell temperature.

A hydraulic model was created of the condensate and feedwater systems to assess the changes in the flows, pressures, and available margins in the feedwater, condensate, and heater drain pumps. The available margin for each component was assessed for a range of SG power levels. Modifications that may be required to implement the power uprate were identified.

20.4 IMPACT OF POWER UPRATE ON PUMPING SYSTEMS

Figures 20.1–20.3 show the available NPSH margin for the MFP and No. 3 and No. 7 heater drain pumps for the indicated SG power levels. Limiting NPSH margins occur in the wintertime when lower main condenser pressures result in lower condensate temperatures, drawing more extraction steam from the LP turbines.

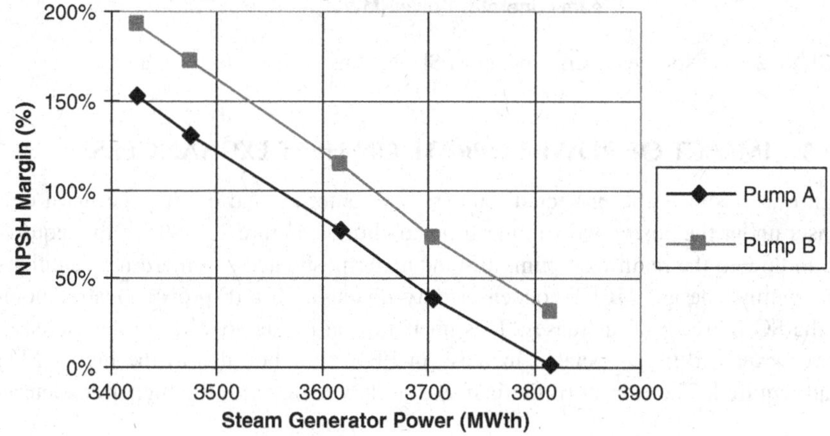

FIGURE 20.1 Main feed pump NPSH margins—wintertime operation.

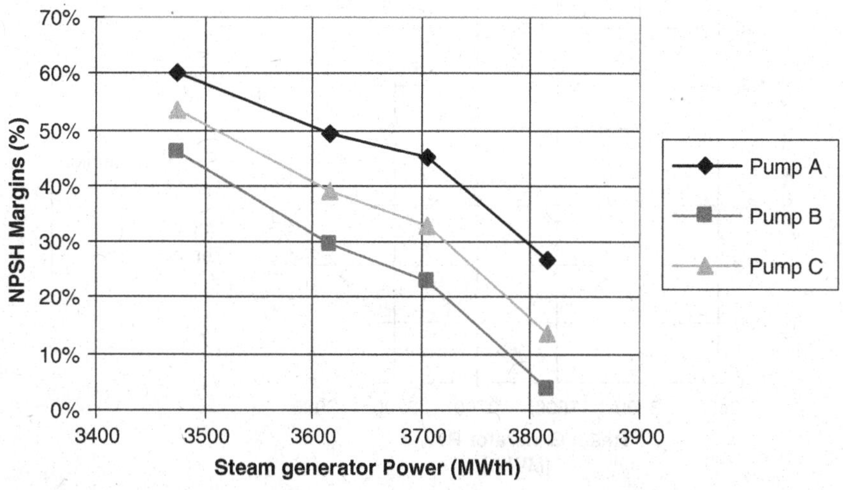

FIGURE 20.2 No. 3 heater drain pump NPSH margins—wintertime operation.

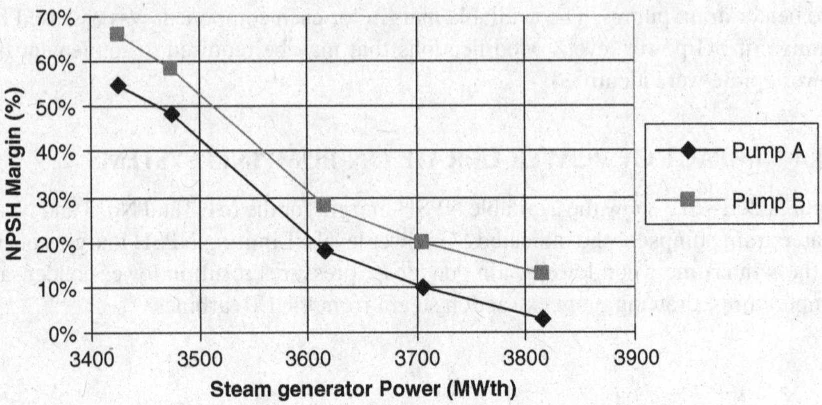

FIGURE 20.3 No.7 heater drain pump NPSH margins—wintertime operation.

20.5 IMPACT OF POWER UPRATE ON HEAT EXCHANGERS

Figure 20.4 shows the projected temperatures entering and exiting the main condenser under the postulated summertime conditions. Figure 20.5 shows the required steam flow to the main feed pump turbine under postulated summertime conditions. The summer design MFPT condenser duty increases in a disproportionate amount as the SG power level increases. This phenomenon is due not only to the increased power level and the associated increase in FFW flow but also to the higher MFP head required. The impact of the increase in duty along with the higher condensate

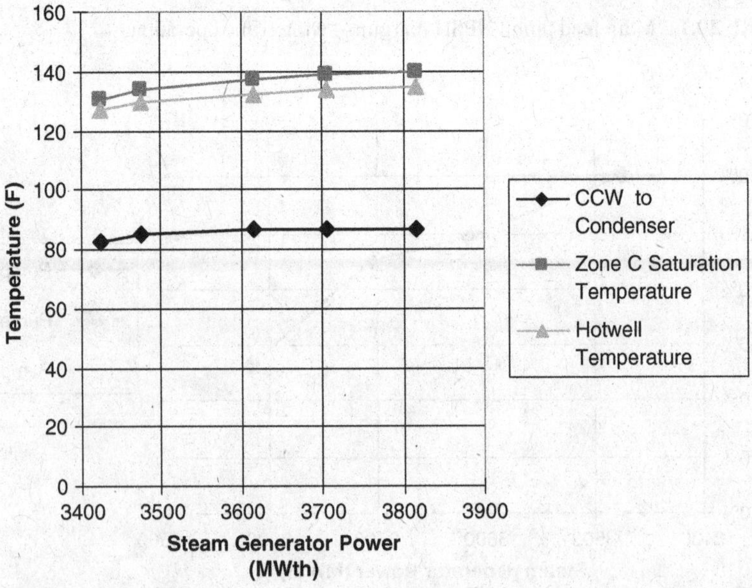

FIGURE 20.4 Main condenser temperatures for summertime operation.

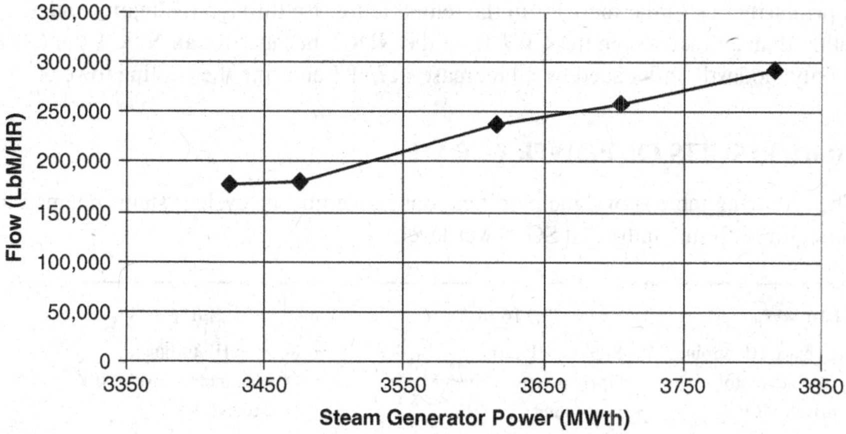

FIGURE 20.5 Main feed pump steam flow for summertime operation.

temperature coming from the main condenser results in an increase in the main feed pump turbine backpressure, requiring more steam flow to the main feed pump turbine. As a result, the source of motive steam must be changed from hot reheat steam (MSR discharge) to main steam at the higher SG power levels.

In general, the required heat transfer rate increases in the MSR and the stages of FWH as the thermal power level is increased, and the TTD and DCA increase slightly. The exceptions to this trend are FWH No. 7 and, to a lesser extent, FWH No. 6, which are affected by the main feed pump condenser performance during summertime operation. The effectiveness of each of the FWH is reduced slightly as thermal power increases because more heat transfer is required from the same total surface area in each FWH (see Chapter 7).

Figure 20.6 shows the main condenser Zone C pressure for summertime operation based on a *PF* of 82%. As Figure 20.6 indicates, this increase in Zone C pressure

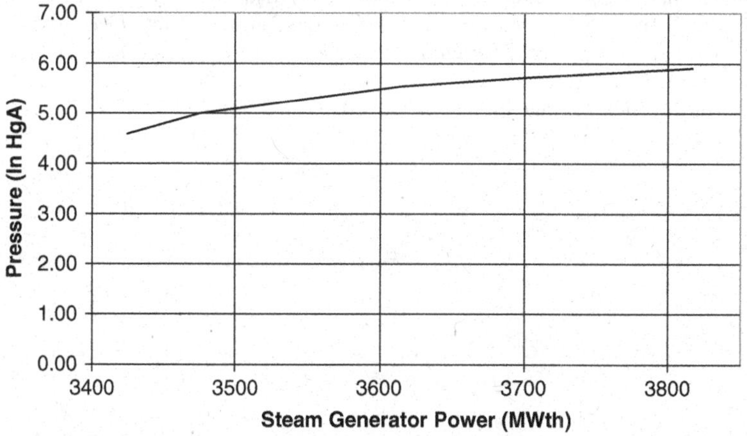

FIGURE 20.6 Main condenser Zone C pressure for summertime operation with 82% *PF*.

is primarily due to the increase in the temperature rise through the main condenser rather than an increase in the *CWT* from the NDCT because for an NDCT the *CWT* is only modestly influenced by an increase in *HWT* entering the cooling tower.

20.6 RESULTS OF POWER UPRATE

The following table shows the modifications in the turbine cycle systems deemed to be required for the indicated SG power level.

3,616 MW_{th}	3,716 MW_{th}	3,816 MW_{th}
• Replace HP turbine • Operate motor-driven MFP year-round • Replace No. 7 heater drain valve	• Replace HP turbine • Operate motor-driven MFP year-round • Replace No. 7 heater drain valve • LP turbine blade monitoring system • Automatic unit load run-back on low turbine-driven MFP NPSH	• Replace HP turbine • Operate motor-driven MFP year-round • New heater drain pumps • New LP turbine rotor blades. • Drive MFP turbine with main steam • Automatic unit load run-back on low turbine-driven MFP NPSH • Replace LP turbine C rotor

The following table shows the increase in T-G output for each SG power level

Reactor Power	3,616 MW_{th}	3,716 MW_{th}	3,816 MW_{th}
Wintertime operation	46.7	74.9	114.2
Summertime operation	43.6	70.8	108.6

REFERENCES

1. Bowman, C. F., et al., Evaluation of Turbine Cycle Systems for a Reactor Power Uprate at the Tennessee Valley Authority's Watts Bar Nuclear Plant, *Proceedings of the Electric Power Research Institute Thermal Performance Improvement Seminar*, 2002.
2. Bowman, C. F., Turbine Cycle Equipment Evaluation (TCEE) Workbook, Electric Power Research Institute Product ID 3002005344, 2015.

Nomenclature

Symbol	Definition	English Units	SI Units
A	reference surface area	ft^2	m^2
A	cross-sectional area of the discharge pipe	ft^2	m^2
a	interface area per unit of volume	ft^2	m^2
A	cross-sectional area of the CV	ft^2	m^2
A	greenhouse floor area	ft^2	m^2
A	total effective area	ft^2	m^2
$A_{annulus}$	annulus area of LP turbine exhaust	ft^2	m^2
A_c	effective cold-side heat exchanger surface area	ft^2	m^2
$A_{c,f}$	total cross-sectional area of plate fins	ft^2	m^2
$A_{c,t}$	total cross-sectional area of tubes	ft^2	m^2
$A_{condenser}$	exterior surface area of condenser	ft^2	m^2
A_{eff}	effective surface area	ft^2	m^2
A_f	total area of plate fins	ft^2	m^2
a_f	area of the fin per fin	ft^2	m^2
A_h	effective hot-side heat exchanger surface area	ft^2	m^2
A_h	total effective area of finned surface	ft^2	m^2
a_h	area of finned side per fin	ft^2	m^2
A_h/A_t	ratio of finned side to inside area	dimensionless	dimensionless
$A_{h\text{-}eff}$	effective hot-side area	ft^2	m^2
$A_{net,f}$	net area of each plate	ft^2	m^2
a_o	area of finned surface per fin	ft^2	m^2
A_p	net outside area of tubes (prime area)	ft^2	m^2
a_p	area of the prime per fin	ft^2	m^2
A_s	shell-side area of the FWH	ft^2	m^2
$A_{s,1}$	sub-cooling section shell-side area	ft^2	m^2
$A_{s,2}$	condensing section shell-side area	ft^2	m^2
AST	average surface temperature of the lake	ft^2	m^2
a_t	cross-sectional area of one tube	ft^2	m^2
A_t	tube-side area	ft^2	m^2
$A_{t,o}$	outside tube area	ft^2	m^2
A_w	tube wall area	ft^2	m^2
B	blowdown	gal/min	L^2/s
B	mass transfer driving potential	dimensionless	dimensionless
B	face height	ft	m
B	spacing between plates	in	m
b_{cal}	uncertainty of instrument calibration	dimensionless	dimensionless
$BHP_{corrected}$	corrected brake horsepower	hp	hp
BHP_{design}	design brake horsepower	hp	hp
Bi	Biot number	dimensionless	dimensionless
$b_{spa\ var}$	uncertainty of spatial variation	dimensionless	dimensionless

(*Continued*)

Symbol	Definition	English Units	SI Units
C	Constant	dimensionless	dimensionless
C	cycles of concentration	dimensionless	dimensionless
C_D	drag coefficient	dimensionless	dimensionless
c_p	specific heat	Btu/lbm-°F	kJ/kg-°C
$c_{p(a)}$	specific heat of moist air	Btu/lbm-°F	kJ/kg-°C
$c_{p,c}$	tube-side specific heat	Btu/lbm-°F	kJ/kg-°C
$c_{p,h}$	hot-side specific heat	Btu/lbm-°F	kJ/kg-°C
$c_{p,mix}$	specific heat of the mixture	Btu/lbm-°F	kJ/kg-°C
$c_{p\text{-}a}$	specific heat of air	Btu/lbm-°F	kJ/kg-°C
$c_{p\text{-}c}$	cold-side specific heat	Btu/lbm-°F	kJ/kg-°C
$c_{p\text{-}ccw}$	specific heat of CCW	Btu/lbm-°F	kJ/kg-°C
$c_{p\text{-}MA}$	specific heat of moist air	Btu/lbm-°F	kJ/kg-°C
$c_{p\text{-}t}$	tube-side specific heat	Btu/lbm-°F	kJ/kg-°C
$c_{p\text{-}w}$	specific heat of the water vapor	Btu/lbm-°F	kJ/kg-°C
$C_{s\text{-}1}$	capacitance of shell-side drains in the sub-cooling	Btu/hr-°F	W/°C
C_t	tube-side capacitance	Btu/hr-°F	W/°C
D	drift	gal/min	L/s
D	drop diameter	ft	m
D	bundle depth	ft	m
dA_i	differential interface area	ft^2	m^2
DCA	drain cooler approach	°F	°C
dE	differential evaporative mass transfer rate	Btu/hr	J/s
D_e	hydraulic diameter	ft	m
d_f	magnitude of the drag force	lbf	N
d_{fin}	effective fin diameter	ft	m
Dh	differential change in enthalpy of moist air	Btu/lbm	kJ/kg
d_i	tube inside diameter	ft	m
d_o	tube outside diameter	ft	m
DOA$_{calc}$	calculated drain cooler approach	°F	°C
DPT	ambient dewpoint temperatures	°F	°C
dQ_e	differential latent heat transfer rate	Btu/hr	J/s
dQ_s	differential sensible heat transfer rate	Btu/hr	J/s
dQ_t	differential heat transfer rate	Btu/hr	J/s
dS	incremental change in entropy	Btu/lbm-°R	kJ/kg-°K
dt_c	tube-side temperature rise	°F	°C
Duty	rate of heat rejection	Btu/lbm-°R	kJ/kg-°K
dV	differential cooling volume	ft^3	m^3
D_x	bulk drag force in the x direction	lbf	N
D_y	bulk drag force in the y direction	lbf	N
E	evaporation	lbm/hr	kg/s
E	equilibrium temperature	°F	°C
e_a	air vapor pressure	mmHg	mmHg
EMTD	effective mean temperature difference	°F	°C
e_s	saturated vapor pressure at lake surface temperature	lbf/in^{2}A	kPa

(Continued)

Symbol	Definition	English Units	SI Units
F	interference factor	dimensionless	dimensionless
F	LMTD correction factor	dimensionless	dimensionless
FA	face area	ft^2	m^2
F_f	fouling factor (performance factor)	dimensionless	dimensionless
F_m	tube material correction factor	dimensionless	dimensionless
FR	fouling ratio	dimensionless	dimensionless
Ft	temperature correction factor	dimensionless	dimensionless
FV	face velocity out of air side	ft/sec	m/s
G	air flow rate	ft^3/min	m^3/s
G	acceleration of gravity	ft/sec^2	m/s^2
G	flow rate	gal/min	L/s
g_c	Newton's proportionality constant	lbm-ft/lbf-sec^2	kg-m/s^2
G_c	service water flow rate	gal/min	L/s
G_t	tube-side mass flux	lbm/ft^2/hr	kg/m^2/s
G_h	hot-side mass flux	lbm/ft^2/hr	kg/m^2/s
G_{CCW}	CCW flow	gal/min	m^3/min
$G_{corrected}$	corrected air flow rate	ft^3/min	m^3/s
G_{design}	design air flow rate	ft^3/min	m^3/s
G_{pump}	required pump flow	gal/min	m^3/min
G_t	tube-side mass flux	lbm/ft^2/hr	kg/m^2/s
$g_{water\text{-}coil}$	service water flow rate per coil	gal/min	m^3/min
H	coefficient of heat transfer	Btu/hr-ft^2-°F	J/s-m^2-°C
H	pump head	ft	m
H	local heat transfer coefficient	Btu/hr-ft^2-°F	J/s-m^2-°C
h_1	enthalpy of steam entering the HP turbine	Btu/lbm	kJ/kg
H_1	peak height of the spray above the nozzle	ft	m
h_2	enthalpy of steam exiting the HP turbine	Btu/lbm	kJ/kg
H_2	height of the nozzle above the water level	ft	m
h_{2s}	isentropic enthalpy of steam exiting the HP turbine	Btu/lbm	kJ/kg
h_a	enthalpy of air	Btu/lbm	kJ/kg
h_c	cold-side convection coefficient	Btu/hr-ft^2-°F	J/s-m^2-°C
$h_{c,design}$	design cold-side convection coefficient	Btu/hr-ft^2-°F	J/s-m^2-°C
h_{CCWin}	enthalpy of CCW entering condenser	Btu/lbm	kJ/kg
h_{CCWout}	enthalpy of CCW exiting condenser	Btu/lbm	kJ/kg
$h_{c\text{-}in}$	cold stream enthalpy in	Btu/lbm	kJ/kg
$h_{condensate}$	enthalpy of the condensate	Btu/lbm	kJ/kg
$h_{c\text{-}out}$	cold stream enthalpy out	Btu/lbm	kJ/kg
h_{CRD}	enthalpy of control rod drive	Btu/lbm	kJ/kg
$H_{discharge}$	discharge head of the CCW pump	ft	m
$h_{drain\ in}$	enthalpy of FWH drains entering the FWH	Btu/lbm	kJ/kg
$h_{drain\ out}$	enthalpy of FWH drains exiting the FWH	Btu/lbm	kJ/kg
h_e	exit specific enthalpy	Btu/lbm	kJ/kg
$h_{exhaust}$	enthalpy of LP turbine exhaust flow	Btu/lbm	kJ/kg
$h_{ex\text{-}m}$	enthalpy of extracted moisture	Btu/lbm	kJ/kg

(Continued)

Symbol	Definition	English Units	SI Units
$h_{extraction}$	enthalpy of extraction steam	Btu/lbm	kJ/kg
h_{fg}	latent heat of the steam	Btu/lbm	kJ/kg
h_g	enthalpy of saturated water vapor	Btu/lbm	kJ/kg
h_h	hot-side convection coefficient	Btu/hr-ft²-°F	J/s-m²-°C
$h_{h,design}$	design hot-side convection coefficient	Btu/hr-ft²-°F	J/s-m²-°C
$h_{h,ideal}$	ideal hot-side convection coefficient	Btu/hr-ft²-°F	J/s-m²-°C
$h_{h\text{-}in}$	hot stream enthalpy in	Btu/lbm	kJ/kg
$h_{h\text{-}out}$	hot stream enthalpy out	Btu/lbm	kJ/kg
h_i	inlet specific enthalpy	Btu/lbm	kJ/kg
h_{in}	enthalpy of extraction entering the stage	Btu/lbm	kJ/kg
$h_{leakage}$	enthalpy of leakage	Btu/lbm	kJ/kg
h_{MS}	enthalpy of main steam	Btu/lbm	kJ/kg
$h_{MS\ drain}$	enthalpy of drains out of MSR	Btu/lbm	kJ/kg
$h_{MS\ stm\ out}$	enthalpy of steam out of MSR	Btu/lbm	kJ/kg
h_{MSin}	enthalpy of steam into MSR	Btu/lbm	kJ/kg
h_{out}	enthalpy of extraction leaving the stage	Btu/lbm	kJ/kg
H_{RJ}	plant heat rejection	Btu/ft²/day	J/m²/day
h_s	enthalpy of saturated air at the water temperature	Btu/lbm	J/kg
$h_{sc,1}$	sub-cooling section coefficient of heat transfer	Btu/hr-ft²-°F	J/s-m²-°C
$h_{sc,2}$	condensing section coefficient of heat transfer	Btu/hr-ft²-°F	J/s-m²-°C
h_{SGB}	enthalpy of steam generator blowdown	Btu/lbm	kJ/kg
h_{shell}	shell-side coefficient of heat transfer	Btu/hr-ft²-°F	J/s-m²-°C
$h_{shell\text{-}in}$	enthalpy of shell-side steam into RH	Btu/lbm	kJ/kg
$h_{shell\text{-}out}$	enthalpy of shell-side steam out of RH	Btu/lbm	kJ/kg
H_{SN}	net solar heating	Btu/ft²/day	J/m²/day
H_{sprays}	spray pump head	ft	kPa
H_{static}	static head	ft	kPa
$H_{suction}$	suction head of the CCW pump	ft	kPa
h_t	tube-side coefficient of heat transfer	Btu/hr-ft²-°F	J/s-m²-°C
$h_{tube\text{-}drain}$	enthalpy of drain from RH tubes	Btu/lbm	kJ/kg
$h_{tube\text{-}in}$	enthalpy of heating steam into RH tubes	Btu/lbm	kJ/kg
h_{UEEP}	enthalpy of the utilized energy end point	Btu/lbm	kJ/kg
$H_{velocity}$	velocity head	ft	kPa
h_{vent}	enthalpy of FWH vent steam	Btu/lbm	kJ/kg
J	number of measurement points for parameters	dimensionless	dimensionless
J_T	hot-side convection coefficient correction factor	dimensionless	dimensionless
K	turbine stage flow coefficient	dimensionless	dimensionless
K	ideal gas law constant for steam	dimensionless	dimensionless
K	mass transfer coefficient	lbm/hr-ft²	kg/s-m²
K	surface heat exchange coefficient	Btu/ft²-day-°F	J/m²-s-°C
k_c	thermal conductivity of water	Btu/hr-ft-°F	kJ/s-m-°C
k_c	tube-side thermal conductivity	Btu/hr-ft-°F	kJ/s-m-°C
k_{cw}	thermal conductivity of the CCW	Btu/hr-ft-°F	kJ/s-m-°C
k_f	shell-side thermal conductivity of condensate film	Btu/hr-ft-°F	kJ/s-m-°C

(*Continued*)

Symbol	Definition	English Units	SI Units
k_h	hot-side thermal conductivity	Btu/hr-ft-°F	kJ/s-m-°C
k_t	thermal conductivity of the tube material	Btu/hr-ft-°F	kJ/s-m-°C
L	water flow rate	gal/min	m^3/min
L	characteristic fin length	ft	m
L	fin height	ft	m
L_c	equivalent fin height	ft	m
L_{CP}	compressed height of plates	in	m
Le	Lewis factor	dimensionless	dimensionless
l_{eff}	effective tube length	ft	m
L_{eff}	effective tube length	ft	m
L_H	height of plates	in	m
LMTD	log mean temperature difference	°F	°C
L_w	width of plates	in	m
m	mass flow-passing capability	lbm/hr	kg/s
m	mass flow rate through LP turbine exhaust	lbm/hr	kg/s
M	moisture of steam at LP turbine exhaust	%	%
m	mass flow rate	lbm/hr	kg/s
M	makeup	lbm/hr	kg/s
m	mass of the drop	lbm	kg
M	air flow through the CV	lbm/hr	kg/s
m_a	mass flow rate of the air	lbm/hr	kg/s
M_a	molecular weight of air	dimensionless	dimensionless
m_c	mass flow rate of cold stream	lbm/hr	kg/s
$m_{c\text{-}coil}$	tube-side flow rate per coil	lbm/hr	kg/s
m_{CCWout}	mass flow rate of CCW exiting condenser	lbm/hr	kg/s
$m_{condensate}$	mass flow rate of the condensate	lbm/hr	kg/s
m_{CRD}	control rod drive mass flow rate	lbm/hr	kg/s
$m_{d,o}$	shell-side drain outlet mass flow rate	lbm/hr	kg/s
m_{drain}	mass flow rate of FWH drains entering condenser	lbm/hr	kg/s
$m_{drain\ in}$	mass flow rate of FWH drains entering the FWH	lbm/hr	kg/s
$m_{drain\ out}$	mass flow rate of FWH drains exiting the FWH	lbm/hr	kg/s
m_e	mass evaporation	lbm/hr	kg/s
$m_{exhaust}$	mass flow rate of LP turbine exhaust	lbm/hr	kg/s
$m_{ex\text{-}m}$	mass flow rate of moisture that is extracted	lbm/hr	kg/s
$m_{extraction}$	mass flow rate of extraction steam	lbm/hr	kg/s
m_{FFW}	final feedwater mass flow rate	lbm/hr	kg/s
m_h	mass flow rate of hot stream	lbm/hr	kg/s
m_h	total mass flow rate on the air side	lbm/hr	kg/s
m_h	hot-side flow measurement per coil	lbm/hr	kg/s
m_{in}	total mass flow rate of extraction entering the stage	lbm/hr	kg/s
$m_{leakage}$	leakage mass flow rate	lbm/hr	kg/s
m_{MS}	main steam mass flow rate	lbm/hr	kg/s
$m_{MS\ drain}$	mass flow rate of drains out of MSR	lbm/hr	kg/s
$m_{MS\ stm\ out}$	mass flow rate of steam out of MSR	lbm/hr	kg/s

(Continued)

Symbol	Definition	English Units	SI Units
m_{MSin}	mass flow rate of steam into MSR	lbm/hr	kg/s
$m_{purge\ stm}$	mass flow rate of purge steam from RH tubes	lbm/hr	kg/s
M_s	mass flow rate of the water sprayed	lbm/hr	kg/s
M_{SGB}	steam generator blowdown mass flow rate	lbm/hr	kg/s
m_{shell}	mass flow rate of shell-side steam through RH	lbm/hr	kg/s
$m_{shell\text{-}in}$	mass flow rate of shell-side steam into RH	lbm/hr	kg/s
$m_{shell\text{-}out}$	mass flow rate of shell-side steam out of RH	lbm/hr	kg/s
m_t	tube-side mass flow rate	lbm/hr	kg/s
$m_{total\text{-}m}$	total moisture entering the stage	lbm/hr	kg/s
$m_{tube\text{-}drain}$	mass flow rate of drain from RH tubes	lbm/hr	kg/s
$m_{tube\text{-}in}$	mass flow rate of heating steam into RH tubes	lbm/hr	kg/s
m_v	mass flow rate of the water vapor	lbm/hr	kg/s
M_v	molecular weight of water vapor	dimensionless	dimensionless
m_{vent}	mass flow rate of FWH vent steam	lbm/hr	kg/s
MW_{th}	megawatt thermal	MW	mW
m_{wv}	mass of the water vapor	lbm/hr	kg/s
N	number of u-tubes	dimensionless	dimensionless
n	empirically derived constant	dimensionless	dimensionless
N_f	number of fins per unit of length	fins/ft	fins/m
NHX	number of PHX in service	dimensionless	dimensionless
N_p	number of plates	dimensionless	dimensionless
N_{pass}	number of passes	dimensionless	dimensionless
N_{pv}	number of measurements taken during the test	dimensionless	dimensionless
N_r	number of rows	dimensionless	dimensionless
N_t	number of tubes per row	dimensionless	dimensionless
NTU	number of transfer units	dimensionless	dimensionless
NTU_1	number of transfer units of the sub-cooling section	dimensionless	dimensionless
NTU_2	number of transfer units of the condensing section	dimensionless	dimensionless
N_{tubes}	number of tubes per pass	dimensionless	dimensionless
Nu	Nusselt number	dimensionless	dimensionless
P	effectiveness	dimensionless	dimensionless
p	pressure	lbf/in^2	kPa
P	total static pressure	lbf/in^2	kPa
P	pressure	in. HgA	kPaa
P	total atmospheric pressure	lbf/in^2	kPa
P^*	upstream pressure	lbf/in^2A	kPaa
P_1	effectiveness of the sub-cooling section	dimensionless	dimensionless
P_2	effectiveness of the condensing section	dimensionless	dimensionless
P_a	partial pressure of dry air	lbf/in^2A	kPaa
PF	performance factor	dimensionless	dimensionless
$P_{first\ stage}$	HP turbine first-stage pressure	lbf/in^2A	kPaa
$P_{in\text{-}sat}$	saturation pressure entering	lbf/in^2A	kPaa
P_o	discharge pressure	lbf/in^2A	kPaa
$P_{cooling\ tower}$	CCW pump power required with a cooling tower	kW	kW

(Continued)

Symbol	Definition	English Units	SI Units
P_{lake}	CCW pump power required with a cooling lake	kW	kW
$Power_{spray}$	spray pond pump power required	kW	kW
Pr	Prandtl number	dimensionless	dimensionless
Pr_t	tube-side Prandtl number	dimensionless	dimensionless
$P_{reactor}$	reactor power	MW	kJ/s
P_{sat}	saturation pressure of the water vapor at DBT	lbf/in²A	kPaa
P_{sat}	saturation pressure at $T_{h,i}$	lbf/in²A	kPaa
$P_{sat\text{-}T}$	saturation pressure at the indicated temperature	lbf/in²A	kPaa
p_t	total pressure	lbf/in²A	kPaa
p_v	partial pressure of the water vapor	lbf/in²A	kPaa
P_w	partial pressure of the water vapor	lbf/in²A	kPaa
P_{wv}	partial pressure of the water vapor in the air	lbf/in²A	kPaa
Q	heat transfer rate	Btu/hr	kJ/s
Q^*	heat transfer rate at accident conditions	Btu/hr	kJ/s
Q_1	sub-cooling section rate of heat transfer	Btu/hr	kJ/s
Q_2	condensing section rate of heat transfer	Btu/hr	kJ/s
q_c	convective heat transfer	Btu/hr	kJ/s
Q_c	tube-side heat transfer rate	Btu/hr	kJ/s
$Q_{c\ test}$	cold-side heat transfer rate	Btu/hr	kJ/s
Q_{calc}	calculated rate of heat transfer	Btu/hr	kJ/s
q_e	evaporative heat transfer	Btu/hr	kJ/s
Q_{glass}	greenhouse glass heat transfer rate	Btu/hr	kJ/s-Ac
Q_h	air-side heat transfer rate	Btu/hr	kJ/s
$Q_{h\ test}$	hot-side heat transfer rate	Btu/hr	kJ/s
Q_{in}	heat added to the turbine cycle	Btu/hr	kJ/s
QL	heat rejected	Btu/hr	kJ/s
q_{solar}	solar heating	Btu/ft²/day	J/s-m²
q_T	total heat transfer from the drop	Btu/hr	J/s
Q_{test}	test rate of heat transfer	Btu/hr	kJ/s
R	capacitance ratio	dimensionless	dimensionless
R	temperature rise through the plant	°F	°C
R_0	magnitude of the total velocity vector	ft/sec	m/s
R_1	sub-cooling section capacitance ratio	dimensionless	dimensionless
r_2	fin radius	Ft	M
r_{2c}	characteristic fin radius	Ft	M
r_{2c}	radius ratio	dimensionless	dimensionless
R_a	gas constant for air	ft-lbf/lbm-°R	J/mol-°K
Re_t	tube-side Reynolds number	dimensionless	dimensionless
$r_{fouling}$	fouling resistance to heat transfer	hr-ft²/Btu	m²-°C/W
RH	relative humidity	%	%
r_{sc}	shell-side convection resistance to heat transfer	hr-ft²/Btu	m²-°C/W
$r_{sc,1}$	sub-cooling section resistance to heat transfer	hr-ft²/Btu	m²-°C/W
$r_{sc,2}$	condensing section resistance to heat transfer	hr-ft²/Btu	m²-°C/W
r_{sf}	shell-side fouling resistance to heat transfer	hr-ft²/Btu	m²-°C/W

(Continued)

Symbol	Definition	English Units	SI Units
r_{shell}	shell-side convection resistance to heat transfer	hr-ft²/Btu	m²-°C/W
$r_{shell\text{-}design}$	design shell-side convection resistance	hr-ft²/Btu	m²-°C/W
r_{tc}	tube-side convection resistance to heat transfer	hr-ft²/Btu	m²-°C/W
r_{tf}	tube-side fouling resistance to heat transfer	hr-ft²/Btu	m²-°C/W
r_{tube}	tube-side convection resistance to heat transfer	hr-ft²/Btu	m²-°C/W
$r_{tube\text{-}design}$	design tube-side convection resistance	hr-ft²/Btu	m²-°C/W
RV	velocity of the drop relative to the air	ft/sec	m/s
r_w	tube wall resistance	hr-ft²/Btu	m²-°C/W
r_{wall}	tube wall resistance to heat transfer	hr-ft²/Btu	m²-°C/W
$r_{wall\text{-}design}$	design tube wall resistance to heat transfer	hr-ft²/Btu	m²-°C/W
S	number of serpentines	dimensionless	dimensionless
S	Entropy	Btu/lbm-°R	kJ/kg-°K
Sc	Schmidt number	dimensionless	dimensionless
$s.g.$	specific gravity	dimensionless	dimensionless
Sh	Sherwood number	dimensionless	dimensionless
T	absolute temperature	°R	°K
T	time of flight of the drop	sec	s
T	student t for the number of measurement points	dimensionless	dimensionless
T	tube wall thickness	In	M
T^*	upstream temperature in degrees absolute	°R	°K
t_1	condensate inlet temperature	°F	°C
t_1	hot water temperature entering the spray nozzle	°F	°C
t_2	condensate temperature exiting the drain cooler	°F	°C
t_2	cold water temperature of the sprayed water	°F	°C
t_3	condensate outlet temperature	°F	°C
TA	tilt angle	rad	rad
T_{amb}	ambient temperature	°F	°C
t_c	cold-side temperature	°F	°C
T_C	CCW temperature rise through the plant	°F	°C
t_c	average water temp	°F	°C
$t_{c,i}$	entering water temp	°F	°C
$t_{c,o}$	exiting water temp	°F	°C
t_{CCWin}	temperature of CCW entering condenser	°F	°C
$T_{CCW\text{-}in}$	CCW temperature entering the plant	°F	°C
t_{CCWout}	temperature of CCW exiting condenser	°F	°C
$T_{CCW\text{-}out}$	CCW temperature exiting the plant	°F	°C
$t_{c\text{-}in}$	cold stream temperature in	°F	°C
$t_{c\text{-}in}$	cold stream temperature out	°F	°C
T_d	drop temperature	°F	°C
$t_{\mathbf{db}}$	local air dry-bulb temperature	°F	°C
T_{GH}	greenhouse temperature	°F	°C
T_h	hot-side temperature	°F	°C
TH	total head of the CCW pump	ft	m
T_h	average air dry bulb temperature	°F	°C

(*Continued*)

Symbol	Definition	English Units	SI Units
$T_{h,i}$	entering air dry bulb temperature	°F	°C
$T_{h,o}$	exiting air dry bulb temperature	°F	°C
$T_{h\text{-}in}$	hot stream temperature in	°F	°C
$T_{h\text{-}out}$	hot stream temperature out	°F	°C
t_i	condensate inlet temperature	°F	°C
t_o	condensate outlet temperature	°F	°C
T_o	shell-side outlet temperature	°F	°C
T_o	discharge temperature in degrees absolute	°R	°K
T_s	average lake surface temperature	°F	°C
t_{sat}	saturation temperature	°F	°C
T_{sat}	shell-side saturation temperature	°F	°C
$t_{shell\text{-}inlet}$	reheater shell-side inlet temperature	°F	°C
$t_{shell\text{-}out}$	reheater shell-side outlet temperature	°F	°C
$t_{surface}$	cooling lake surface temperature	°F	°C
TTD	terminal temperature difference	°F	°C
TTD_{calc}	calculated terminal temperature difference	°F	°C
t_{tube}	RH tube temperature	°F	°C
t_w	tube wall thickness	in	m
t_w	water temperature	°F	°C
t_{WB}	ambient wet-bulb temperature	°F	°C
$t_{WB\text{-}amb}$	wet-bulb temperature outside the spray region	°F	°C
$t_{WB\text{-}local}$	wet-bulb temperature at the spray nozzle	°F	°C
$T_{wh\text{-}ave}$	average waste heat temperature	°F	°C
t_β	average of AST and DPT	°F	°C
U	overall heat transfer coefficient	Btu/hr-ft^2-°F	W/m^2-°C
U	air velocity	ft/sec	m/s
u^*	uncertainty of heat transfer rate at accident conditions	Btu/hr	kJ/s
U_1	sub-cooling section overall heat transfer coefficient	Btu/hr-ft^2-°F	W/m^2-°C
U_2	condensing section overall heat transfer coefficient	Btu/hr-ft^2-°F	W/m^2-°C
U_C	covering heat transfer coefficient	Btu/hr-ft^2-°F	W/m^2-°C
U_{clean}	overall heat transfer coefficient while clean	Btu/hr-ft^2-°F	W/m^2-°C
$U_{clean\ design}$	overall design heat transfer coefficient while clean	Btu/hr-ft^2-°F	W/m^2-°C
U_{design}	design overall heat transfer coefficient	Btu/hr-ft^2-°F	W/m^2-°C
u_{dP}	uncertainty in measured dP	lbf/in^2	kPa
U_e	horizontal components of exit air velocity	ft/sec	m/s
U_F	floor heat transfer coefficient	Btu/hr-ft^2-°F	W/m^2-°C
u_G	uncertainty in measured flow	lbm/hr	kg/s
U_i	horizontal components of inlet air velocity	ft/sec	m/s
$U_{operating}$	overall heat transfer coefficient while operating	Btu/hr-ft^2-°F	W/m^2-°C
u_{pv}	uncertainty of process variation	°F	°C
u_{Qc}	uncertainty of the cold-side rate of heat transfer	Btu/hr	J/s
u_{Qh}	uncertainty of the hot-side rate of heat transfer	Btu/hr	J/s
U_{test}	overall heat transfer as determined by test	Btu/hr-ft^2-°F	W/m^2-°C
V	specific volume at LP turbine exhaust	ft^3/lbm	m^3/kg

(Continued)

Symbol	Definition	English Units	SI Units
V	velocity	ft/sec	m/s
V	cooling volume	ft^3	m^3
V	flow rate of air through the OSCS control volume	ft^3/sec	m^3/s
V	volumetric flow rate	ft^3/min	m^3/s
v_{0x}	horizontal velocity vector	ft/sec	m/s
v_{0y}	vertical velocity vector	ft/sec	m/s
$v_{first\ stage}$	first-stage specific volume	ft^3/lbm	m^3/kg
v_t	tube velocity	ft/sec	m/s
W	Work	Btu/hr	kJ/s
W	wind speed	Mi/hr	m/s
w	humidity ratio	dimensionless	dimensionless
w_{actual}	actual work	Btu/hr	kJ/s
$w_{isentropic}$	isentropic work	Btu/hr	kJ/s
x	quality of the steam entering the stage	%	%
x	mass fraction of the water in the bulk air stream	dimensionless	dimensionless
x	radial distance of the outer edge of the spray	ft	m
ΔX	thickness of plate	ft	m
x_{air}	tube spacing in line with air flow	ft	m
X_{ave}	average measured value at each location during test		
$X_{bulk\ ave}$	bulk average temperature at flow cross section	°F	°C
X_i	measured values		
x_s	mass fraction of the water at water temperature	%	%
Y_0	elevation of the top nozzle on the OSCS tree,	ft	m
Y_{max}	top of the control volume	ft	m
Y_{max1}	maximum height of drops from the side of the nozzle closest to the central axis of the OSCS	ft	m
Y_{max2}	maximum height of drops from the side of the nozzle away from the central axis of the OSCS	ft	m
ρ_a	density of air-water vapor mixture	lbm/ft^3	kg/m^3
ρ_w	density of water	lbm/ft^3	kg/m^3
Δ	density	lbm/ft^3	kg/m^3
ΔP	pressure drop	lbf/in^2	kPa
ΔT_1	greater terminal temperature difference	°F	°C
ΔT_2	lesser terminal temperature difference	°F	°C
T	absolute humidity	dimensionless	dimensionless
α	minimum angle of inclination of droplets	rad	rad
δ	fin thickness	ft	m
β	slope of the saturated vapor pressure curve	mmHg/°F	
β	maximum angle of inclination of droplets	rad	rad
δQ	incremental change in heat transfer	Btu/sec	kJ/s
η	spray efficiency of one spray nozzle	dimensionless	dimensionless
η	fin efficiency	dimensionless	dimensionless
η_h	surface efficiency	dimensionless	dimensionless
η_{th}	thermal efficiency	dimensionless	dimensionless

(Continued)

Symbol	Definition	English Units	SI Units
η_{urbine} –	turbine efficiency	dimensionless	dimensionless
μ	moisture removal effectiveness	dimensionless	dimensionless
μ_c	tube-side absolute viscosity	lbm/ft-hr	kg/m-s
μ_f	viscosity of the condensate film	lbm/ft-hr	kg/m-s
μ_h	hot-side viscosity	lbm/ft-hr	kg/m-s
μ_h	hot-side absolute viscosity	lbm/ft-hr	kg/m-s
μ_s	shell-side viscosity	lbm/ft-hr	kg/m-s
μ_t	tube-side viscosity	lbm/ft-hr	kg/m-s
μ_w	viscosity at the wall surface temperature	lbm/ft-hr	kg/m-s
θ	amount entering CCW temperature exceeds E	°F	°C
θ	angle of the velocity vector	rad	rad
Δ	density	lbm/ft^3	kg/m^3
ρ	density of the steam in the discharge plain	lbm/ft^3	kg/m^3
ρ_a	density of air	lbm/ft^3	kg/m^3
ρ_c	tube-side density	lbm/ft^3	kg/m^3
ρ_f	density of the steam	lbm/ft^3	kg/m^3
$\rho_{a\infty}$	inlet air density	lbm/ft^3	kg/m^3
ρ_a	density of air-water vapor mixture	lbm/ft^3	kg/m^3
ρ_w	density of water	lbm/ft^3	kg/m^3

Appendix A
Example Problem in English Units

CHAPTER 1

Refer to Figure 1.1.

In a Rankine cycle, steam leaves the SG and enters a turbine at 1,000 lbm/in^{2}A saturated. The steam expands in an isentropic process to a condenser pressure of 1.0 lbm/in^{2}A. The enthalpy increase across the main feed pump is 1.8 Btu/lbm. Determine the thermal efficiency of the cycle. Hint: because the mass flow rate is not given, compute work and added heat on a per pound basis.

$$\eta_{th} = \frac{W}{Q_H} = \frac{w}{q_H} = \frac{(h_4 - h_5) - (h_2 - h_1)}{h_4 - h_2}$$

From the *ASME Steam Tables*[1]

$$h_4 = 1192.9\left(\text{Btu}/_{\text{lbm}}\right);\ s_4 = 1.391\left(\text{Btu}/_{\text{lbm-°R}}\right)$$

$$s_5 = s_4 = 1.391\left(\text{Btu}/_{\text{lbm-°R}}\right)$$

$$s_1 = s_f \text{ at } 1.0\left(\text{lbf}/_{\text{in}^2\text{A}}\right) = 0.1326\left(\text{Btu}/_{\text{lbm-°R}}\right)$$

$$x = \frac{s_5 - s_1}{s_{fg}} = \frac{1.391 - 0.1326}{1.8455} = 0.6819$$

$$h_5 = h_f + xh_{fg}$$

$$h_5 = 69.73 + (0.6819)1036.1 = 776.2\left(\text{Btu}/_{\text{lbm}}\right)$$

$$w_T = h_4 - h_5 = 1192.9 - 776.2 = 416.7\left(\text{Btu}/_{\text{lbm}}\right)$$

$$w_P = h_2 - h_1 = 1.8\left(\text{Btu}/_{\text{lbm}}\right)$$

$$h_1 = h_f = 69.7\left(\text{Btu}/_{\text{lbm}}\right)$$

$$h_2 = 69.7 + 1.8 = 71.5\left(\text{Btu}/_{\text{lbm}}\right)$$

$$q_H = h_4 - h_2 = 1192.9 - 71.5 = 1121.4\left(\text{Btu}/_{\text{lbm}}\right)$$

$$\eta_{th} = \frac{w_T - w_p}{q_H} = \frac{416.7 - 1.8}{1121.4} = 37.0\%$$

CHAPTER 2

Refer to Figure 2.3.

The FFW flow to a PWR SG is 15,000,000 lbm/hr, at a pressure of 1,100 lbf/in^2 and a temperature of 442°F. The SG is operating at a pressure of 1,000 lbf/in^2 and discharging saturated steam with a quality of 99.75%. The blowdown flow is 175,000 lbm/hr. The reactor coolant pumps increase the heat going from the reactor to the SG by 15 MW. Determine how much heat the reactor is producing.

$$Q_{in} = m_{MS}\, h_{MS} + m_{SGB}\, h_{SGB} - m_{FFW}\, h_{FFW}$$

$$m_{MS} = m_{FFW} - m_{SGB} = 15{,}000{,}000 - 175{,}000 = 14{,}825{,}000 \left({}^{\text{lbm}}\!/_{\text{hr}} \right)$$

From the *ASME Steam Tables*[1]

$$h_{MS} = 1191.3 (\text{Btu/lbm})$$

$$h_{SGB} = 542.6 (\text{Btu/lbm})$$

$$h_{FFW} = 421.7 (\text{Btu/lbm})$$

$$Q_{in} = m_{MS}\, h_{MS} + m_{SGB}\, h_{SGB} - m_{FFW}\, h_{FFW}$$

$$14{,}825{,}000\ (1191.3) + 175{,}000\ (542.6) - 15{,}000{,}000\ (421.7)$$

$$1.143 \times 10^{10} \left({}^{\text{Btu}}\!/_{\text{hr}} \right)$$

Converting to MW_{th}

$$MW_{th} = \left(\frac{1.143 \times 10^{10} \left({}^{\text{Btu}}\!/_{\text{hr}} \right)}{3{,}412.14 \frac{{}^{\text{Btu}}\!/_{\text{hr}}}{\text{kW}}} \right) 0.001 \left({}^{\text{MW}}\!/_{\text{kW}} \right) = 3{,}350 \left(MW_{th} \right)$$

Subtracting the heat added by the reactor coolant pumps yields the reactor power.

$$P_{Reactor} = 3{,}350 - 15 = 3{,}335\ MW_{th}$$

CHAPTER 3

Refer to Figure 3.4.

The pressure at the HP turbine throttle valve inlet is 1,000 lbf/in^2 saturated vapor, and the exhaust pressure is 170 lbf/in^2. The turbine passes 15,000,000 lbm/hr with an efficiency of 80%. There are no steam extractions from the turbine. Calculate the power out of the turbine.

$$W = m_{HP\text{-}in} h_{HP\text{-}in} - m_{HP\text{-}out}\, h_{HP\text{-}out} = m_{HP}\left(h_{HP\text{-}in} - h_{HP\text{-}out}\right)$$

$$h_{HP\text{-}in} - h_{HP\text{-}out} = \eta_{HP}\left(h_{HP\text{-}in} - h_{HP\text{-}out\,(s)}\right)$$

$$W = m_{HP}\,\eta_{HP}\left(h_{HP\text{-}in} - h_{HP\text{-}out\,(s)}\right)$$

From the *ASME Steam Tables*[1]

$$\text{with } P_{in\text{-}sat} = 1{,}000\left({}^{\text{lbf}}\!/\!_{\text{in}^2}\right)$$

$$h_{HP\text{-}in} = 1192.9\left({}^{\text{Btu}}\!/\!_{\text{lbm}}\right);\; s_{HP\text{-}in} = 1.391\left({}^{\text{Btu}}\!/\!_{\text{lbm-}^\circ\text{R}}\right)$$

$$s_{HP\text{-}out} = s_{HP\text{-}in} = 1.391\left({}^{\text{Btu}}\!/\!_{\text{lbm-}^\circ\text{R}}\right)$$

$$\text{with } P_{HP\text{-}out} = 170 \text{ lbf/in}^2\text{A}$$

$$x = \frac{s_{HP\text{-}out} - s_f}{s_{fg}} = \frac{1.391 - 0.5269}{1.0322} = 0.8371$$

$$h_f = 341.2 \text{ and } h_{fg} = 854.8\left({}^{\text{Btu}}\!/\!_{\text{lbm}}\right)$$

$$h_{HP\text{-}out\,(s)} = h_f + x h_{fg} = 341.2 + (0.8371)\,854.8 = 1056.8\left({}^{\text{Btu}}\!/\!_{\text{lbm}}\right)$$

$$W = \frac{m_{HP}\left({}^{\text{lbm}}\!/\!_{\text{hr}}\right)\eta_{HP}\left(h_{HP\text{-}in} - h_{HP\text{-}out\,(s)}\right)\left({}^{\text{Btu}}\!/\!_{\text{lbm}}\right)}{3412.14\left(\frac{{}^{\text{Btu}}\!/\!_{\text{hr}}}{\text{kW}}\right)}$$

$$= \frac{(15{,}000{,}000)\left({}^{\text{lbm}}\!/\!_{\text{hr}}\right)(0.8)(1192.9 - 1056.8)\left({}^{\text{Btu}}\!/\!_{\text{lbm}}\right)}{3412.14\left(\frac{{}^{\text{Btu}}\!/\!_{\text{hr}}}{\text{kW}}\right)} = 479{,}000\,(\text{kW})$$

CHAPTER 4

Refer to Figure 4.10.

Given a single-stage reheater operating with a shell-side mass flow rate of 10,200,000 lbm/hr, inlet and outlet shell pressures of 168 and 163 lbf/in^2, respectively, and an inlet quality of 99.74%. Reheat steam on the tube side is from the main steam with an enthalpy of 1,192 Btu/lbm and a pressure of 945 lbf/in^2 saturated. The purge steam flow is 2% of the reheat steam. The tube-side drain flow is measured to be 1,325,000 lbm/hr, and the drain temperature is 536°F. The hot reheat temperature is not available. Calculate the TTD of the reheater.

$$m_{tube\text{-}in} = \frac{m_{tube\text{-}drain}}{1-0.02} = 1{,}352{,}000 \text{ lbm/hr}$$

$$m_{purge\ stm} = m_{tube\text{-}in} - m_{tube\text{-}drain} = 27{,}000 \text{ lbm/hr}$$

Because the shell-side inlet pressure is 168 lbf/in^2 and the inlet quality is 99.74%, from the *ASME Steam Tables*[1]

$$h_{shell\text{-}in} = 1193.3 \text{ Btu/lbm}$$

The tube-side drain is saturated, so from the *ASME Steam Tables*[1]

$$T_{tube\text{-}in} = 537.8°\text{F},\ h_{tube\text{-}drain} = 531.75 \text{ Btu/lbm}$$

Because the shell-side mass flow rate out is equal to the shell-side mass flow rate in, from the first law of thermodynamics

$$m_{shell\text{-}in}h_{shell\text{-}in} + m_{tube\text{-}in}h_{tube\text{-}in} = m_{shell\text{-}out}\, h_{shell\text{-}out} + m_{tube\text{-}drain}\, h_{tube\text{-}drain} + m_{purge\ steam}\, h_{tubeside\ in}$$

$$h_{shell\text{-}out} = \frac{m_{shell\text{-}in}\, h_{shell\text{-}in} + m_{tube\text{-}in}h_{tube\text{-}in} - m_{tube\text{-}drain}\, h_{tube\text{-}drain} - m_{purge\ steam}\, h_{tube\text{-}in}}{m_{shell\text{-}out}}$$

$$= 1279.1 \text{ Btu/lbm}$$

Because the shell-side inlet pressure is 163 lbf/in^2 and the enthalpy is 1279.1 Btu/lbm, from the *ASME Steam Tables*, $t_{shell\text{-}out} = 511.5°\text{F}$ and

$$TTD = T_{tube\text{-}in} - t_{shell\text{-}out} = 537.8 - 511.5 = 26.3°\text{F}$$

CHAPTER 5

Refer to Figure 5.5.

The total steam flow entering a moisture removal stage is 9,058,000 lbm/hr, with a quality of 95.7%. The pressure downstream of the stage is 16 lbf/in^{2}A. Calculate the quality downstream of the moisture removal stage. From Figure 5.6, μ = 16.2%.

$$m_{ex-m} = m_{in}\left(1 - x_{in}\right)\mu = 63{,}100 \text{ lbm/hr}$$

The moisture is removed at 16 lbf/in^{2}A saturated with a quality of 95.7%. From the *ASME Steam Tables*[1]

$$h_{in} = 1110.5 \text{ Btu/lbm.}$$

The drain exiting is saturated liquid at 16 lbf/in^{2}A, so from the *ASME Steam Tables*[1]

$$h_{ex-m} = 184.5 \text{ Btu/lbm.}$$

From the first law of thermodynamics,

$$h_{out} = \frac{m_{in}\, h_{in} - m_{ex-m}\, h_{ex-m}}{m_{in} - m_{ex-m}} = 1117.0 \text{ Btu/lbm}$$

The steam leaves the stage at a pressure of 16 lbf/in^{2}A and an enthalpy of 1117.0 Btu/lbm, so from the *ASME Steam Tables*,

$$x_{out} = 96.37\%$$

CHAPTER 6

The following example problem is based on actual plant data from the Sequoyah Nuclear Plant

Condenser Physical Data		**Units**
Number of tubes, N	77,592	
Tube outside diameter, D_o	0.750	in
Tube outside diameter, $d_o[D_o/12]$	0.063	ft
Tube wall thickness, t	0.022	in
Tube inside diameter, $D_i[D_o - 2t]$	0.706	in
Tube inside diameter, $d_i[D_i/12]$	0.059	ft
Effective straight length of tubes per zone	49.840	ft
Surface area, A	759,310	ft^2
Tube material	Titanium	
Tube material thermal conductivity	12.5	Btu/hr-ft-°F

Condenser Operating Boundary Conditions	**Design**	**Test**	**Units**
Tube-side flowrate, G [test data]	524,000	537,524	gal/min
Tube-side inlet temperature, t_i [test data]	61.00	81.55	°F
Tube-side outlet temperature, t_o [test data]	90.50	109.10	°F
Specific volume, v [steam tables]	0.01600	0.01607	ft^3/lbm
Tube-side flowrate, m_{ccw} [$8.0208 \times G/v$]	2.627E+08	2.827E+08	lbm/hr
Tube-side velocity, V [$0.408 \times (G/N)/D_i^2$]	5.528	5.671	ft/s
Tube-side flow area, A_t [$(3.1416/4) \times d_i^2 \times N$]	210.94	210.94	ft^2
Tube-side specific heat at inlet temperature, $c_{p\text{-}in}$	1.002	1.000	Btu/(lbm-°F)
Condensing duty, Q [$m_{ccw}c_{p\text{-}in}(t_o - t_i)$]	7.769E+09	7.791E+09	Btu/hr
Thermal capacity rate at inlet temp., C_t [$m_{ccw}c_{p\text{-}in}$]	2.633E+08	2.828E+08	Btu/(hr-°F)
Tube-side average temperature, t_{ave} [$(t_o - t_i)/2$]	75.75	95.33	°F
Tube-side specific heat, $c_{p\text{-}ave}$	1.001	0.999	Btu/(lbm-°F)
Tube-side thermal conductivity, k_t	0.351	0.360	Btu/(hr-ft-°F)
Tube-side viscosity, μ_t	2.240	1.713	lbm/(ft-hr)

Tube-side Convection Resistance	**Design**	**Test**	**Units**
Reynolds number, Re [$m_t d_o/A_t\mu_t$]	32,712	46,043	
Tube-side Prandtl number, Pr [$\mu_t c_{p\text{-}ave}/k_t$]	6.39	4.75	
Fanning friction factor, f [$(1.58 \ln \text{Re}_t - 3.28)^{-2}$]	0.00579	0.00534	
Nusselt number, Nu_t (from Petukhov and Kirillov)[a]	221	257	
Calculated tube-side film coefficient, h_{tube} [$k_t\text{Nu}_t/d_i$]	1,317	1,575	Btu/hr-ft^2-°F
Calculated tube-side film resistance, r_{tube}[$(1/h_{tube}$]	0.000760	0.000635	hr-ft^2-°F/Btu

[a] $$\text{Nu} = \frac{\left(\frac{f}{2}\right)\text{Re}_t\,\text{Pr}_t}{1.07 + 12.7\sqrt{\frac{f}{2}}\left(\text{Pr}_t^{2/3} - 1\right)}$$

Tube Wall Resistance	Design	Test	Units
Tube wall resistance, r_{wall} $[(d_o/2k_w)(\ln(d_o/d_i))]$	0.000151	0.000151	hr-ft²-°F/Btu

Shell-side Convection Resistance	Design	Test	Units
Shell-side pressure, P [design/test data]	1.935	3.170	HgA
Shell-side temperature, t_{sat} [steam tables]	100.27	116.84	°F
Tube-side outlet temperature, t_o [design/test data]	90.50	109.10	°F
TTD $[t_{sat} - t_o]$	9.77	7.74	°F
Condensing LMTD $[(t_o - t_i)/\ln\{(t_{sat} - t_i)/(t_{sat} - t_o)\}]$	21.20	18.15	°F

Additive Resistance Method	Design	Test	Units
Shell-side condensate temperature, t_f $[(t_{sat} + t_i)/2]$	80.63	99.19	°F
Shell-side condensate thermal conductivity, k_f	0.353	0.362	Btu/hr-ft-°F
Shell-side condensate film density, ρ_f	62.213	62.007	lbm/ft³
Shell-side latent heat of steam, h_{fg} [steam tables]	1048.01	1037.54	Btu/lbm
Shell-side condensate film viscosity, μ_f	0.000572	0.000466	lbm/(ft-hr)
$t_{sat} - t_i$	39.27	35.29	°F
Nusselt parameter, Nu_{shell}' [a]	53.128	58.236	
Nusselt correction, $Nu_{correct}$ $[Nu_{shell\text{-}design}'/Nu_{shell\text{-}test}']$		0.912	
Calculated shell-side film coefficient, h_{shell} [b]	990	1,085	Btu/hr-ft²-°F
Calculated shell-side film resistance, r_{shell} [c]	0.001010	0.000922	hr-ft²-°F/Btu
Condensing heat trans. rate, clean, U_{clean} [d]	508	572	Btu/hr-ft²-°F
Condensing heat trans. rate, operating, $U_{operating}$ [e]	483	565	Btu/hr-ft²-°F
Condenser performance factor by ARM, PF [f]	95%	99%	
Tube-side fouling resistance [g]	0.000104	0.000022	hr-ft²-°F/Btu

a $$Nu_{shell}' = \left[\frac{k_f^3\,\rho_f^2\,g\,h_{fg}}{\mu_f\left(t_{sat} - t_{CCWin}\right)d_o}\right]^{1/4}$$

b $$h_{shell\text{-}design} = \frac{1}{r_{shell\text{-}design}}$$

$$r_{shell\text{-}design} = \frac{1}{U_{clean\text{-}design}} - r_{wall\text{-}design} - \left(\frac{d_o}{d_i}\right) r_{tube\text{-}design}$$

c $$h_{shell\text{-}test} = \frac{h_{shell\text{-}design}}{Nu_{correct}}$$

$$r_{shell\text{-}test} = \frac{1}{h_{shell\text{-}test}}$$

(Continued)

d $U_{clean\text{-}design} = \dfrac{U_{operating}}{PF_{design}}$ $U_{clean\text{-}test} = \dfrac{1}{r_{shell\text{-}test} + r_{wall} + \left(\dfrac{d_o}{d_i}\right) r_{tube\text{-}test}}$

e $U = \dfrac{Q}{A\,LMTD}$

f $PF = \dfrac{U_{operating}}{U_{clean}}$

g $r_{fouling} = \dfrac{1}{U_{operating}} - \dfrac{1}{U_{clean}}$

Heat Exchange Institute Method	**Design**	**Test**	**Units**
HEI constant, C	267	267	
Material correction factor – 22 BWG Titanium, F_m	0.870	0.870	
Tube-side inlet temperature	61.00	81.55	°F
HEI temperature correction factor, F_t	0.930	1.050	
Condenser performance factor by HEI, PF[a]	95%	97%	

a $PF = \dfrac{U}{C\sqrt{V}\,F_t\,F_m}$

CHAPTER 7

Refer to Figure 7.4.

An FWH is operating at a shell pressure of 41.3 lbf/in^{2}A. Approximately 9,580,000 lbm/hr of condensate passing through the tubes is heated from 210°F to 264°F by extraction steam having an enthalpy of 1176.5 Btu/lbm and a drain flow of 400,000 lbm/hr coming from a higher pressure FWH at a temperature of 262°F. Approximately 0.5% of the extraction steam is vented to the main condenser. The temperature of the drain coming out of the drain cooler is 222°F. Calculate the extraction steam flow rate.

The enthalpy of drains and condensate (i.e., FWH in the cycle below the main feed pump) is taken as a saturated liquid because the amount of sub-cooling is insignificant. Therefore, the enthalpies of the condensate and drains entering and exiting the FWH are as follows from the *ASME Steam Tables*[1]:

$$h_{condensate\text{-}i} = 178.2 \text{ Btu/lbm}$$

$$h_{condensate\text{-}o} = 232.8 \text{ Btu/lbm}$$

$$h_{drains\text{-}i} = 230.8 \text{ Btu/lbm}$$

$$h_{drains\text{-}o} = 190.2 \text{ Btu/lbm}$$

The enthalpy of the vent steam corresponds to saturated vapor at the FWH operating pressure. Therefore, from the *ASME Steam Tables*[1]:

$$h_{vent} = 1170.3 \text{ Btu/lbm}$$

Therefore, from the first law of thermodynamics,

$$m_{drain\text{-}i}\,h_{drain\text{-}i} + m_{ex}\,h_{ex} + m_{condensate}\,h_{condensate\text{-}i} = m_{drain\text{-}o}\,h_{drain\text{-}o} + m_{vent}\,h_{vent} + m_{condensate}\,h_{condensate\text{-}o}$$

$$m_{drain\text{-}o} = m_{drain\text{-}i} + m_{ex} - m_{vent} \qquad m_{vent} = 0.005\, m_{ex}$$

$$m_{drain\text{-}o} = m_{drain\text{-}i} + m_{ex}\left(1 - 0.005\right)$$

$$m_{ex}(h_{ex} - 0.995\, h_{drain\text{-}o} - 0.005\, h_{vent}) = m_{condensate}(h_{condensate\text{-}o} - h_{condensate\text{-}i}) - m_{drain\text{-}i}(h_{drain\text{-}i} - h_{drain\text{-}o})$$

$$m_{ex} = \frac{m_{condensate}\,(h_{condensate\text{-}o} - h_{condensate\text{-}i}) - m_{drain\text{-}i}\,(h_{drain\text{-}i} - h_{drain\text{-}o})}{h_{ex} - 0.995\, h_{drain\text{-}o} - 0.005\, h_{vent}}$$

$$m_{ex} = \frac{9{,}580{,}000\,(232.8 - 178.2) - 400{,}000\,(230.8 - 190.2)}{1176.5 - 0.995\,(190.2) - 0.005\,(1170.3)} = 516{,}000 \ {}^{\text{lbm}}\!/_{\text{hr}}$$

CHAPTER 8

Assume a 12-in extraction line bypass to the condenser. The pipe surface temperature measured just upstream of the main condenser is 188.6°F (648.6°R). This corresponds to a saturation pressure of 9.1 lbf/in^{2}A. Because the valve discharges into the condenser operating at a pressure of 1.5 inHg (0.732 lbf/in^{2}A), a critical pressure ratio exists. Therefore,

$$T* = 0.859\, T_O = 557.1°\text{R}$$

$$k = {}^{V^2}\!\!/_{g_c RT*} \Rightarrow V = \sqrt{k\, g_c\, RT*} = \sqrt{(1.33)(32.17)(85.76)(557.1)} = 1{,}430\ (\text{ft/sec})$$

From the *ASME Steam Tables*[1], the density of the steam at 0.732 lbf/in^{2}A is

$$\rho = \frac{1}{v} = 0.00223\left({}^{\text{ft}^3}\!/_{\text{lbm}}\right)$$

The area is

$$A = \frac{\pi}{4}\left(\frac{D_o}{12}\right)^2 = 0.7854\ (\text{ft}^2)$$

And, the mass flow rate is

$$m = V\,\rho\,A = (1{,}430)(0.00223)(0.7854)(3{,}600) = 9{,}016\left({}^{\text{lbm}}\!/_{\text{hr}}\right)$$

Alternately, use Figure 14 in the *ASME Steam Tables*[1], Critical, (Choking), Mass Flow Rate for Isentropic Process and Equilibrium Conditions. This figure provides the mass flow rate per square inch of flow area per lbf/in^2 of pressure, m', as a function of upstream enthalpy and pressure. In the instant case, the extraction enthalpy is 1,106 Btu/lbm. Based on the critical pressure ratio, the pressure upstream of the diffuser is

$$P_O = \frac{P*}{0.54} = \frac{0.732}{0.54} = 1.36\ \text{lbf/in}^2$$

From Figure 14, m' is equal to 57 [lbm/(hr-in^2-psi)].

Therefore,

$$m = m'\,a\,P_o = (57)\left(\frac{\pi}{4}\right)(12^2)(1.36) = 8{,}739\left({}^{\text{lbm}}\!/_{\text{hr}}\right)$$

CHAPTER 9

The following water chemistry data were taken from the Tennessee River, which provides makeup to the Watts Bar Nuclear Plant:

Methyl orange alkalinity: 51.8
Calcium hardness: 53.2 PPM $CaCO_3$
Total dissolved solids: 89.8 PPM.

Assuming a maximum operating temperature of 130°F, the following values may be determined from Figure A9.1 below:

pAK = 3.00
pCa = 3.28
$C = 1.58$
pH = 7.50.

The Ryznar stability index (*RSI*) is defined as follows:

$$RSI = 2\left(\text{pH}_s\right) - \text{pH}$$

where

$$\text{pH}_s = \text{pCa} + \text{pAlk} + C$$

Table A9.1 is a tabulation of the *RSI* for an array of cycles of concentration:

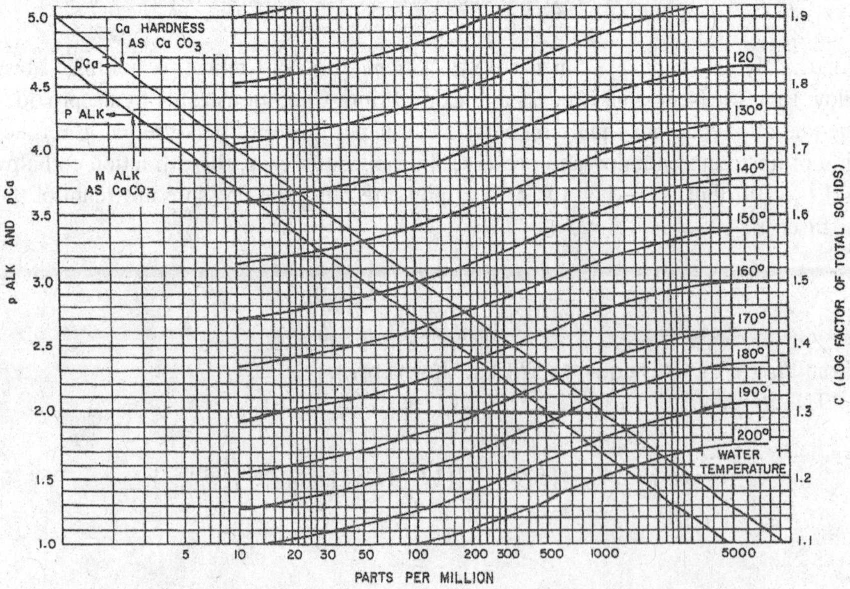

FIGURE A9.1 Water chemistry data.

TABLE A9.1
Tabulation of the RSI

Cycles of concentration	1	2	3
Methyl orange alkalinity	51.8	103.6	207.2
Calcium hardness	53.2	106.4	159.6
Total dissolved solids, PPM	89.8	179.6	269.4
pAK	3.00	2.68	2.55
pCa	3.28	2.96	2.82
C	1.58	1.60	1.62
pH_s	7.86	7.24	6.99
pH	7.50	7.85	8.15
RSI	8.22	6.63	5.83

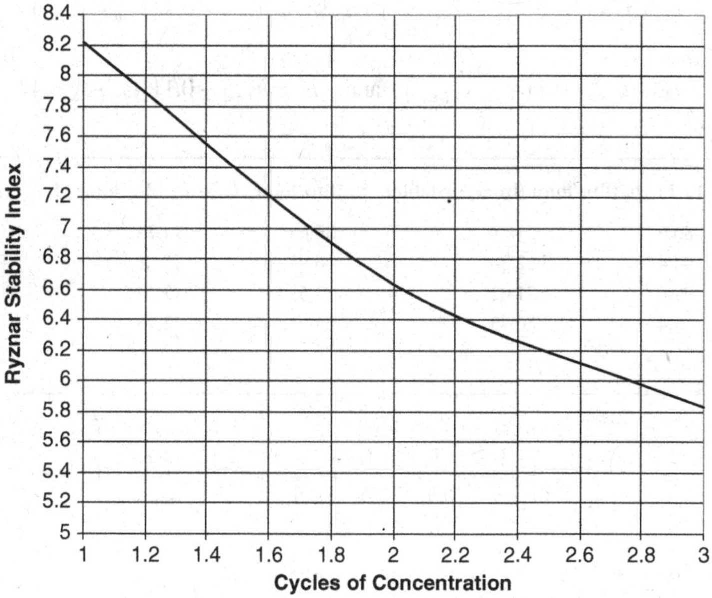

FIGURE A9.2 Ryznar stability index.

Figure A9.2 shows the *RSI* as a function of the cycles of concentration. Because an *RSI* of less than 6 may result in scale formation at the outlet of the main condenser and an *RSI* of greater than 7 may result in corrosion, an optimum cycle of concentration would be approximately 2.2.

CHAPTER 10

Refer to Figure 10.9.

Calculate the integral of the area under the curve using the Tchebycheff method.

$$\frac{KaV}{L} = \int_{t_{cold}}^{t_{hot}} \frac{dt}{h_S - h_a} = \frac{t_{S\text{-}in} - t_{S\text{-}out}}{4}\left(\frac{1}{\Delta h_1} + \frac{1}{\Delta h_2} + \frac{1}{\Delta h_3} + \frac{1}{\Delta h_4}\right)$$

where $t_{S\text{-}in} = 110°F$, $t_{S\text{-}out} = 84°F$,

$$t_1 = t_{S\text{-}out} + 0.1\left(t_{S\text{-}in} - t_{S\text{-}out}\right) \text{ and } h_1 = h_{a\text{-}in} + 0.1\left(t_{S\text{-}in} - t_{S\text{-}out}\right)$$

$$t_2 = t_{S\text{-}out} + 0.4\left(t_{S\text{-}in} - t_{S\text{-}out}\right) \text{ and } h_2 = h_{a\text{-}in} + 0.4\left(t_{S\text{-}in} - t_{S\text{-}out}\right)$$

$$t_3 = t_{S\text{-}in} - 0.4\left(t_{S\text{-}in} - t_{S\text{-}out}\right) \text{ and } h_3 = h_{a\text{-}out} - 0.4\left(t_{S\text{-}in} - t_{S\text{-}out}\right)$$

$$t_4 = t_{S\text{-}in} - 0.1\left(t_{S\text{-}in} - t_{S\text{-}out}\right) \text{ and } h_4 = h_{a\text{-}out} - 0.1\left(t_{S\text{-}in} - t_{S\text{-}out}\right)$$

Interval	t (°F)	h_S (Btu/lbm) (from Air Table)	h_A (Btu/lbm)	$h_W - h_A$ (Btu/lbm)	$1/\Delta h$ (lbm/Btu)
1	86.6	51.41	36.63	14.78	0.0676
2	94.4	62.38	46.77	15.61	0.0640
3	99.6	71.02	53.53	17.49	0.0571
4	107.4	86.43	63.67	22.76	0.0441
Total					0.2328

$$\frac{KaV}{L} = \frac{t_{S\text{-}in} - t_{S\text{-}out}}{4}\left(\frac{1}{\Delta h_1} + \frac{1}{\Delta h_2} + \frac{1}{\Delta h_3} + \frac{1}{\Delta h_4}\right) = \left(\frac{110 - 84}{4}\right)0.2328 = 1.51$$

Consider the following cooling tower test results:

	Design	Test
Water Flow, L, gal/min	10,000	8,950
Airflow, ft³/min	850,000	
Hot water temperature, HWT, °F	110	110
Coldwater temperature, CWT, °F	90	84
Cooling range, °F	20	26
Wet-bulb temperature, °F	80	69
Approach to wet-bulb temperature, °F	10	15
Fan power, bhp	136	115

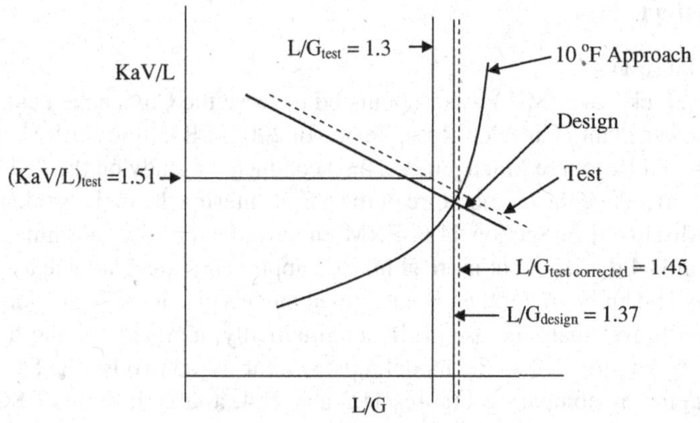

$$Tower\ Capability = \frac{\left(L/G\right)_{test\ corrected}}{\left(L/G\right)_{design}} = \frac{1.45}{1.37} = 105.8\ \%$$

FIGURE A10.1 Results of the cooling tower test.

$$L/G_{design} = 116.5\left(\frac{10{,}000}{850{,}000}\right) = 1.37 \quad L/G_{test} = 1.37\left(\frac{8{,}950}{10{,}000}\right)\left(\frac{136}{115}\right)^{1/3} = 1.30$$

Because the value of *KaV/L* is determined to be 1.51, the test results establish a different point on the *KaV/L* vs *L/G* plot, as shown in Figure A10.1.

By drawing a second straight line parallel to the vendor-supplied line through the intersection of *KaV/L* = 1.51 and *L/G* = 1.30, one may see that because the line representing the test results is above the line representing the vendor's design, the test line intersects the 10°F approach curve slightly to the right of that representing the design value of *L/G*, which is 1.37. The results of the test indicate an *L/G* of 1.45. The ratio of these two values represents the cooling tower capability of 105.8% of the design value.

CHAPTER 11

Refer to Figure 11.4.

Squaw Creek Lake (SCL) was impounded to serve the Comanche Peak two-unit Nuclear Power Plant near Glen Rose, Texas. In 2005, ERM, Inc., (ERM) (formerly Environmental Resource Management, Inc.) conducted a study on the SCL to investigate ways to reduce the temperature of the CCW entering the main condenser of the plant. As discussed in Section 11.6, ERM employed sophisticated numerical techniques to model the lake, but more simplistic approaches are available as discussed in Sections 11.3 and 11.4. Of the two analytical models discussed in Section 11.4, this example will investigate the use of the longitudinally unmixed cooling lake model illustrated by Figure 11.3 as the model most similar to the two for the SCL. As one might imagine by comparing Figures 11.3 and 11.4, a large portion of SCL would be ineffective in rejecting heat to the environment due to the irregular shape of the lake. In this example problem, the effective surface area will be determined for the Comanche Peak HRS parameters and annual approximate average meteorological conditions at the site as follows:

SCL area: 3,287 acres
CCW flow rate for two nuclear units: 2,200,000 gal/min
Temperature rise through the plant: 15.3°F
Waste heat rejected to the lake: 1.68×10^{10} Btu/hr
Annual average CCW temperature entering the plant: 81.0°F
Annual average solar radiation, 455 Langleys
Annual average dew point temperature: 50.0°F
Annual average ambient wind velocity: 15 mi/hr.

The solar radiation is first converted to Btu/hr as follows:

$$q_{solar} = 3.688 \times 455 = 1{,}678 \text{ Btu/ft}^2\text{/day}$$

The lake area is converted to square feet:

$$A = 43{,}560 \times 3{,}287 = 1.432 \times 10^8 \text{ft}^2$$

The total heat that must be rejected to the atmosphere is the sum of the heat rejected by the plant plus the solar radiation.

$$Q_{total} = A \times q_{solar} / 24 + \dot{Q}_{plant} = 1.432 \times 10^8 \times 69.9 + 1.68 \times 10^{10} = 2.68 \times 10^{10} \text{ Btu/hr}$$

As discussed in Section 11.3, the average lake surface temperature is first computed by iteration as follows:

First, assume $t_{surface} = 80.3$°F.
Then,

$$t_\beta = \frac{t_{surface} + t_{dew\ point}}{2} = 65.25\ °\text{F}$$

$$\beta = 0.255 - 0.0085\, t_\beta + 0.000204\, {t_\beta}^2 = 0.567 \text{ mmHg/°F}$$

$$f(W) = 70 + 0.7(W^2) = 227.5$$

where:

W = wind velocity in mi/hr.

Then,

$$K = 15.7 + (\beta + 0.26)\, f(W) = 203.9 \text{ Btu/ft}^2\text{/day/°F}$$

where:

K = coefficient of surface heat exchange.

Then,

$$t_{equilibrium} = t_{dew\ point} + \frac{q_{solar}}{K} = 58.2\text{°F}$$

where q_{solar} is the solar heating in Btu/ft²/day.

Then,

$$H_{rj} = \frac{24 \times Q_{total}}{A} = 4{,}494 \text{ Btu/ft}^2\text{/day}$$

where H_{rj} is the heat rejected by the lake in Btu/ft²/day and Q_{total} is in Btu/hr.

Then,

$$t_{surface} = t_{equilibrium} + \frac{H_{rj}}{K} = 80.3\text{ °F} \quad \text{(by iteration)}$$

Then, knowing K

$$NTU = \frac{G_{CCW}\ \text{(gal/min)}\ \rho\ \text{(lbm/ft}^3\text{)}\ c_p\ \text{(Btu/lbm-°F)}\ 1{,}440\ \text{(min/day)}}{K\ \text{(Btu/ft}^2\text{-day-°F)}\ A\ \text{(ft}^2\text{)}\ 7.48\ \text{(gal/ft}^3\text{)}} = 0.905$$

where:

G_{CCW} = plant CCW flow rate.

As indicated previously, because the SCL is relatively shallow, the longitudinally unmixed cooling lake model will be used with plug flow in which the temperature of the water is the same in a cross-section perpendicular to the direction of flow to estimate the temperature of the CCW entering the plant. For the latter case, the temperature of the water leaving the lake may be calculated as follows:

$$X = e^{-\frac{1}{NTU}} = 0.331$$

$$\theta = R\left(\frac{X}{1 - X}\right) = 12.1$$

where

$$R = \frac{Q_{total}}{500G_{CCW}} = 24.37°\text{F}$$

Finally,

$$t_{out} = t_{equilibrium} + \theta = 70.3°\text{F}$$

Because the observed temperature of the CCW entering the plant is 81.0°F, as expected, the effective surface area is somewhat less than the actual 3,287 acres. The area corresponding to an inlet temperature of 81.0°F may be found by performing the above calculation for an array of areas. As may be seen in Figure A11.1, the effective area of the lake is 50% of the actual lake.

One of the alternatives considered by ERM in 2005 was to increase the surface area by 5% by raising the lake level slightly. The tool developed herein could be utilized to determine the anticipated reduction in the CCW inlet temperature by doing the same as follows:

Assume $t_{surface} = 89.1°\text{F}$.

Then,

$$t_\beta = \frac{t_{surface} + t_{dew\ point}}{2} = 69.6\ °\text{F}$$

$$\beta = 0.255 - 0.0085\, t_\beta + 0.000204\, t_\beta^{\ 2} = 0.651\ \text{mmHg/°F}$$

$$f(W) = 70 + 0.7\left(W^2\right) = 227.5$$

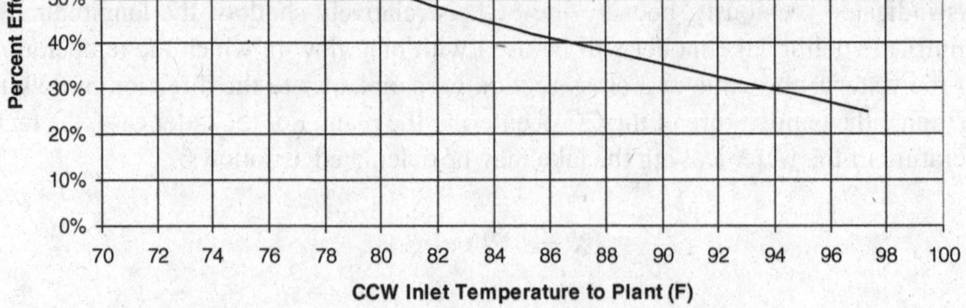

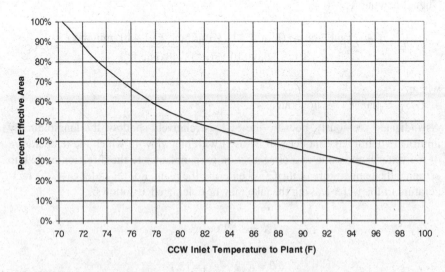

FIGURE A11.1 Percent effective lake area vs. CCW inlet temperature.

$$K = 15.7 + (\beta + 0.26)\, f(W) = 222.9 \text{ Btu/ ft}^2\text{/ day/°F}$$

$$t_{equilibrium} = t_{dew\ point} + \frac{q_{solar}}{K} = 57.5\,°F$$

$$H_{rj} = \frac{24 \times Q_{total}}{A} = 7{,}042 \text{ Btu/ft}^2\text{/day}$$

$$t_{surface} = t_{equilibrium} + \frac{H_{rj}}{K} = 89.1°F$$

$$t_{surface} = t_{equilibrium} + \frac{q_{heat\ rejection}}{K} = 80.3°F \quad \text{(by iteration)}$$

$$NTU = \frac{G_{CCW}\ (\text{gal/ min})\ \rho\ (\text{lbm/ft}^3)\ c_p\ (\text{Btu/lbm-° F})\ 1{,}440\ (\text{min/day})}{K\ (\text{Btu/ft}^2\text{-day- °F})\ \ A\ (\text{ft}^2)\ 7.48\ (\text{gal/ ft}^3)} = 1.578$$

$$X = e^{-\frac{1}{NTU}} = 0.531$$

$$R = \frac{Q_{total}}{500 G_{CCW}} = 22.7°F$$

$$\theta = R\left(\frac{X}{1-X}\right) = 12.1°F$$

$$t_{out} = t_{equilibrium} + \theta = 80.2°F$$

The indicated reduction in CCW temperature entering the plant (exiting the lake) would be 81–80.2 or 0.8°F. The ERM study using the numerical GLLVHT model concluded that the annual average reduction in the CCW temperature entering the plant would be 0.75°F.

CHAPTER 12

Refer to Figures 12.6 and 12.7.

Calculate the reduction in the CCW temperature entering the main condenser if an FBSP sprays a portion of the water prior to entering the plant and the power required to spray the water.

In 2005, a study was conducted to consider alternatives to reduce the temperature of the CCW entering the main condenser at the Comanche Nuclear Power Plant near Glen Rose, Texas. Among the alternatives considered was to spray approximately 25% of 2,200,000 gal/min of CCW that was circulated through the plant and the SCR (see Chapter 11).

Assuming that flatbed sprays located 5ft above the SCR and similar to those utilized at Rancho Seco that each spray 50 gal/min at a nozzle pressure of 7 lbf/in^2 were used, the following number of spray nozzles would be required:

$$N_{nozzles} = \frac{0.25 \times 2,200,000}{50} = 11,000$$

Assuming each nozzle requires a space of 5′ × 5′ or 25 ft^2, the spray field would require approximately 6.3 acres. Such a large spray area would require the maximum interference factor of 0.6, as shown in Figure 12.7.

The LWBT may be calculated as follows:

$$t_{WB\text{-}local} = t_{WB\text{-}amb} + f\left(t_1 - t_{WB\text{-}amb}\right)$$

where:

$t_{WB\text{-}local}$ = wet-bulb temperature at the spray nozzle
$t_{WB\text{-}amb}$ = wet-bulb temperature outside the spray region
t_1 = temperature of the CCW entering the spray nozzle
f = interference factor.

Refer to the Chapter 11 example in Appendix A. Assuming an ambient WBT of 50°F and a CCW temperature of 81°F entering the spray nozzle from SCR, the LWBT at the spray nozzle is as follows:

$$t_{WB\text{-}local} = 50 + 0.6(81 - 50) = 68.6\text{°F}$$

From Figure 12.6, assume a spray efficiency of 50% at a wind speed of 15 mi/hr. Calculate the CWT from the sprays as follows:

$$\eta = \frac{t_1 - t_2}{t_1 - t_{WB\text{-}local}} \Rightarrow t_2 = t_1 - \eta\left(t_1 - t_{WB\text{-}local}\right) = 74.8\text{ °F}$$

where:

η = spray efficiency of one spray nozzle

t_2 = CWT of the sprayed water.

Because only 25% of the 2,200,000 gal/min of CCW is sprayed, the temperature of the CCW entering the main condenser is calculated as follows:

$$t_{CCW\text{-}in} = \frac{0.75 \times G_{CCW} \times t_1 + 0.25 \times G_{CCW} \times t_2}{G_{CCW}} = 79.5\ °F$$

Calculate the power required to spray the water as follows:

The required pump flow is as follows:

$$G_{pump} = 0.25 \times 2{,}200{,}000 = 550{,}000\ \text{gal/min}$$

The required pump head referenced to the lake level and neglecting piping friction losses where the pump and motor efficiencies are 80% and 95%, respectively, is as follows:

$$H_{sprays} = 5\ Ft + 2.31 \times 7 \left({}^{\text{lbf}}\!/_{\text{in}^2} \right) = 21.2\ \text{ft}$$

$$Power_{spray} = 0.7457 \frac{G_{sprays} \times H_{sprays}}{3955 \times \eta_{pump}\ \eta_{motor}} = 2{,}889\ \text{kW}$$

CHAPTER 13

Refer to Figures 13.14.

Calculate the number of spray trees identical to those used at the CGS that would be required to achieve the same reduction in the temperature of the CCW entering the main condenser as in the Chapter 12 example and the power required to spray the water.

Continuing with Example 12 in Appendix A, in 2005, a study was conducted to consider alternatives to reduce the temperature of the CCW entering the main condenser at the Comanche Nuclear Power Plant near Glen Rose, Texas. Among the alternatives considered was to spray a portion of the 2,200,000 gal/min of CCW that was circulated through the plant and the SCR (see Chapter 11).

Assume an OSCS with spray trees identical to those described in Section 13.1 with seven 1.5 in Spraying Systems Whirljet nozzles on each tree delivering a total of 325 gal/min. From Figure 13.15, the efficiency of the OSCS at a wind velocity of 15 mi/hr is 62%, and due to the design of the OSCS, there is no interference to the ambient air reaching the nozzles. The operating pressure of the spray trees referenced to the SCR surface is 20.7 lbf/in^{2}g. Assuming an ambient WBT of 50°F and a CCW temperature of 81°F entering the spray nozzle from SCR, the CWT of the CCW leaving the OSCS may be calculated as follows:

$$\eta = \frac{t_1 - t_2}{t_1 - t_{WB\text{-}amb}} \Rightarrow t_2 = t_1 - \eta\left(t_1 - t_{WB\text{-}amb}\right) = 61.8\,°\text{F}$$

where:

η = spray efficiency of the OSCS
t_1 = temperature of the CCW entering the spray nozzle
t_2 = cold-water temperature of the sprayed water
$t_{WB\text{-}amb}$ = WBT outside the spray region.

The percentage of the CCW that must be sprayed in order to achieve a reduction of the CCW entering the main condenser from 81°F to 79.5°F may be calculated as follows:

$$G_{CCW}\, t_{CCW\text{-}in} = \left(1 - X\right) G_{CCW}\, t_1 + X\, G_{CCW}\, t_2$$

$$t_{CCW\text{-}in} = t_1 + X\left(t_2 - t_1\right)$$

$$X = \frac{t_{CCW\text{-}in} - t_1}{t_2 - t_1} = 7.8\%$$

where:

X = percent of CCW sprayed.

The following number of spray trees would be required:

$$N_{nozzles} = \frac{0.078 \times 2{,}200{,}000}{325} = 528$$

Calculate the power required to spray the water as follows:
The required pump flow is as follows:

$$G_{pump} = 0.078 \times 2{,}200{,}000 = 171{,}700 \text{ gal/min}$$

The required pump head referenced to the lake level, neglecting piping friction losses where the pump and motor efficiencies are 80% and 95%, respectively, is as follows:

$$H_{sprays} = 2.31 \times 20.7 \left({}^{\text{lbf}}\!/_{\text{in}^2} \right) = 47.8 \text{ ft}$$

$$Power_{spray} = 0.7457 \frac{G_{sprays} \times H_{sprays}}{3{,}955 \times \eta_{pump}\, \eta_{motor}} = 2{,}037 \text{ kW}$$

CHAPTER 14

Determine the reduction in CWT achievable by converting a counter-flow NDCT to an OSACT.

The main condensers at the Watts Bar Nuclear Plant in Tennessee are of the single-pass, multi-pressure, multi-shell main condenser sign with three pressure zones matching the three LP turbine exhausts (see Section 6.2). During periods of high WBT, the staff at Unit 1 has experienced shell pressures in the highest pressure zone exceeding the manufacturer's limit on LP turbine backpressure. In one instance, this resulted in a turbine blade failure, leading to an extended outage (see Section 5.3). In 2005, a study was conducted to investigate ways to reduce the CCW temperature entering the main condenser. Alternatives considered included improving the distribution of CCW over the fill material and adding fill material around the perimeter of counter-flow NDCT, adding an MDCT parallel to the NDCT, and converting the NDCT to an OSACT.

The following table shows a comparison between the critical performance parameters with a hot-water temperature of 128°F and an ambient WBT of 80°F:

Parameter	w/o OSA		w/ OSA	
Dry air flow	155,962,793	lbm/hr	175,161,511	lbm/hr
Average water loading	1,870	lbm/(hr-ft²)	1,630	lbm/(hr-ft²)
Liquid/gas ratio	1.35	dimensionless	1.05	dimensionless
Spray CWT			96.01	°F
Basin CWT			90.19	°F
Tower CWT	91.80	°F	90.90	°F

As one may see, although the CWT coming off the sprays is higher than that from the NDCT without the OSA, the increase in the airflow coupled with the reduction in the water loading on the NDCT with the OSA dramatically reduces the *L*/*G* ratio and thus increases the value of *KaV*/*L* (see Figure 10.11). The result is a mixed CWT that is lower than the NDCT without the OSA.

Figure A14.1 illustrates the reduction in CWT that may be achieved by converting the Watts Bar NDCT to an OSACT. Applying the equations in the figure, one may determine the annual average reduction in the CWT by knowing the monthly average WBT as follows:

	Jan	Feb	Mar	Apr	May	Jun	Jul	Aug	Sep	Oct	Nov	Dec	Ave
WBT, °F	*35.1*	*38.0*	*42.7*	*51.8*	*60.2*	*67.1*	*69.6*	*69.1*	*63.5*	*52.9*	*42.9*	*36.7*	
CWT w/o OSA	67.1	68.5	70.9	75.7	80.3	84.2	85.7	85.4	82.2	76.3	71.0	67.9	
CWT w/OSA	66.4	67.8	70	74.6	79.1	83	84.5	84.2	81	75.2	70.1	67.2	
Difference	0.7	0.8	0.9	1.1	1.2	1.2	1.2	1.2	1.2	1.1	0.9	0.7	1.0

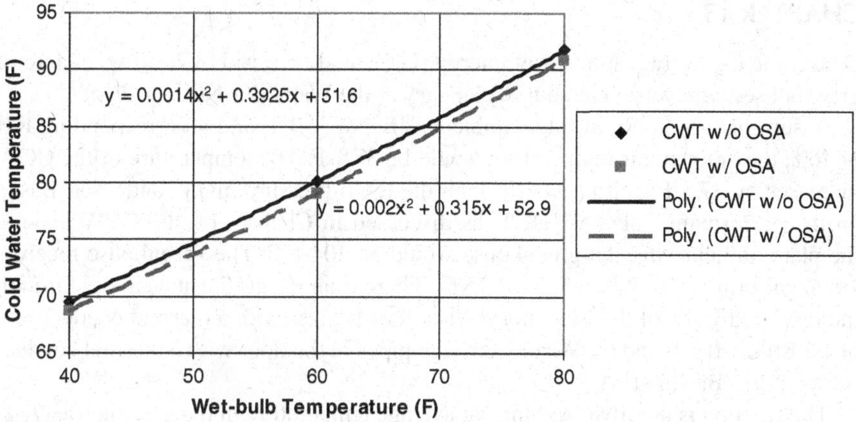

FIGURE A14.1 Reduction in CWT achievable by converting the Watts Bar NDCT to an OSACT.

Note that the improvement in CWT is the greatest during the periods of the highest WBT.

It should be noted that Watts Bar did not modify their NDCT to become an OSACT but instead improved the distribution of CCW over the fill material and added fill material around the perimeter of the NDCT and subsequently procured an LP turbine spindle with ruggedized last stage blades that could withstand higher backpressures.

CHAPTER 15

Determine the average amount of energy that would be saved by heating an acre of greenhouses with waste heat during January at the Watts Bar Nuclear Plant.

Assuming a monthly average ambient WBT of 35.1°F and a corresponding RH of 76%, the ambient air temperature would be 37.8°F. The temperature of the CCW increases by 37.5°F as it passes through the multi-pressure main condenser. Based on the performance of the NDCT, as discussed in Chapter 14, the CCW leaving the plant and entering the greenhouse would be 104.6°F. The greenhouse receives 1,000 gal/min of CCW (see Section 15.6). There is an insignificant heat loss through the insulated walls of the greenhouse. The roof is glass with a thermal conductivity of 1.2 Btu/hr-ft^2-°F and CCW runs through pipes in the floor with a thermal conductivity of 0.73 Btu/hr-ft^2-°F.

The solution is iterative. Assume an average temperature in the greenhouse (T_{GH}) and calculate the heat loss through the roof and the heat added by the floor until the two values are equal as follows:

$$\text{Assume } T_{GH} = 62.6\,°\text{F} \quad \text{(by iteration)}$$

$$Q_{glass} = U_{glass}\,(T_{GH} - t_{amb})\,A = 1{,}296{,}000 \text{ Btu/hr-acre}$$

$$\Delta t_{ccw} = \frac{Q_{glass}}{500 \times G_{ccw}} = 2.6\,°\text{F}$$

$$t_{ccw\text{-}out} = t_{CCW\text{-}in} - \Delta t_{ccw} = 102.0\,°\text{F}$$

$$t_{ccw\text{-}ave} = \frac{t_{ccw\text{-}out} + t_{CCW\text{-}in}}{2} = 103.3\,°\text{F}$$

$$Q_{glass} = U_{floor}\,(T_{GH} - t_{ccw\text{-}ave})\,A = 1{,}294{,}000 \text{ Btu/hr-acre}$$

Therefore, each acre of greenhouse heated by waste heat in January would save 1.3 million Btu/hr of energy that would otherwise have been derived from oil or natural gas.

CHAPTER 16

Calculate the heat balance error and the uncertainty of the heat balance error.

In 2005, the staff at the Callaway Nuclear Generating Station in Missouri conducted a heat transfer test on their Train A component cooling system (CCS) HX. The Callaway CCS HX is a shell-and-tube HX similar to a TEMA Type G design with two passes on the tube side and split flow on the shell side, furnished by Struthers Wells Corporation. The table below is a listing of the physical data for each bundle.

CCS HX Physical Data

Total number of tubes	N_{tubes}	4464.0	
Percentage of tubes plugged		0.0%	
Total number of active tubes	$N_{active} = N_{tubes}(1\%)$	4464.0	
Number of passes	*Passes*	2	
Effective tube length	L_{eff}	36.41	ft
Outside tube diameter	d_o	0.0625	ft
Tube wall thickness	t	0.049	in
Inside tube diameter	$d_i = d_o - 2(t/12)$	0.0543	ft
Total effective hot-side area	$A_h = \pi d_o L_{eff} N_t$	31,917	ft^2
Tube-side area	$A_c = (d_i/d_o)A_h$	27,747	ft^2
Tube wall area	$A_w = (A_h - A_c)/\ln(A_h/A_c)$	29,783	ft^2
Tube material		90-10	
Hot-side fouling resistance	$r_{f,h}$	0.0005	hr-ft^2/Btu
Tube-side fouling resistance	$r_{f,c}$	0.0020	hr-ft^2/Btu
Ratio of hot-side to tube side area	A_h/A_c	1.15	
Tube material thermal conductivity	k_t	26	Btu/hr-ft-°F
Tube wall resistance	$r_w = (d_o - d_i)/2k_t$	0.00016	hr-ft^2/Btu

The following table is a listing of the design basis conditions for operating in the normal full power operation with maximum cool-down mode. The properties of the hot-side and cold-side fluids are based on the average temperatures on each side of the HX except for the density that is based on the inlet and outlet temperatures for the cold and hot-side fluids, respectively.

Design Basis Conditions for Normal Full Power Operation with Maximum Letdown

Heat transfer rate	Q	77,180,000	Btu/hr
Required hot-side flow	g_h	10,501	gal/min
Entering hot-side temperature	$T_{h,i}$	119.7	°F
Leaving hot-side temperature	$T_{h,o}$	105.0	°F
Average hot-side temperature	T_h	*112.4*	°F
Hot-side thermal conductivity	k_h	0.3675	Btu/hr-ft-°F
Hot-side absolute viscosity	μ_h	1.47	lbm/ft-hr
Hot-side specific heat	$c_{p,h}$	0.998	Btu/lbm-°F

(Continued)

Design Basis Conditions for Normal Full Power Operation with Maximum Letdown

Hot-side density	ρ_h	61.8	lbm/ft³
Required tube-side flow	g_t	13,500	gal/min
Entering water temperature	$t_{c,i}$	95.0	°F
Exiting water temperature	$t_{c,o}$	106.4	°F
Average water temperature	t_c	*100.7*	°F
Tube-side thermal conductivity	k_c	0.364	Btu/hr-ft-°F
Tube-side absolute viscosity	μ_c	1.65	lbm/ft-hr
Tube-side specific heat	$c_{p,c}$	0.997	Btu/lbm-°F
Tube-side density	ρ_c	62.0	lbm/ft³
Water temperature rise	dT_c	11.4	°F
LMTD correction factor	F	0.9455	
Tube-side flow rate	m_c	6,790,547	lbm/hr
Hot-side flow rate	m_h	5,260,862	lbm/hr

The overall heat transfer coefficient at design conditions, U_{design}, is calculated for the normal mode of operation in the following table.

Verification of U_{design} at for Normal Mode of Operation

Greater terminal temperature difference	$\Delta t_1 = T_{h,I} - t_{c,o}$	13.3	°F
Lesser terminal temperature difference	$\Delta t_2 = T_{h,o} - t_{c,i}$	10.0	°F
Log mean temperature difference	$LMTD = (\Delta t_1 - \Delta t_2)/\ln(\Delta t_1/\Delta t_2)$	11.6	°F
Effective mean temperature difference	$EMTD = F(LMTD)$	10.9	°F
Design overall heat transfer coefficient	$U_{design} = Q/(A_h EMTD)$	221	Btu/hr-ft²-°F

The tube-side convection coefficient may be computed from the Petukhov correlation as shown in the following table.

Tube-side Convection Coefficient

Mass flow rate/tube	$m_t = Passes(m_c/N_t)$	3042.4	lbm/hr
Volumetric flowrate/tube	$V_t = m_t/\rho_\tau$	49.1	ft³/hr
Tube area	$a_t = (\pi/4)d_i^2$	0.00232	ft²
Tube velocity	$v_t = V_t/a_t$	21,167	ft/hr
Tube velocity	$v_t = V_t/3{,}600a_t$	5.88	ft/s
Prandtl number	$\mathrm{Pr}_c = c_{p,c}\mu_c/k_c$	4.5	
Reynolds number	$\mathrm{Re}_c = \rho v_t d_i/\mu_c$	4.32E+04	
Fanning friction factor	$f = (1.58 \ln \mathrm{Re}_c - 3.28)^{-2}$	0.00542	
Nusselt number	$\mathrm{Nu} = ((f/2)\,\mathrm{Re}_c\mathrm{Pr}_c)/$ $(1.07 + 12.7(f/2)^{1/2}(\mathrm{Pr}_c^{\ 2/3} - 1))$	238.66	
Tube-side film coefficient	$h_c = \mathrm{Nu}(k_c/D_i)$	1599	Btu/hr-ft²-°F

The hot-side convection coefficient may be computed as follows:

$$h_{h,\,design} = \frac{1}{\dfrac{1}{U_{design}} - r_{f,h} - \left(\dfrac{A_h}{A_w}\right) r_w - \left(\dfrac{A_h}{A_c}\right) r_{f,c} - \left(\dfrac{A_h}{A_c}\right)\dfrac{1}{h_c}} = 1{,}195.8 \text{ Btu/(hr-ft}^2\text{-°F)}$$

The following table shows the results of the CCS test.

CCS Test Results

Required hot-side flow	g_h	12,686	gal/min
Hot-side inlet temperature	$T_{h,i}$	92.8	°F
Hot-side outlet temperature	$T_{h,o}$	78.1	°F
Average hot-side temperature	T_h	85.5	°F
Hot-side thermal conductivity	k_h	0.3575	Btu/hr-ft-°F
Hot-side absolute viscosity	μ_h	1.92	lbm/ft-hr
Hot-side specific heat	$c_{p,h}$	0.998	Btu/lbm-°F
Hot-side density	ρ_h	62.2	lbm/ft³
Hot-side flow rate	$m_h = g_h\rho_h(60/7.48)$	6,329,009	gal/min
Water flow rate	g_t	8850	°F
Tube-side Inlet temperature	$t_{c,i}$	66.7	°F
Tube-side outlet temperature	$t_{c,o}$	85.5	°F
Average water temperature	t_c	76.1	°F
Tube-side thermal conductivity	k_c	0.352	Btu/hr-ft-°F
Tube-side absolute viscosity	μ_c	2.20	lbm/ft-hr
Tube-side specific heat	$c_{p,c}$	0.999	Btu/lbm-°F
Tube-side density	ρ_t	62.3	lbm/ft³
Tube-side flow rate	$m_c = g_t\rho_t(60/7.48)$	4,425,000	gal/min
Greater terminal temperature difference	$\Delta t_1 = T_{h,i} - t_{c,i}$	7.3	°F
Lesser terminal temperature difference	$\Delta t_2 = T_{h,o} - t_{c,o}$	11.4	°F
Log mean temperature difference	$LMTD = (\Delta t_1 - \Delta t_2)/\ln(\Delta t_1/\Delta t_2)$	9.2	°F
Ratio of temperature change	$R = (T_{h,i} - T_{h,o})/(t_{c,o} - t_{c,i})$	0.78	
Effectiveness	$P = (t_{c,o} - t_{c,i})/(T_{h,i}\, t_{c,o})$	0.72	
LMTD correction factor	F	0.840	
LMTD (corrected)	$EMTD = F(LMTD)$	7.7	°F
Test heat transfer (hot-side)	$Q_{test,h} = m_h c_p(T_{h,i} - T_{h,o})$	93.166	MBtu/hr
Test heat transfer (tube side)	$Q_{test,c} = m_c c_p(t_{c,o} - t_{c,i})$	83.046	MBtu/hr
Average test heat balance	$Q_{test} = (Q_{test,h} + Q_{test,c})/2$	88.106	MBtu/hr
Test overall heat transfer coefficient	$U_{test} = dQ_{test}/A_h(EMTD)$	357.49	Btu/hr-ft-°F
Mass flow rate/tube	$m_t = m_c/\text{Passes } N_t$	1,982.5	lbm/hr
Volumetric flowrate/tube	$V_t = m_c/\rho_t$	31.8	ft³/hr
Tube velocity	$v_t = V_t/a_t$	13,725	ft/hr
Tube velocity	$v_t = v_t/3{,}600$	3.81	ft/s
Prandtl number	$\mathrm{Pr}_c = c_{p,c}\mu_c/k_c$	6.24	
Reynolds number	$\mathrm{Re}_c = \rho v_t d_i/\mu_c$	21,100	

(Continued)

CCS Test Results			
Fanning friction factor	$f = (1.58 \ln Re_c - 3.28)^{-2}$	0.00645	
Nusselt number	Nu^a	152	
Tube-side film coefficient	$h_c = Nu(k_c/d_i)$	985	Btu/hr-ft-°F
Hot-side flow rate	$m_h = m_c(c_{p,c}/c_{p,h})(t_{c,o} - t_{c,i})/(T_{h,i} - T_{h,o})$	5,641,519	lbm/hr
Hot-side film coefficient	$h_{h,test}{}^b$	1140.0	Btu/hr-ft-°F
Resistances Referenced to Hot-side Area			
Total resistance	$r = 1/U_{test}$	0.00280	hr-ft-°F /Btu
Hot-side convection resistance	$r_h = 1/(\eta_h h_h)$	0.00088	hr-ft-°F /Btu
Wall resistance	$r_w = (A_h/A_w)(d_o - d_i)/(2k_t)$	0.00017	hr-ft-°F /Btu
Cold-side convection resistance	$r_c = (A_h/A_c)(1/h_c)$	0.00117	hr-ft-°F /Btu
Fouling referenced to hot-side	$r_{f,h,test}{}^c$	0.00058	hr-ft-°F /Btu
Tube-side fouling resistance	$r_{f,c} = (A_c/A_h)r_{f,h}$	0.00051	hr-ft-°F /Btu

a $Nu = \left(\left(f/2\right)Re_c\,Pr_c\right)/\left(1.07 + 12.7\left(f/2\right)^{1/2}\left(Pr_c^{2/3} - 1\right)\right)$

b $$h_{h,test} = \frac{\left[\left(\frac{m_h}{\mu_h}\right)^{0.6}\left(\frac{\mu_h c_{p,h}}{k_h}\right)^{0.333} k_h\right]_{test}}{\left[\left(\frac{m_h}{\mu_h}\right)^{0.6}\left(\frac{\mu_h c_{p,h}}{k_h}\right)^{0.333} k_h\right]_{design}} h_{h,design}$$

c $$r_{f,h,test} = \frac{1}{\frac{1}{U_{test}} - \left(\frac{1}{h_{h,test}}\right) - \left(\frac{A_h}{A_w}\right) r_w - \left(\frac{A_h}{A_c}\right)\frac{1}{h_{c,test}}}$$

Therefore, the total fouling resistance as measured by the test is only approximately 20% of the design value.

Uncertainty Analysis

Refer to Section 16.9.

The following is the frequency and the number of each temperature measuring instrument with the associated Student's "t" for each (see Table 16.1):

Number of temperature measurements	N_t	31
Temperature measurement Student's "t"	t	2
Number of SW inlet RTDs	J_t	4
Number of sensors Student's "t"	t	3.182
Number of SW outlet RTDs	J_t	8
Number of sensors Student's "t"	t	2.306
Number of CCS inlet RTDs	J_t	4
Number of sensors Student's "t"	t	3.182
Number of CCS outlet RTDs	J_t	8
Number of sensors Student's "t"	t	2.306

The following tables are the temperature measurements. The table numbers refer to those in Appendix K of Reference 1.

Temperature Measurements

Calibration bias	b_{cal}	0.20				°F
Installation bias (estimated)	$b_{install}$	0.05				°F
Table 1 – Average, *t(i)*		X^a	$S_x{}^b$	$b_{spat\ var}{}^c$	$S_{pv}{}^e$	
Service water inlet temperature, top	$t_{i\text{-}0}$	66.83	0.06	0.0008	0.00005	°F
Service water inlet temperature, right	$t_{i\text{-}90}$	66.63	0.06	0.0009	0.00006	°F
Service water inlet temperature, bottom	$t_{i\text{-}180}$	66.78	0.03	0.0002	0.00001	°F
Service water inlet temperature, left	$t_{i\text{-}270}$	*66.69*	0.02	0.0002	*0.00001*	°F
	Ave	66.73		Sum	0.00013	°F
	$b_{spat\ var,\ ti}{}^d$			0.07		
	$u_{pv}{}^f$				0.023	
Table 2 – Average, *T(i)*		X^a	$S_x{}^b$	$b_{spat\ var}{}^c$	$S_{pv}{}^e$	
Hot-side inlet temperature No. 1	$T_{i\text{-}0}$	92.74	0.07	0.0013	0.00009	°F
Hot-side inlet temperature No. 2	$T_{i\text{-}90}$	92.94	0.04	0.0005	0.00003	°F
Hot-side inlet temperature No. 3	$T_{i\text{-}180}$	92.90	0.02	0.0001	0.00001	°F
Hot-side inlet temperature No. 4	$T_{i\text{-}270}$	*92.88*	0.01	0.0000	0.00000	°F
	Ave	92.87		Sum	0.00013	°F
	$b_{spat\ var,\ Ti}{}^d$			0.07		°F
	$u_{pv}{}^f$				0.022	°F
Table 3 – Average, *t(o)*		X^a	$S_x{}^b$	$b_{spatvar}{}^c$	$S_{pv}{}^e$	
Service water outlet temperature, top	$t_{o\text{-}0}$	85.42	0.04	0.0004	0.00001	°F
Service water outlet temperature, top-right	$t_{o\text{-}45}$	85.62	0.04	0.0005	0.00001	°F
Service water outlet temperature, right	$t_{o\text{-}90}$	85.50	0.01	0.0000	0.00000	°F
Service water outlet temperature, bottom-right	$t_{o\text{-}135}$	85.54	0.01	0.0000	0.00000	°F
Service water outlet temperature, bottom	$t_{o\text{-}180}$	85.52	0.00	0.0000	0.00000	°F
Service water outlet temperature, bottom-left	$t_{o\text{-}225}$	85.30	0.08	0.0019	0.00006	°F
Service water outlet temperature, left	$t_{o\text{-}275}$	85.72	0.08	0.0018	0.00005	°F
Service water outlet temperature, top-left	$t_{o\text{-}315}$	*85.50*	0.01	0.0000	*0.00000*	°F
	Ave	85.52		Sum	0.00013	°F
	$b_{spat\ var,to}{}^d$			0.06		°F
	$u_{pv}{}^f$				0.023	°F
Table 4 – T_o		X^a	$S_x{}^b$	$b_{spat\ var}{}^c$	$S_{pv}{}^e$	
Hot-side outlet temperature, top	$T_{o\text{-}0}$	78.00	0.03	0.0004	0.00001	°F
Hot-side outlet temperature, top-right	$T_{o\text{-}45}$	78.18	0.03	0.0003	0.00001	°F
Hot-side outlet temperature, right	$T_{o\text{-}90}$	78.05	0.02	0.0001	0.00000	°F

(Continued)

Temperature Measurements						
Hot-side outlet temperature, bottom-right	$T_{o\text{-}135}$	78.14	0.02	0.0001	0.00000	°F
Hot-side outlet temperature, bottom	$T_{o\text{-}180}$	78.20	0.04	0.0005	0.00001	°F
Hot-side outlet temperature, bottom-left	$T_{o\text{-}225}$	77.99	0.04	0.0004	0.00001	°F
Hot-side outlet temperature, left	$T_{o\text{-}275}$	78.10	0.00	0.0000	0.00000	°F
Hot-side outlet temperature, top-left	$T_{o\text{-}315}$	*78.08*	0.00	0.0000	*0.00000*	°F
	Ave	78.09		Sum	0.00005	°F
Bias uncertainty, *b*(spat var)	$b_{spat\ var,\ To}$[d]			0.03		
	u_{pv}[f]				0.014	°F
Table 6 – Average Temperature Measurements		**Avg. Temp.**	$b_{spat\ var}$			
Service water inlet temperature	t_i	66.73	0.07			°F
Service water outlet temperature	t_o	92.87	0.08			°F
Hot-side inlet temperature	T_i	85.52	0.07			°F
Hot-side outlet temperature	T_o	78.09	0.05			°F
Table 7 – Temperature Uncertainty due to Process Variation						°F
Service water inlet temperature	$u_{pv,ti}$	0.023				°F
Service water outlet temperature	$u_{pv,to}$	0.023				°F
Hot-side inlet temperature	$u_{pv,Ti}$	0.022				°F
Hot-side outlet temperature	$u_{pv,To}$	0.014				°F
Table 8 – Overall Measurement Uncertainty for Temperature	b_{cal}	$b_{install}$	$b_{spat\ var}$	u_{pv}	$u_{overall}$[g]	°F
Service water inlet temperature	0.20	0.05	0.07	0.023	0.22	°F
Service water outlet temperature	0.20	0.05	0.08	0.023	0.22	°F
Hot-side inlet temperature	0.20	0.05	0.07	0.022	0.22	°F
Hot-side outlet temperature	0.20	0.05	0.05	0.014	0.21	°F

[a] $X = \frac{1}{N_t}\sum_{i=1}^{N} X_i$

[b] $S_X = \sqrt{\sum_{i=1}^{N_t} \frac{(X_k - X)^2}{N_t - 1}}$

[c] $\left(\frac{1}{J_t}\right)\frac{(T - T_{ave})^2}{J_t - 1}$

[d] $b_{spat\,var} = t\sqrt{\left(\frac{1}{J_t}\right)\sum_{i=1}^{J} \frac{(t - t_{ave})^2}{J_t - 1}}$

[e] $\frac{1}{J_t}\frac{s_X{}^2}{N_{pv}}$

[f] $u_{pv} = 2\sqrt{\sum_{i=1}^{J} \frac{1}{J_t}\frac{s_X{}^2}{N_{pv}}}$

[g] $u_{overall} = \sqrt{b_{cal}{}^2 + b_{spat\,var}{}^2 + u_{pv}{}^2}$

The following tables are the average standard deviations for and uncertainties in flow measurements.

Number of flow measurements	J_f	1		
Hot-side flow measurement calibration uncertainty	$u_{f\text{-}cal\text{-}H}$	2.0		%
Cold-side flow measurement calibration uncertainty	$u_{f\text{-}cal\text{-}C}$	2.0		%
Table 5 – Average Flow Measurement			S_X	
Cold stream flow rate	G_c	8,850	300	gal/min
Hot stream flow rate	G_H	12,686	400	gal/min
Mass flow rate, m_c per coil	m_c	4,425,000[a]	149,164	lbm/hr
Mass flow rate	m_H	6,329,009[a]	198,276	lbm/hr
Table 9 – Flow Measurement Uncertainty, Calibration, etc.				
Service water flow measurement	$u_{m,c\text{-}cal}$	88,500[b]		lbm/hr
Hot-side flow measurement	$u_{m,h\text{-}cal}$	126,580[b]		lbm/hr
Table 10 – Flow Measurement Uncertainty due to Process Variation				
Service water flow measurement	$u_{m,c\text{-}pv}$	77,028[c]		lbm/hr
Hot-side flow measurement	$u_{m,h\text{-}pv}$	102,389[c]		lbm/hr
Table 11 – Overall Flow Measurement Uncertainty				
Service water flow measurement	$u_{m,\ c\text{-}overall}$	117,237[d]		lbm/hr
Hot-side flow measurement	$u_{m,\ h\text{-}overall}$	162,807[d]		lbm/hr

[a] $m = G\rho(60/7.48)$

[b] $u_{cal} = u_{f\text{-}cal} \times m$

[c] $s_{pv} = 2\sqrt{\sum_{i=1}^{J} \frac{1}{J_t} \frac{s_X^{\ 2}}{N_{pv}}}$

[d] $u_{m\text{-}overall} = \sqrt{u_{m\text{-}cal}^{\ 2} + u_{m\text{-}pv}^{\ 2}}$

The following table is an assessment of the uncertainty in the rate of heat transfer:

Table 12 – Uncertainty in Q_H	$u_{Q\text{-}H}$	**Units**	$\theta_{Q\text{-}H}$	**Units**	$(u_{Q\text{-}H} \times \theta_{Q\text{-}H})^2$	**Units**
Hot stream inlet temperature, T_i	0.22	°F	−6,316,351[a]	Btu/(hr-°F)	1.907E + 12	(Btu/hr)^2
Hot stream outlet temperature, T_o	0.21	°F	6,316,351[b]	Btu/(hr-°F)	1.751E + 12	(Btu/hr)^2
Hot stream mass flow rate, m_H	162,807	lbm/hr	7.424[c]	Btu/lbm	1.461E + 12	(Btu/hr)^2
Hot stream specific heat, $C_{p\text{-}H}$	0.010	Btu/(LbM-°F)	46,977,069[d]	(LbM-°F)/hr	2.207E + 11	(Btu/hr)^2
Total uncertainty in the hot stream heat transfer rate, $u_{Q\text{-}H}$					2,2310,663[e]	Btu/hr

Table 13 – Uncertainty in Q_C	$u_{Q\text{-}C}$	Units	$\theta_{Q\text{-}C}$	Units	$(u_{Q\text{-}C} \times \theta_{Q\text{-}C})^2$	Units
Cold stream inlet temperature, t_i	0.22	°F	−4,411,725[a]	Btu/(hr-°F)	9.364E + 11	(Btu/hr)^2
Cold stream outlet temperature, t_o	0.21	°F	4,411,725[b]	Btu/(hr-°F)	8.963E + 11	(Btu/hr)^2
Cold stream mass flow rate, m_C	117,327	lbm/hr	26.144[c]	Btu/lbm	9.409E + 12	(Btu/hr)^2
Cold stream specific heat, $C_{p\text{-}C}$	0.0100	Btu/(LbM-°F)	115,636,313[d]	(LbM-°F)/hr	1.337E + 12	(Btu/hr)^2
Total uncertainty in the cold stream heat transfer rate, $u_{Q\text{-}C}$					3,546,697[e]	Btu/hr

a $\theta_{Q,T_{in}} = -m\, c_p$
b $\theta_{Q,T_{out}} = m\, c_p$
c $\theta_{Q,M} = c_p\left(T_{in} - T_{out}\right)$
d $\theta_{Q,c_p} = m\left(T_{in} - T_{out}\right)$
e $u_Q = \sqrt{\sum (u_Q \times \theta)^2}$

The following table is an assessment of the heat balance between the hot and cold streams:

Hot Stream Range			
Upper limit of the heat transfer rate	$Q_h + u_{Q,h}$	95,476,841	Btu/hr
Heat transfer rate	Q_h	93,166,177	Btu/hr
Lower limit of the heat transfer rate	$Q_h - u_{Q,h}$	90,855,514	Btu/hr
Cold Stream Range			
Upper limit of the heat transfer rate	$Q_c + u_{Q,c}$	86,592,672	Btu/hr
Heat transfer rate	Q_c	83,045,975	Btu/hr
Lower limit of the heat transfer rate	$Q_c - u_{Q,c}$	79,499,279	Btu/hr
Weighted Average			
Weighted average heat transfer rate	Q_{ave}	90,150,619	Btu/hr
Uncertainty in the average heat transfer rate	u_{Qave}	1,936,035	Btu/hr
Upper limit of the heat transfer rate	$Q_{ave} + u_{Qave}$	92,086,655	Btu/hr
Lower limit of the heat transfer rate	$Q_{ave} - u_{Qave}$	88,214,584	Btu/hr

One may note that there is no overlap in the ranges of uncertainty as $Q_c + u_{Q,c} < Q_h - u_{Q,h}$.

The following table is an assessment of the uncertainty of the rate of heat transfer:

Hot-side heat transfer rate	Q_H	93,166,177	Btu/hr
Cold-side heat transfer rate	Q_c	83,045,975	Btu/hr
Total uncertainty in Q_H	$u_{Q,h}$	2,310,663	Btu/hr
Total uncertainty in Q_C	$u_{Q,c}$	3,546,697	Btu/hr
Heat balance error (referenced to hot side)	HBE	10.9%	
Uncertainty in heat balance error	u_{HBE}	4.4%	

Although the uncertainty of the test is very good, the test fails to meet the criteria of $u_{HBE} \geq HBE$, which is an indication of a good test. However, the imbalance in the computed heat transfer rate only slightly exceeds the anticipated range of 3%–10% as indicated in Reference 1.

CHAPTER 17

In December 2009, the staff at the Virgil C. Summer Nuclear Generating Station conducted two heat transfer tests on one of their four RBCU. The tests were conducted one day apart with the SW flow rate during the second test being half that of the first test. The following is the physical data for the RBCU:

RBCU Physical Data			**Units**
Number of coil sections in parallel	C	8	
Number of rows	N_r	8	
Number of serpentines	S	2	
Effective tube length	L_{eff}	11.00	ft
Number of tubes per row	N_t	16	
Outside tube diameter	d_o	0.0533	ft
Tube wall thickness	t	0.0490	in
Fin thickness	δ	0.0070	ft
Number of fins per unit of length	N_f	74	fins/ft
Bundle depth	D	1.00	ft
Face height	B	2	ft
Hot-side fouling resistance	$r_{f,h}$	0.0000	hr-ft^2-°F /Btu
Tube-side fouling resistance	$r_{f,c}$	0.0005	hr-ft^2-°F /Btu

The following physical parameters may be calculated.

Additional RBCU Physical Data			**Units**
Face area	$FA = L_{eff}B$	22.00	ft^2
Inside tube diameter	$d_i = d_o - 2(t/12)$	0.04517	ft
Tube-side area	$A_c = \pi d_i L_{eff} N_r N_t$	199.8	ft^2
Total cross-sectional area of plate fins	$A_{c,f} = BD$	2.00	ft^2
Total cross-sectional area of tubes	$A_{c,t} = (\pi/4)d_o^2\, N_t N_{rows}$	0.29	ft^2
Net area of each plate	$A_{net,f} = 2(A_{c,f} - A_{c,t})$	3.43	ft^2
Number of plates	$NP = L_{eff}N_r$	814	
Total area of plate fins	$A_f = A_{net,f}NP$	2,790	ft^2
Net outside area of tubes (prime area)	$A_p = \pi d_o N_t N_r L_{eff}(1 - N_f\delta)$	114	ft^2
Total effective area of the finned surface	$A_h = A_f + A_p$	2,904	ft^2
Ratio of the finned side to inside area	A_h/A_t	14.54	
Area of the finned surface per fin	$a_o = A_f/N_r N_t NP$	0.02678	ft^2/fin
Effective fin diameter	$d_{fin} = ((2/\pi)a_o + d_o^2)^{0.5}$	0.14105	ft
Area of the finned side per fin	$a_h = A_h/N_r N_t N_f L_{eff}$	0.02787	ft^2/fin
Area of the prime per fin	$a_p = (d_o(1/N_f - \delta)$	0.00109	ft^2/fin
Area of the fin per fin	$a_f = a_h - a_p$	0.02678	ft^2/fin
Outside tube area	$A_{t,o} = (d_o/d_i)A_t$	235.9	ft^2
Tube wall area	$A_w = (A_{t,o} - A_c)/\ln(A_{t,o}/A_c)$	217.4	ft^2
Tube and fin thermal conductivity (Cu 122)	k_t	196	Btu/hr-ft-°F
Tube wall resistance	$r_w = [d_o \ln(A_h/A_c)]/2k_t$	0.000364	hr-ft^2-°F /Btu

The fin efficiency may be calculated as follows:

Fin Efficiency			**Units**
Characteristic fin length	$l = \delta/2$	0.00350	ft
Assumed air-side film coefficient	h_h	13.3	Btu/hr-ft²-°F
Biot number	$\text{Bi} = h_h l/k_{fin}$	0.0002	
Fin radius	$r_2 = d_{fin}/2$	0.0705	ft
Characteristic fin radius	$r_{2c} = r_2 + \delta/2$	0.0740	ft
Fin height	$L = r_2 - d_o/2$	0.0439	ft
Equivalent fin height	$L_c = L + \delta/2$	0.0474	ft
Radius ratio (Consult Figure 17.2 below[2])	$r_{2c} = (d_o/2r_1)$	2.78	
x-axis	$L_c^{1.5}[h_h/k_{fin}L_c\delta]^{0.5}$	0.147	
Fin efficiency	$\eta_{fin,h}$	0.952	
Surface efficiency	$\eta_h = (a_p + a_f\eta_{fin,h})/a_h$	0.952	

The following parameters were either taken from the HX datasheet or calculated as shown:

Design HX Data			**Units**
Total service water flow rate	G	650	gal/min
Service water flow rate per coil	$g_c = G/C$	81.25	gal/min
Entering water temperature	$t_{c,i}$	85.00	°F
Exiting water temperature	$t_{c,o}$	89.60	°F
Average water temperature	$t_c = (t_{c,i} + t_{c,o})/2$	87.30	°F
Tube-side thermal conductivity	k_c	0.357	Btu/hr-ft-°F
Tube-side absolute viscosity	μ_c	1.91	lbm/ft-hr
Tube-side specific heat	$c_{p,c}$	0.997	Btu/lbm-°F
Tube-side density	ρ_c	62.1	lbm/ft³
Tube-side temperature rise	$dt_c = t_{c,o} - t_{c,i}$	4.60	°F
Tube-side heat transfer rate	$Q_c = (60/7.48)g_c c_{p,c}\rho_c dt_c$	185,682	Btu/hr
Entering air dry bulb temperature	$T_{h,i}$	100.00	°F
Leaving air dry bulb temperature	$T_{h,o}$	88.75	°F
Average air dry bulb temperature	T_h	94.38	°F
Hot-side thermal conductivity	k_h	0.0156	Btu/hr-ft-°F
Hot-side absolute viscosity	μ_h	0.0460	lbm/ft-hr
Hot-side specific heat	$c_{p,h}$	0.240	Btu/lbm-°F
LMTD correction factor	F	0.9890	
Tube-side flow rate per coil	$m_{c\text{-}coil} = (60/7.48)g_c\rho_c$	40,500	lbm/hr
Face velocity out of air side	FV	739.20	ft/min
Relative humidity	RH	40.00%	%
Total atmospheric pressure	P	14.60	lbf/in²A

The overall heat transfer coefficient at design conditions, U_{design}, is calculated as follows:

Design Heat Transfer Coefficient			**Units**
Greater terminal temperature difference	$\Delta T_1 = T_{h,i} - t_{c,o}$	10.40	°F
Lesser terminal temperature difference	$\Delta T_2 = T_{h,o} - t_{c,i}$	3.75	°F
Log mean temperature difference	$LMTD = (\Delta T_1 - \Delta T_2)/\ln(\Delta T_1/\Delta T_2)$	6.52	°F
Effective mean temperature difference	$EMTD = F(LMTD)$	6.45	°F
Design overall heat transfer coefficient	$U_{design} = Q/(A_h EMTD)$	9.92	Btu/hr-ft²-°F

The tube-side convection coefficient may be computed from the Petukhov correlation in Chapter 16 as follows:

Mass flow rate/tube	$m_t = m_{c\text{-}coil}/N_t$	1265.6	lbm/hr
Volumetric flow rate/tube	$V_t = m_t/\rho$	20.4	ft³/hr
Tube area	$a_t = (\pi/4)d_i^2$	0.00160	ft²
Tube velocity	$v_t = V_t/a_t$	12,711	ft/hr
Prandtl number	$\mathrm{Pr}_c = c_{p,c}\mu_c/k_c$	5.3	
Reynolds number	$\mathrm{Re}_c = \rho v_t d_i/\mu_c$	18,698	
Fanning friction factor	$f = (1.58 \ln \mathrm{Re}_c - 3.28)^{-2}$	0.00665	
Nusselt number	$\mathrm{Nu} = ((f/2)\mathrm{Re}_c \mathrm{Pr}_c)/(1.07 + 12.7(f/2)^{1/2} (\mathrm{Pr}_c^{2/3}-1))$	128.82	
Tube-side film coefficient	$h_c = \mathrm{Nu}(k_c/d_i)$	1017	Btu/hr-ft²-°F

The hot-side convection coefficient may be computed as follows:

$$h_{h,\,design} = \frac{1/\eta_h}{\dfrac{1}{U_{design}} - r_{f,h} - r_w - \left(\dfrac{A_h}{A_c}\right) r_{f,c} - \left(\dfrac{A_h}{A_c}\right)\dfrac{1}{h_c}} = 13.3 \text{ Btu/hr-ft}^2\text{-}°\text{F}$$

The design air side mass flow rate may be calculated as follows:

Volumetric flow rate per coil	V	16,262	ft³/min
Saturation pressure at $T_{h,i}$	*ASME Steam Tables*	0.96	lbf/in²
Partial pressure of the water vapor in the air	$P_{wv} = \mathrm{RH}\, P_{sat}$	0.38	lbf/in²
Partial pressure of the air	$P_a = P - P_{wv}$	14.22	lbf/in²
Humidity ratio	$w = (M_{wv}/M_a)(P_{wv}/P_a)$	0.017	
Mass flow rate of air	$m_a = [P_a\, 60V/R(T + 460)]$	66,879	lbm/hr
Mass of the water vapor	$m_{wv} = w m_a$	1,116	lbm/hr
Total mass flow rate on the air side	$m_h = m_a + m_{wv}$	67,995	lbm/hr
Specific heat of moist air	$c_{p\text{-}MA} = (c_{p\text{-}DA} m_a + c_{p\text{-}WV}\, m_{wv})/m_h$	0.2529	Btu/lbm-°F
Density of air	$\rho_a = m_h/V$	0.0697	lbm/ft³
Hot-side flow measurement per coil	$m_h = V\rho_a$	67,995	lbm/hr
Air-side heat transfer rate	$Q_h = m_h c_p (T_{h,i} - T_{h,o})$	183,958	Btu/hr

The HX test data is as follows:

HX Test Data		Test 1	Test 2	Units
Total service water flow rate	G	1315	631	gal/min
Service water flow rate per coil	$g_c = G/C$	164.38	78.88	gal/min
Entering water temperature	$t_{c,i}$	59.76	82.39	°F
Exiting water temperature	$t_{c,o}$	64.30	88.28	°F
Average water temperature	$t_c = (t_{c,i} + t_{c,o})/2$	62.03	85.34	°F
Tube-side thermal conductivity	k_c	0.344	0.356	Btu/hr-ft²-°F
Tube-side absolute viscosity	μ_c	2.72	1.96	lbf/ft-hr
Tube-side specific heat	$c_{p,c}$	1.002	1.000	Btu/lbm-°F
Tube-side density	ρ_c	62.4	62.2	lbm/ ft³
Tube-side flow rate per coil	$m_{c\text{-}coil} = (60/7.48)g_c\rho_c$	82,222	39,332	lbm/hr
Tube-side heat transfer rate	$Q_c = (60/7.48)g_c c_{p,c}\rho_c dt_c$	374,130	231,638	Btu/hr
Inlet relative humidity	RH	22.19%	11.12%	%
Total volumetric flow rate	V_{total}	114,902	115,513	ft³/min
Volumetric flow rate per coil, Cu-ft/hr	$V_{coil} = 60\ V_{total}/C$	861,765	866,348	F³/hr
Entering air dry bulb temperature	$T_{h,i}$	92.46	104.23	°F
Leaving air dry bulb temperature	$T_{h,o}$	64.97	86.81	°F
Average air dry bulb temperature	$T_h = (T_{h,i} + T_{h,o})/2$	78.72	95.52	°F
Hot-side thermal conductivity	k_h	0.0152	0.0156	Btu/hr-ft²-°F
Hot-side absolute viscosity	μ_h	0.0451	0.0461	lbm/ft-hr
Hot-side specific heat	$c_{p,h}$	0.240	0.240	Btu/lbm-°F
Hot-side density	ρ_h	0.0723	0.0727	lbm/ ft³
Total pressure	P	14.85	15.23	lbf/in²A
Saturation pressure at inlet temperature	P_{sat}	0.76	1.08	lbf/in²A
Partial pressure of water	$P_{wv} = RH \times P_{sat}$	0.17	0.12	lbf/in²A
Molecular weight of water	M_w	18	18	
Molecular weight of air	M_a	29	29	
Specific heat of air	$c_{p\text{-}a}$	0.240	0.240	Btu/lb-F
Specific heat of water	$c_{p\text{-}w}$	1.002	1.000	Btu/lb-F
Partial pressure of air	$P_a = P - P_{wv}$	14.68	15.11	lbf/in²A
Humidity ratio	$W = (M_{wv}/M_a)(P_{wv}/P_a)$	0.0071	0.0050	
Mass flow rate of air per coil	$m_a = [P_a\ 60V_{coil}/R(T{+}460)]$	61,827	62,632	lbm/hr
Mass flow rate of water vapor per coil	$m_{w\text{-}coil} = wm_a$	440	310	lbm/hr
Total mass flow rate per coil	$m_h = m_a + m_{w\text{-}coil}$	62,266	62,942	lbm/hr
Specific heat of moist air	$c_{p\text{-}ma} = (c_{p\text{-}DA}m_a + c_{p\text{-}WV}m_{wv})/m_h$	0.2454	0.2437	Btu/lbm-°F
density of air, lbm/Cuft	$\rho_a = m_h/V$	0.0723	0.0727	lbm/ft³
Test heat transfer (hot side)	$Q_h = m_h c_p(T_{h,i}T_{h,o})$	420,025	267,255	Btu/hr
Average heat transfer rate	$Q_{test,ave} = (Q_c + Q_h)/2$	397,077	249,446	Btu/hr

(Continued)

HX Test Data		Test 1	Test 2	Units
Greater terminal temperature difference	$\Delta T_1 = T_{h,i} - t_{c,o}$	28.16	15.95	°F
Lesser terminal temperature difference	$\Delta T_2 = T_{h,o} - t_{c,i}$	5.21	4.42	°F
Log mean temperature difference	$LMTD = (\Delta T_1 - \Delta T_2)/\ln(\Delta T_1/\Delta T_2)$	13.60	8.98	°F
Ratio of temperature change	$R = (T_{h,i} - T_{h,o})/(t_{c,o} - t_{c,i})$	6.06	2.96	
Effectiveness	$P = (t_{c,o} - t_{c,i})/(T_{h,i} - t_{c,i})$	0.14	0.27	
LMTD correction factor	F	0.975	0.954	
LMTD (corrected)	$EMTD = F(LMTD)$	13.26	8.57	°F
Test overall heat transfer coefficient	$U_{test} = Q_{ave}/(A_h EMTD)$	10.05	9.56	Btu/hr-ft^2-°F
Average tube temperature	$t_{tube} = (T_h + t_c)/2$	70.37	90.43	°F
Wall resistance, $R(w)$	$r_w = [d_o \ln(A_h/A_t)]/2k_t$	0.00036	0.00036	hr-ft^2-°F /Btu
Mass flow rate/tube	$m_t = m_{c\text{-}coil}/SN_t$	2,569	1,229	lbm/hr
Volumetric flowrate/tube	$V_t = m_t/\rho_t$	41.2	19.8	ft^2/hr
Tube velocity	$v_t = V_t/a_t$	25,716	12,340	ft/hr
Tube velocity	$v_t = v_t/3{,}600$	7.14	3.43	ft/s
Prandtl number	$\mathrm{Pr}_c = c_{p,c}\mu_c/k_c$	7.9	5.5	
Reynolds number	$\mathrm{Re}_c = \rho v_t D_i/\mu_c$	26,592	17,676	
Fanning friction factor	$f = (1.58 \ln \mathrm{Re}_c - 3.28)^{-2}$	0.00609	0.00675	
Nusselt number	Nu^a	203.4	124.7	
Tube-side film coefficient	$h_c = \mathrm{Nu}(k_c/D_i)$	1548	982	Btu/hr-ft^2-°F
Hot-side flow rate	$m_{h\text{-}coil} = (60/7.48)g_h\rho_h$	62,266	62,942	lbm/hr
Hot-side film coefficient	h_h^b	12.4	12.6	Btu/hr-ft^2-°F
Total resistance	$r = 1/U_{test}$	0.09948	0.10460	hr-ft^2-°F /Btu
Hot-side convection resistance	$r_h = 1/(\eta_h h_h)$	0.08451	0.08313	hr-ft^2-°F /Btu
Wall resistance	$r_w = [d_o \ln(A_h/A_t)]/2k_t$	0.00036	0.00036	hr-ft^2-°F /Btu
Cold-side convection resistance	$R_c = (A_h/A_c)h_c$	0.00939	0.01480	hr-ft^2-°F /Btu
Fouling referenced to hot side	$r_{f,h}^c$	0.00522	0.00631	hr-ft^2-°F /Btu
Fouling referenced to tube side	$r_{f,c} = (A_c/A_h)r_{f,h}$	0.00036	0.00043	hr-ft^2-°F /Btu

The test fouling resistance may be calculated as follows:

[a] $\mathrm{Nu} = \left(\left(f/2\right)\mathrm{Re}_c\,\mathrm{Pr}_c\right)/\left(1.07 + 12.7\left(f/2\right)^{1/2}\left(\mathrm{Pr}_c^{2/3} - 1\right)\right)$

[b] $$h_{h,test} = \frac{\left[\left(\frac{m_h}{\mu_h}\right)^{0.681}\left(\frac{\mu_h c_{p,h}}{k_h}\right)^{0.333} k_h\right]_{test}}{\left[\left(\frac{m_h}{\mu_h}\right)^{0.681}\left(\frac{\mu_h c_{p,h}}{k_h}\right)^{0.333} k_h\right]_{design}} h_{h,design}$$

[c] $$r_{f,test} = \frac{1}{\frac{1}{U_{test}} - \left(\frac{1}{\eta_h h_{h,test}}\right) - r_w - \left(\frac{A_h}{A_c}\right)\frac{1}{h_{c,test}}}$$

The uncertainty of the fouling resistance for Test 1 may be calculated as follows:

The test measurement data using the procedure described in Chapter 16 is summarized as follows:

Uncertainty of the Fouling Resistance for Test 1

Parameter	Value	Uncertainty
t_i	59.76°F	±0.17°F
t_o	64.30°F	±0.12°F
T_i	92.46°F	±0.15°F
T_o	64.97°F	±0.18°F
m_c	82,440 lbm/hr	2,519 lbm/hr
m_h	62,266 lbm/hr	1,384 lbm/hr

The sensitivity coefficients, θ, as described in Appendix B of Reference 4, are a measure of how much a dependent variable such as Q varies with an independent variable such as t (i.e., the derivative of Q with respect to t). For the heat transfer rate, Q, the independent variables are the mass flow rate, the specific heat, and the inlet and outlet temperatures.

$$\theta_{Q,t_i} = -m_c\, c_{p,c} = -8.263\times 10^4$$

$$\theta_{Q,t_o} = m_c\, c_{p,c} = 8.263\times 10^4$$

$$\theta_{Q,m_c} = c_{p,c}\left(t_o - t_i\right) = 4.45$$

$$\theta_{Q,c_p c} = m_c\left(t_o - t_i\right) = 3.743\times 10^5$$

$$\theta_{Q,T_i} = -m_h\, c_{p,h} = -1.494\times 10^4$$

$$\theta_{Q,T_o} = m_h\, c_{p,h} = 1.494\times 10^4$$

$$\theta_{Q,m_h} = c_{p,h}\left(T_i - T_o\right) = 6.598$$

$$\theta_{Q,c_p h} = m_h\left(T_i - T_o\right) = 1.712\times 10^6$$

Following the procedure in Tables K.12 and K.13 of Reference 4, the uncertainty of the hot and cold stream heat transfer rates may now be calculated. Recalling that

$$u_{Q,c} = \left[\left(u_{t_i}\theta_{t_i}\right)^2 + \left(u_{t_o}\theta_{t_o}\right)^2 + \left(u_{m_c}\theta_{m_c}\right)^2 + \left(u_{c_{p,c}}\theta_{c_{p,c}}\right)^2 \right]^{1/2}$$

Uncertainty in the Cold Stream Heat Transfer Rate

Contributing Factor	Uncertainty	Sensitivity	Uncertainty Contribution
t_i	0.17	-8.263×10^4 (BTU-hr-°F)	1.947×10^8
t_o	0.12	8.263×10^4 (BTU-hr-°F)	9.036×10^7
m_c	2,519	4.45 (Btu/lbm)	1.313×10^8
$c_{p,c}$	0.01	3.743×10^5 (lbm-°F /hr)	1.401×10^7
$u_{Q,c}$			2.075×10^4

Note: The uncertainty of specific heat of 10% is from Reference 4, Appendix I.

Uncertainty in Hot Stream Heat Transfer Rate

Contributing Factor	Uncertainty	Sensitivity	Uncertainty Contribution
T_i	0.15	1.495×10^4 (Btu-hr-°F)	5.293×10^6
T_o	0.18	-1.495×10^4 (Btu-hr-°F)	6.968×10^6
m_h	1,384	6.598 (Btu/lbm)	8.335×10^7
$c_{p,h}$	0.01	1.712×10^6(lbm-°F /hr)	2.930×10^8
$u_{Q,h}$			1.971×10^4

The weighted average heat transfer rate and the uncertainty of the average heat transfer rate may now be calculated as follows:

$$Q_{ave} = \left(\frac{u_{Q,h}^2}{u_{Q,c}^2 + u_{Q,h}^2} \right) Q_c + \left(\frac{u_c^2}{u_{Q,c}^2 + u_{Q,h}^2} \right) Q_h = 393{,}877 \left(\mathrm{Btu}/_{\mathrm{hr}} \right)$$

$$u_{Q_{ave}} = \frac{\left(u_{Q,c}^4 \, u_{Q,h}^2 + u_{Q,h}^4 \, u_{Q,c}^2 \right)^{1/2}}{u_{Q,c}^2 + u_{Q,h}^2} = 14{,}290 \left(\mathrm{Btu}/_{\mathrm{hr}} \right)$$

The heat balance error with respect to the hot stream may be computed as

$$HBE = \frac{Q_c - Q_h}{Q_{hot}} = 8.7\%$$

The uncertainty of the heat balance error may be computed as follows:

$$u_{HBE} = \left(\frac{Q_c}{Q_h}\right)\left[\left(\frac{u_{Q,h}}{Q_h}\right)^2 + \left(\frac{u_{Q,c}}{Q_c}\right)^2\right]^{1/2} = 6.7\%$$

The heat balance error between Q_h and Q_c such that $HBE \geq u_{HBE}$ is an indication of a poor test.

The effective mean temperature difference, *EMTD*, is defined as $EMTD = F(LMTD)$ (where for the purposes of uncertainty analysis $F = 1$).

The *EMTD* may be computed from

$$EMTD = \frac{(T_i - t_o) - (T_o - t_i)}{\ln\left[\frac{(T_i - t_o)}{(T_o - t_i)}\right]}$$

One may calculate the uncertainty of the *EMTD* as a function of the uncertainties of the temperatures in the equation above. The sensitivity coefficients for *EMTD* to temperature as found in Appendix B of Reference 4 may be computed as follows:

$$\theta_{EMTD,T_i} = EMTD\frac{\left(1 - EMTD/\Delta T_1\right)}{(\Delta T_1 - \Delta T_2)} = 0.31$$

$$\theta_{EMTD,To} = EMTD\frac{\left(1 - EMTD/\Delta T_2\right)}{(\Delta T_1 - \Delta T_2)} = 0.93$$

$$\theta_{EMTD,t_i} = -EMTD\frac{\left(1 - EMTD/\Delta T_2\right)}{(\Delta T_1 - \Delta T_2)} = -0.93$$

$$\theta_{EMTD,t_o} = -EMTD\frac{\left(1 - EMTD/\Delta T_1\right)}{(\Delta T_1 - \Delta T_2)} = -0.31$$

In addition, Example K.1 in Reference 4 includes additional uncertainty to account for the uncertainty associated with possible variation in the heat exchanger fouling by assuming a Biot number of 0.5 for a water-to-water heat exchanger (see Appendix G of Reference 4).

$$\frac{b_{EMTD,u}}{EMTD} = 1 - \frac{2\left(\mathrm{Bi} - \Delta T_1/\Delta T_2\right)\ln\left(\Delta T_1/\Delta T_2\right)}{(1+\mathrm{Bi})\left(\left(\Delta T_1/\Delta T_2\right) - 1\right)\ln\left(\frac{\mathrm{Bi}}{\Delta T_1/\Delta T_2}\right)}$$

$$b_{EMTD,u} = 0.0091 \times EMTD = -0.011°F$$

where

$$\Delta T_1 = T_i - t_o$$

$$\Delta T_2 = T_o - t_i$$

An additional uncertainty in the *EMTD* of 2% is assumed for incomplete mixing in the heat exchanger.

The sensitivity coefficients for the variation in fouling and incomplete mixing are 1.0 and 0.27, respectively.

Therefore, the total uncertainty in the *EMTD* is as follows:

Uncertainty in EMTD at Test Conditions

Contributing Factor	Uncertainty	Sensitivity	Uncertainty Contribution
T_i	0.15	0.31	0.002
T_o	0.18	0.93	0.027
t_i	0.17	−0.93	0.025
t_o	0.12	−0.31	0.001
Variation in fouling	−0.11	1.0	0.011
Incomplete mixing	0.27	1.0	0.070
u_{EMTD}			0.36

The overall coefficient of heat transfer indicated by the test conditions, U_{test}, may be calculated.

$$U_{test} = \frac{Q_{ave}}{A \times EMTD} = 10.23 \text{ Btu/hr-ft}^2\text{-°F}$$

The uncertainty in U_{test} may be determined by first calculating the sensitivity coefficients for the variables Q_{ave} and EMTD (see Table B.3 in Reference 4). The area, A, is assumed to be constant.

$$\theta_{Q_{ave}} = \frac{1}{A \times EMTD} = 2.54 \times 10^{-5}\left(\frac{1}{\text{ft}^2\ °\text{F}}\right)$$

$$\theta_{EMTD} = \frac{Q_{ave}}{A \times EMTD} = 0.74\left(\frac{\text{Btu}}{\text{hr-ft}^2\text{-° F}}\right)$$

Therefore, the total uncertainty in the U_{test} is as follows:

Uncertainty in U_{test} at Test Conditions

Contributing Factor	Uncertainty	Sensitivity	Uncertainty Contribution
Q_{ave}	1.43×10^4 (Btu/hr)	2.58×10^{-5} (ft^{-2}-°F^{-1})	0.14
EMTD	0.36 (°F)	0.77 (Btu/hr-ft^2-°F)	0.08
u_{UTest}			0.46

Recalling that the fouling resistance measured during a test is defined as

$$r_{f,\,test} = \frac{1}{U_{test}} - \frac{1}{U_{clean}}$$

The uncertainty in the fouling resistance is a function of the uncertainty in U_{test}, h_h, and h_c. The sensitivity coefficients, θ, and uncertainty of the fouling are determined numerically by perturbing each parameter plus and minus by the amount of the uncertainty to determine the impact on the fouling. The sensitivity coefficients are as follows:

$$\theta_{U,test} = \frac{\Delta r_{f,test}}{\Delta U_{test}}$$

and

$$\theta_{h,c} = \frac{\Delta r_{f,test}}{\Delta h_c} \qquad \theta_{h,h} = \frac{\Delta r_{f,test}}{\Delta h_h}$$

and

$$u_{f,c} = \left[\left(u_{U_{test}} \theta_{U_{test}} \right)^2 + \left(u_{h_c} \theta_{h_c} \right)^2 + \left(u_{h_h} \theta_{h_h} \right)^2 \right]^{1/2}$$

The following table shows the calculated uncertainty in the apparent fouling resistance where the analytical uncertainties in the equations for the cold and hot-side convection coefficients of heat transfer are 10% and 15%, respectively:

Uncertainty of Fouling Resistance Referenced to the Hot Side

Parameter	Uncertainty	Value	Overall Fouling	Sensitivity Coefficient	Coefficient Uncertainty	Overall Fouling
U(test) + U	4.5%	10.54	0.003333	−6.782E-04	0.46	−0.00031
U(test) + U	−4.5%	9.63	0.003959			
h(c + d)test	10.0%	1,703	0.003637	3.706E-08	154.82	0.00001
h(c − d)test	−10.0%	1,393	0.003625			
h(h + d)test	15.0%	32.97	0.003960	1.058E-04	1.86	0.00020
h(h − d)test	−15.0%	20.71	0.002663			
Composite						0.00037

Therefore, calculation of the range of apparent fouling for both tests where the results of Test 2 are calculated in a similar fashion is done as follows:

Parameter	Test 1	Test 2	Units
Referenced to hot side			
$r_{f,h} + u_{f,h}$	0.00559	0.00725	hr-ft^2-°F /Btu
$r_{f,h}$	0.00522	0.00631	hr-ft^2-°F /Btu
$r_{f,h} - u_{f,h}$	0.00485	0.00537	hr-ft^2-°F /Btu
Referenced to tube side			
$r_{f,c} + u_{f,c}$	0.00038	0.00050	hr-ft^2-°F /Btu
$r_{f,c}$	0.00036	0.00043	hr-ft^2-°F /Btu
$r_{f,c} - u_{f,c}$	0.00033	0.00037	hr-ft^2-°F /Btu

Because the two tests were conducted only one day apart, one would expect to be able to calculate similar values for fouling resistance, which is the case. Because $r_{f,c} + u_{f,c}$ for Test 1 slightly exceeds $r_{f,c} - u_{f,c}$ for Test 2, there is a slight overlap of the estimated fouling resistance when uncertainty is considered.

CHAPTER 18

Refer to Section 18.4. Calculate the constant C in the Colburn analogy equation.
A PHX has the following dimensions:

$$L_H = 67.50 \text{ in}$$

$$L_w = 33.25 \text{ in}$$

$$L_{cp} = 31.375 \text{ in}$$

$$N_p = 191$$

$$\Delta X = 0.02 \text{ in}$$

$$\phi = 1.3$$

$$N_{cp} = (N_p - 1)/2 = (191 - 1)/2 = 95$$

$$A_{eff} = \phi\, L_H L_W N_p = 1.3 \times 67.5 \times 33.25 \times 191/144 = 3{,}870 \text{ ft}^2$$

$$b = \left(L_{cp} / N_p \right) - \Delta X = \left(31.375/191 \right) - 0.02 = 0.1443 \text{ in}$$

$$D_e = \frac{4L_w b}{(2L_w + 2\phi b)} = \frac{4 \times 33.25 \times 0.1443}{(2 \times 33.25 + 2 \times 1.3 \times 0.1443)} = 0.0239 \text{ ft}$$

The design operating conditions are as follows:

CWT entering, $t_i = 82.0°\text{F}$
CWT exiting, $t_o = 91.0°\text{F}$
HWT entering, $T_i = 100.0°\text{F}$
HWT exiting, $T_o = 83.5°\text{F}$
Coldwater flow rate, $G_c = 4{,}250$ gal/min
Hot water flow rate, $G_h = 2{,}100$ gal/min
Plate wall resistance, $r_w = 0.000134$ hr-ft^2-°F /Btu

The properties of the fluids are as follows:

Cold-side density, $\rho_c = 62.2$ lbm/ft^3
Cold-side specific heat, $c_{pc\text{-}ave} = 0.976$ Btu/lbm-°F
Cold-side viscosity, $\mu_c = 1.929$ lbm/ft-hr
Cold-side thermal conductivity, $k_c = 0.356$ Btu/hr-ft-°F
Hot-side density, $\rho_h = 62.1$ lbm/ft^3

Hot-side specific heat, $c_{ph\text{-}ave} = 0.996$ Btu/lbm-°F
Hot-side viscosity, $\mu_h = 1.796$ lbm/ft-hr
Hot-side thermal conductivity, $k_h = 0.359$ Btu/hr-ft-°F

Therefore,

$$\Delta t_1 = T_i - t_o = 9.0$$

$$\Delta t_2 = T_o - t_i = 1.51$$

$$LMTD = \frac{\Delta t_1 - \Delta t_2}{\ln\left(\frac{\Delta t_1}{\Delta t_2}\right)} = 4.20 \text{ °F}$$

$$m_c = 0.1337 \times 60 \times g_c \times \rho_c = 2{,}119{,}100 \text{ lbm/hı}$$

$$m_h = 0.1337 \times 60 \times g_h \times \rho_h = 1{,}046{,}100 \text{ lbm/hr}$$

$$Q_c = m_c c_p \left(t_o - t_i\right) = 18{,}621{,}000 \text{ Btu/hr}$$

$$\mathrm{Q}_h = m_h c_p \left(T_i - T_o\right) = 17{,}185{,}000 \text{ Btu/hr}$$

$$Q_{ave} = \left(Q_{c\text{-}c} + Q_h\right)/2 = 17{,}903{,}000 \text{ Btu/hr}$$

$$U_{req} = \frac{Q_{ave}}{A_{eff}\, LMTD} = 1{,}103 \text{ Btu/hr-ft}^2\text{- °F}$$

Calculating the mass flux, G, as follows:

$$G_c = \frac{m_c}{N_{cp}\left(\frac{b}{12}\right)\left(\frac{L_w}{12}\right)\phi} = 515{,}080 \text{ lbm/ft}^2\text{-hr}$$

$$G_h = \frac{m_h}{N_{cp}\left(\frac{b}{12}\right)\left(\frac{L_w}{12}\right)\phi} = 254{,}270 \text{ lbm/ft}^2\text{-hr}$$

The Reynolds numbers are

$$\mathrm{Re}_c = D_e G_c / \mu_c = 6{,}384$$

$$\mathrm{Re}_h = D_e G_h / \mu_h = 3{,}385$$

The Prandtl numbers are

$$\Pr_c = \frac{\mu_c c_{pc}}{k_c} = 5.29$$

$$\Pr_h = \frac{\mu_h c_{ph}}{k_h} = 4.99$$

Therefore,

$$C = \frac{\dfrac{D_e}{\mathrm{Re}_c^{3/4}\,\Pr_c^{1/3}\,k_c} + \dfrac{D_e}{\mathrm{Re}_h^{3/4}\,\Pr_h^{1/3}\,k_h}}{\dfrac{A\,(LMTD)}{Q} - r_w} = 0.184$$

CHAPTER 19

From June 2002 to May 2014, the staff at the V. C. Summer Nuclear Generating Station conducted heat transfer tests on all four of their RBCU. The physical data for the RBCU may be found in the example problem of Chapter 17.

The following tables provide the results of the RBCU heat transfer tests on the indicated days:

RBCU 1A Heat Transfer Test Results

		06/03/02	05/28/05	06/06/08	05/22/14	Units
Air volumetric flow rate	V	115,012	123,130	120,092	111,731	ft³/min
Service water flow measurement	G_c	668	1,273	1,273	1,269	gal/min
Service water inlet temperature	t_i	73.58	71.34	74.75	75.39	°F
Service water outlet temperature	t_o	81.26	75.54	78.25	81.02	°F
Hot-side inlet temperature	T_i	102.09	97.2	96.51	112.45	°F
Hot-side outlet temperature	T_o	79.73	75.96	78.71	82.09	°F

RBCU 2A Heat Transfer Test Results

		06/03/02	05/28/05	12/07/09	05/22/14	Units
Air volumetric flow rate	V	116,494	120,642	108,539	112,191	ft³/min
Service water flow measurement	G_c	689	1,233	1,269	1,252	gal/min
Service water inlet temperature	t_i	73.36	71.44	74.76	74.87	°F
Service water outlet temperature	t_o	81.09	75.52	78.22	80.9	°F
Hot-side inlet temperature	T_i	102.96	97.61	97.84	114.52	°F
Hot-side outlet temperature	T_o	79.72	75.82	78.38	81.69	°F

RBCU 1B Heat Transfer Test Results

		6/3/02	5/28/05	12/7/09	12/8/09	5/22/14	Units
Air volumetric flow rate	V	116,100	117,395	114,902	115,513	104,625	ft³/min
Service water flow measurement	G_c	609	1,266	1,315	631	1,269	gal/min
Service water inlet temperature	t_i	73.41	71.25	59.76	82.39	76.36	°F
Service water outlet temperature	t_o	82.14	74.98	64.30	88.28	82.25	°F
Hot-side inlet temperature	T_i	101.97	94.87	92.46	104.23	115.96	°F
Hot-side outlet temperature	T_o	80.36	75.45	64.97	86.81	83.62	°F

RBCU 2B Heat Transfer Test Results

		06/03/02	05/28/05	06/06/08	05/22/14	Units
Air volumetric flow rate	V	119,570	118,459	111,443	109,816	ft³/min
Service water flow measurement	G_c	587	1,245	1,279	1,314	gal/min
Service water inlet temperature	t_i	73.37	71.24	74.45	76.41	°F
Service water outlet temperature	t_o	82.45	75.15	78.82	83.03	°F
Hot-side inlet temperature	T_i	102.46	96.83	103.21	117.55	°F
Hot-side outlet temperature	T_o	80.64	75.57	78.87	84.21	°F

The following tables show the calculated results using the procedure described in Chapter 17:

RBCU 1A Calculated Results

		06/03/02	05/28/05	06/06/08	05/22/14	Units
Mass flow rate, m_c per coil	m_c	41753.20	79607.14	79554.07	79256.26	lbm/hr
Mass flow rate, $m(h)$	m_h	60855.85	66457.71	64445.59	58187.22	lbm/hr
Heat transfer rate	Q_c	3.209×10^5	3.347×10^5	2.787×10^5	4.465×10^5	Btu/hr
Weighted average heat transfer rate	Q_{ave}	3.228×10^5	3.364×10^5	2.784×10^5	4.317×10^5	Btu/hr
Heat balance error (referenced to hot side)	*HBE*	1.74%	1.20%	−1.22%	−5.31%	%
Effective mean temperature difference	*EMTD*	11.90	10.91	9.25	15.82	°F
Overall test heat transfer coefficient	U_{test}	9.3	10.5	10.3	9.3	Btu/hr-ft-°F

RBCU 2A Calculated Results

		06/03/02	05/28/05	12/07/09	05/22/14	Units
Mass flow rate, m_c per coil	m_c	43067.85	77104.07	79326.44	78234.15	lbm/hr
Mass flow rate, $m(h)$	m_h	61583.10	65062.68	58125.08	57942.40	lbm/hr
Heat transfer rate	Q_c	3.332×10^5	3.149×10^5	2.747×10^5	4.721×10^5	Btu/hr
Weighted average heat transfer rate	Q_{ave}	3.375×10^5	3.113×10^5	2.744×10^5	4.593×10^5	Btu/hr
Heat balance error (referenced to hot side)	*HBE*	3.01%	−2.20%	−1.19%	−3.40%	%
Effective mean temperature difference	*EMTD*	12.42	10.18	9.36	16.61	°F
Overall test heat transfer coefficient	U_{test}	9.3	10.5	10.0	9.5	Btu/hr-ft-°F

RBCU 1B Calculated Results

		6/3/02	5/28/05	12/7/09	12/8/09	5/22/14	Units
Mass flow rate, m_c per coil	m_c	38066.81	79170.94	82440.44	39365.10	79289.58	lbm/hr
Mass flow rate, $m(h)$	m_h	64125.96	63592.17	62266.38	62941.78	54812.22	lbm/hr
Heat transfer rate	Q_c	3.325×10^5	2.956×10^5	3.741×10^5	2.316×10^5	4.673×10^5	Btu/hr
Weighted average heat transfer rate	Q_{ave}	3.326×10^5	2.960×10^5	3.939×10^5	2.470×10^5	4.358×10^5	Btu/hr
Heat balance error (referenced to hot side)	*HBE*	0.01%	0.25%	8.69%	11.90%	–9.83%	%
Effective mean temperature difference	*EMTD*	12.15	9.98	13.45	8.89	17.04	°F
Overall test heat transfer coefficient	U_{test}	9.4	10.1	10.0	9.5	8.7	Btu/hr-ft-°F

RBCU 2B Calculated Results

		06/03/02	05/28/05	06/06/08	05/22/14	Units
Mass flow rate, m_c per coil	m_c	36691.98	77857.845	79902.95	82062.23	lbm/hr
Mass flow rate, $m(h)$	m_h	65980.12	63936.99	58855.20	57484.30	lbm/hr
Heat transfer rate	Q_c	3.334×10^5	3.048×10^5	3.494×10^5	5.435×10^5	Btu/hr
Weighted average heat transfer rate	Q_{ave}	3.375×10^5	3.165×10^5	3.463×10^5	4.928×10^5	Btu/hr
Heat balance error (referenced to hot side)	*HBE*	3.52%	6.58%	−1.64%	−18.16%	%
Effective mean temperature difference	*EMTD*	12.44	10.65	11.56	17.77	°F
Overall test heat transfer coefficient	U_{test}	9.3	10.2	10.2	9.5	Btu/hr-ft-°F

As stated in Section 19.8 and shown in Figure 19.5, previous heat transfer tests have demonstrated that the *EMTD* is a linear function of the cold-stream heat transfer rate. Consider the RBCU 1A test conducted on May 22, 2014. From Figure 19.5,

$$EMTD = 0.000035 \times Q_c$$
$$= 0.000035 \times 4.465 \times 10^5 = 15.63\ °F$$

It is noted that the *EMTD* thus calculated is slightly different from that calculated from the May 22, 2014 RBCU 1A heat transfer test data. This difference should be immaterial as long as the same equation is used going forward for future tests.

The last piece of required information to determine the NTU is the effective heat transfer surface area, $A_{h\text{-}eff}$, which is obtained by conducting a flow and *dP* test along with the final heat transfer test.

From Figure 19.6,

$$\%_{unplugged} = 100 \times (0.000280 \times dP^2 - 0.033340 \times dP + 1.693)$$
$$= 100 \times (0.000280 \times 33.67^2 - 0.033340 \times 33.67 + 1.693) = 88.8\%$$

Then,

$$A_{h\text{-}eff} = \%_{unplugged} \times A_h = 0.888 \times 2{,}925 = 2{,}597\ \text{ft}^2$$

Therefore,

$$U = \frac{q_c}{A_h\ EMTD} = \frac{4.465 \times 10^5\ \text{Btu/hr}}{2{,}925\left(\text{ft}^2\right) \times 15.63\,(\ \text{F})} = 9.77\left(\text{Btu}/\text{hr-ft}^2\text{-°F}\right)$$

and

$$NTU = \frac{UA_{h-eff}}{m_c\ c_{p_c}} = \frac{9.77\left(\text{Btu}/\text{hr-ft}^2\text{-}\ \text{F}\right) \times 2{,}597\ \text{ft}^2}{79{,}256\left(\text{lbm}/\text{hr}\right) 1.0006\left(\text{Btu}/\text{lbm-°F}\right)} = 0.32$$

Therefore, the heat transfer rate for some future test may be estimated as follows:

$$(Q^* - u^*)_{future} = \left(\frac{NTU_{future}}{NTU_{test}}\right)(Q^* - u^*)_{test}$$

For the RBCU 1A test conducted on May 22, 2014, the heat transfer rate was calculated for design-basis LOCA conditions less the uncertainty of the test to be 133.2 MBtu/hr, whereas the required value was 90.67 MBtu/hr, a margin of 47%. The corresponding calculated heat transfer rate for design-basis LOCA conditions less the uncertainty of the test conducted on June 6, 2006 was only 101.8 MBtu/hr, so obviously timely cleaning of the HX was performed. It would be useful to compare the calculated heat transfer rate for the test conducted on June 6, 2006 with that calculated on May 22, 2014 by using the procedure discussed herein. Unfortunately, no *dP* test was conducted during the June 6, 2006 test. Therefore, for illustrative purposes, the table below shows a hypothetical comparison between the results of the RBCU 1A test conducted on May 22, 2014 and some future tests, where the extrapolated heat transfer rate is calculated as shown above.

Comparison between Calculated Efficiency Results for RBCU 1A on May 22, 2014 and a Hypothetical Future Test

		RBCU 1A Conducted on May 22, 2014	Future Test	
Mass flow rate, m_c per coil	m_c	79,256	79,200	lbm/hr
Tube-side specific heat	$c_{p,c}$	1.0006	1.0006	Btu/lbm-°F
Service water inlet temperature	t_i	75.39	76.80	°F
Service water outlet temperature	t_o	81.02	80.70	°F
Test heat transfer (cold side)	$q_{test,c}$	4.465E + 05	3.091E + 05	Btu/hr
Hot-side heat transfer area	A_h	2925	2925	ft²
Effective mean temperature difference	*EMTD*	15.63	10.82	°F
Overall test heat transfer coefficient	U_{test}	9.77	9.77	Btu/hr-ft-°F
dP corrected to 2,000 gal/min	*dP*	33.67	35.21	lbf/in²
Percentage of tubes unplugged		88.80%	86.64%	%
Effective surface area	$A_{h\text{-}eff}$	2,597	2,534	ft²
Number of transfer units	*NTU*	0.32	0.312	
Extrapolated LOCA heat transfer rate-uncertainty	$Q^* - u$	133.2	130.0	MBtu/hr

CHAPTER 20

Refer to Section 20.5. Calculate the required MFP turbine steam flow and the MFP turbine condenser condensate outlet temperature. Enthalpies are taken from PEPSE (Figure A20.1).

$$Work\ (\text{kW}) = 0.7457\ \frac{G\ (\text{gpm})\ H\ (\text{ft})\ s.g.}{3955\ \eta_{pump}} = 0.7457\ \frac{(39{,}133)\ (2{,}227)\ (0.853)}{3955\ (0.9)}$$

$$= 15{,}574\ (\text{kW})$$

$$Work\left(\text{BTU}/\text{hr}\right) = 3412.14\left(\frac{\text{Btu}/\text{hr}}{\text{kW}}\right) Work\ (\text{kW}) = 3412.14\ (15{,}574)$$

$$= 53{,}140{,}000\left(\text{Btu}/\text{hr}\right)$$

$$m_{HRH} = \frac{Work\left(\text{Btu}/\text{hr}\right)}{(h_{HRH} - h_{MFPT-exh})\left(\text{Btu}/\text{lbm}\right)}\ \frac{53{,}072{,}000}{1283.5 - 1096.5} = 284{,}100\left(\text{lbm}/\text{hr}\right)$$

$$Duty\left(\text{Btu}/\text{hr}\right) = m_{steam}\left(\text{lbm}/\text{hr}\right)\left(h_{exhaust} - h_{drain}\right)\left(\text{Btu}/\text{lbm}\right)$$

$$= 284{,}100\left(1096.5 - 138.1\right) = 2.72 \times 10^{8}\ \left(\text{Btu}/\text{hr}\right)$$

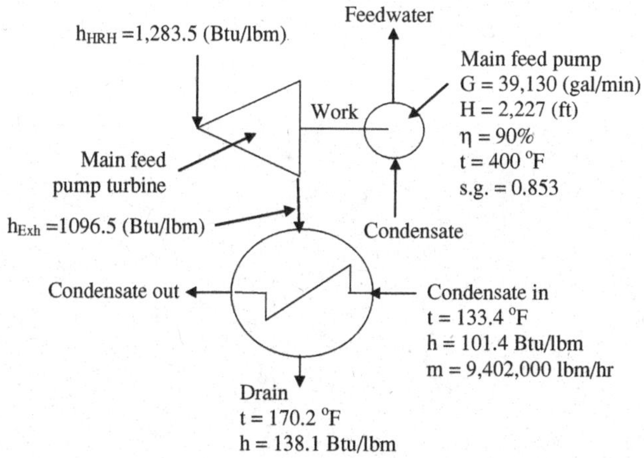

FIGURE A20.1 Main feed pump turbine and condenser.

$$Duty\left(\mathrm{Btu}/_{\mathrm{hr}}\right) = m_{cond}\left(\mathrm{lbm}/_{\mathrm{hr}}\right)\left(h_{cond\text{-}out} - h_{cond\text{-}in}\right)\left(\mathrm{Btu}/_{\mathrm{lbm}}\right)$$

$$h_{cond\text{-}out} = \frac{m_{cond}\ h_{cond\text{-}in} + Duty}{m_{cond}}$$

$$= \frac{(9{,}402{,}000)\ (\mathrm{lbm/hr})\ (101.4)\ (\mathrm{Btu/lbm}) + 2.72 \times 10^{8}\ \left(\mathrm{Btu}/_{\mathrm{hr}}\right)}{9{,}402{,}000\ (\mathrm{lbm/hr})}$$

$$= 130.4\left(\mathrm{Btu}/_{\mathrm{lbm}}\right)$$

$$t_{cond\text{-}out} = 162.4\ ^{\circ}\mathrm{F}$$

REFERENCE

1. Meyer, C. A., et al., *ASME Steam Tables*, 6th ed., American Society of Mechanical Engineers, New York, 1993.

Appendix B
Example Problems in Standard International Units

CHAPTER 1

Refer to Figure 1.1.

In a Rankine cycle, steam leaves the SG and enters a turbine at 4,000 kPaA saturated. The steam expands in an isentropic process to a condenser pressure of 3.0 kPaA. The enthalpy increase across the main feed pump is 1.8 kJ. Determine the thermal efficiency of the cycle. Hint: because the mass flow rate is not given, compute work and added heat on a per pound basis.

$$\eta_{th} = \frac{W}{Q_H} = \frac{w}{q_H} = \frac{(h_4 - h_5) - (h_2 - h_1)}{h_4 - h_2}$$

From the *ASME Steam Tables, Compact Edition*[2]

$$h_4 = 2800.0\left(\text{kJ}/\text{kg}\right);\ s_4 = 6.0685\left(\text{kJ}/\text{kg-K}\right)$$

$$s_5 = s_4 = 6.0685\left(\text{kJ}/\text{kg-K}\right)$$

$$s_1 = s_f \text{ at } 3.0\left(\text{kPaA}\right) = 0.3544\left(\text{kJ}/\text{kg-K}\right)$$

$$x = \frac{s_5 - s_1}{s_{fg}} = \frac{6.0685 - 0.3544}{8.2241} = 0.6948$$

$$h_5 = h_f + xh_{fg}$$

$$h_5 = 101.0 + \left(0.6948\right)2444.7 = 1799.6\left(\text{kJ}/\text{kg}\right)$$

$$w_T = h_4 - h_5 = 2,800 - 1,799 = 1,001\left(\text{kJ}/\text{kg}\right)$$

$$w_P = h_2 - h_1 = 1.8\left(\text{kJ}/\text{kg}\right)$$

$$h_1 = h_f = 101.0\left(\text{kJ}/\text{kg}\right)$$

$$h_2 = h_1 + w_p = 101 + 1.8 = 102.8\left(\text{kJ}/\text{kg}\right)$$

$$q_H = h_4 - h_2 = 2800.0 - 102.8 = 2697.2\left(\text{kJ}/\text{kg}\right)$$

$$\eta_{th} = \frac{w_T - w_p}{q_H} = \frac{1,001 - 1.8}{2697.2} = 37.0\%$$

CHAPTER 2

Refer to Figure 2.3.

The FFW flow to a PWR SG is 1,200 kg/s, at a pressure of 5,000 kPaA and a temperature of 175.0°C. The SG is operating at a pressure of 4,500 kPaA and discharging saturated steam with a quality of 99.98%. The blowdown flow is 2 kg/s. The reactor coolant pumps increase the heat going from the reactor to the SG by 15,000 kJ/s. Determine how much heat the reactor is producing.

From the *ASME Steam Tables, Compact Edition*[2]

$$Q_{in} = m_{MS}\, h_{MS} + m_{SGB}\, h_{SGB} - m_{FFW}\, h_{FFW}$$

$$m_{MS} = m_{FFW} - m_{SGB} = 1{,}200 - 2 = 1{,}198 \left({}^{\text{kg}}\!/\!_{\text{s}} \right)$$

$$h_{MS} = 2797.4\ (\text{kJ/kg})$$

$$h_{SGB} = 1122.1\ (\text{kJ/kg})$$

$$h_{FFW} = 747.0\ (\text{kJ/kg})$$

$$Q_{in} = m_{MS}\, h_{MS} + m_{SGB}\, h_{SGB} - m_{FFW}\, h_{FFW}$$

$$1{,}198\,(2797.4) + 2\,(1122.1) - 1{,}200\,(747.0)$$

$$2.457 \times 10^6 \left({}^{\text{kJ}}\!/\!_{\text{s}} \right)$$

Subtracting the heat added by the reactor coolant pumps yields the reactor power:

$$P_{Reactor} = 2.457 \times 10^6 - 15{,}000 = 2.442 \times 10^6 \left({}^{\text{kJ}}\!/\!_{\text{s}} \right)$$

CHAPTER 3

Refer to Figure 3.4.

The pressure at the HP turbine throttle valve inlet is 6,900 kPaA saturated vapour, and the exhaust pressure is 1180 kPaA. The turbine passes 1,890 kg/s with an efficiency of 80%. There are no steam extractions from the turbine. Calculate the power out of the turbine.

$$W = m_{HP\text{-}in}h_{HP\text{-}in} - m_{HP\text{-}out}\, h_{HP\text{-}out} = m_{HP}\left(h_{HP\text{-}in} - h_{HP\text{-}out}\right)$$

$$h_{HP\text{-}in} - h_{HP\text{-}out} = \eta_{HP}\left(h_{HP\text{-}in} - h_{HP\text{-}out\,(s)}\right)$$

$$W = m_{HP}\,\eta_{HP}\left(h_{HP\text{-}in} - h_{HP\text{-}out\,(s)}\right)$$

From the *ASME Steam Tables, Compact Edition*[2]

$$\text{with } P_{HP\text{-}in(s)} = 6{,}900 \text{ kPaA}$$

$$h_{HP\text{-}in} = 2774.7\left(\text{kJ}/\text{kg}\right);\ s_{HP\text{-}in} = 5.8234\left(\text{kJ}/\text{kg-K}\right)$$

$$s_{HP\text{-}out} = s_{HP\text{-}in} = 5.8234\left(\text{kJ}/\text{kg-K}\right)$$

From the *ASME Steam Tables, Compact Edition*[2],

$$\text{with } P_{HP\text{-}out} = 1{,}180 \text{ kPaA}$$

$$x = \frac{s_{HP\text{-}out} - s_f}{s_{fg}} = \frac{5.8234 - 2.2088}{4.3165} = 0.8374$$

$$h_f = 795.1 \text{ and } h_{fg} = 1987.1\left(\text{kJ}/\text{kg}\right)$$

$$h_{HP\text{-}out\,(s)} = h_f + x h_{fg} = 795.1 + (0.8374)1987.1 = 2459.1\left(\text{kJ}/\text{kg}\right)$$

$$W = m_{HP}\left(\text{lbm}/\text{hr}\right)\eta_{HP}\left(h_{HP\text{-}in} - h_{HP\text{-}out\,(s)}\right)$$

$$= 1890\left(\text{kg}/\text{s}\right)0.80\left(2774.7 - 2459.1\right)\left(\text{kJ}/\text{kg}\right) = 477.000\left(\text{kJ}/\text{s}\right)$$

CHAPTER 4

Refer to Figure 4.10.

Given a single-stage reheater operating with a shell-side mass flow rate of 1285.2 kg/s, inlet and outlet shell pressures of 1,158 and 1,124 kPaA, respectively, and an inlet quality of 99.74%. Reheat steam on the tube side is from main steam with an enthalpy of 2772.6 kJ/kg and a pressure of 6,516 kPaA saturated. Purge steam flow is 2% of the reheat steam. The tube-side drain flow is measured to be 166.9 kg/s, and the drain temperature is 281.1°C. The hot reheat temperature is not available. Calculate the TTD of the reheater.

$$m_{tube\text{-}in} = \frac{m_{tube\text{-}drain}}{1-0.02} = 170.3\left(\mathrm{kg}/\mathrm{s}\right)$$

$$m_{purgest\ steam} = m_{tube\text{-}in} - m_{tube\text{-}drain} = 3.4\left(\mathrm{kg}/\mathrm{s}\right)$$

Because the shell-side inlet pressure is 1158.0 kPaA and the inlet quality is 99.74%, from the *ASME Steam Tables Compact Edition*[2]

$$h_{shell\text{-}in} = 2775.5\ \mathrm{kJ/kg}$$

The tube-side drain is saturated, so from the *ASME Steam Tables Compact Edition*[2]

$$T_{tube\text{-}in} = 281.1°\mathrm{C},\ h_{tube\text{-}drain} = 1242.6\ \mathrm{kJ/kg}$$

From the first law of thermodynamics,

$$m_{\text{shell-in}} h_{\text{shell-in}} + m_{\text{tube-in}} h_{\text{tube-in}} = m_{\text{shell-out}}\ h_{\text{shell-out}} + m_{\text{tube-drain}}\ h_{\text{tube-drain}} + m_{\text{purge steam}}\ h_{\text{tubeside in}}$$

$$h_{\text{shell-out}} = \frac{m_{\text{shell-in}}\ h_{\text{shell-in}} + m_{\text{tube-in}}\ h_{\text{tube-in}} - m_{\text{tube-drain}}\ h_{\text{tube-drain}} - m_{\text{purge steam}}\ h_{\text{tube-in}}}{m_{\text{shell-out}}}$$

$$= 2974.2\ \mathrm{kJ}/\mathrm{kg}$$

Because the shell-side inlet pressure is 1158.0 kPaA and the enthalpy is 2974.2 kJ/kg, from the *ASME Steam Tables Compact Edition*[2] $t_{\text{shell-out}} = 266.4°\mathrm{C}$ and

$$TTD = T_{tube\text{-}in} - t_{shell\text{-}out} = 281.1 - 266.4 = 14.7°\mathrm{C}$$

CHAPTER 5

Refer to Figure 5.5.

The total steam flow entering a moisture removal stage is 1140.3 kg/s with a quality of 95.7%. The pressure downstream of the stage is 110.3 kPaA. Calculate the quality downstream of the moisture removal stage. From Figure 5.6, $\mu = 16.2\%$.

$$m_{ex\text{-}m} = m_{in}\left(1 - x_{in}\right)\mu = 7.9 \text{ kg/s}$$

The moisture is removed at 110.3 kPaA saturated with a quality of 95.7%. From the *ASME Steam Tables Compact Edition*[2]

$$h_{in} = 2583.0 \text{ Btu/lbm}$$

The drain exiting is saturated liquid at 110.3 kPaA, so from the *ASME Steam Tables Compact Edition*[2]

$$h_{ex\text{-}m} = 429.2 \text{ kJ/kg}$$

From the first law of thermodynamics,

$$h_{out} = \frac{m_{in}\ h_{in} - m_{ex\text{-}m}\ h_{ex\text{-}m}}{m_{in} - m_{ex\text{-}m}} = 2598.1 \, \text{kJ}/\text{kg}$$

The steam leaves the stage at a pressure of 110.3 kPaA and an enthalpy of 2598.1 kJ/kg, so from the *ASME Steam Tables Compact Edition*,[2]

$$x_{out} = 96.37\%$$

CHAPTER 6

The following example problem is based on actual plant data from the Bruce Nuclear Station

Condenser Physical Data		Units
Number of tubes, N	43,200	
Tube outside diameter, D_o	0.0254	m
Tube outside diameter, $d_o[D_o/12]$	1.0	in
Tube wall thickness, t	0.0007	m
Tube inside diameter, $D_i[D_o - 2t]$	0.023978	m
Effective straight length of tubes per zone	12.13	m
Surface area, A	41,808	m^2
Tube material	Stainless steel	
Tube material thermal conductivity	15.0	kJ/(s-m-°C)

Condenser Operating Boundary Conditions	Design	Test	Units
Tube-side flowrate, G [test data]	41,739	37,265	kg/s
Tube-side inlet temperature, t_i [test data]	8.33	9.14	°C
Tube-side outlet temperature, t_o [test data]	17.29	19.32	°C
Specific volume, v [steam tables]	0.00090	0.00090	m^3/kg
Tube-side velocity, V $[0.001273 \times (G/N)/D_i^2]$	2.139	1.910	m/s
Tube-side flow area, A_t $[(3.1416/4) \times D_i^2 \times N]$	19.51	19.51	m^2
Tube-side specific heat at inlet temperature, $c_{p\text{-}in}$	4.198	4.196	kJ/(kg-°C)
Condensing duty, $Q[m_{ccw}c_{p\text{-}in}(t_o - t_i)]$	1,569,844	1,591,969	W/(m^2-°C)
Thermal capacity rate at inlet temperature, $C_t[m_{ccw}c_{p\text{-}in}]$	175,206	156,382	m^2-°C/W
Tube-side average temperature, $t_{ave}[(t_o - t_i)/2]$	12.81	14.23	°C
Tube-side specific heat, $c_{p\text{-}ave}$	4.192	4.190	kJ/(kg-°C)
Tube-side thermal conductivity, k_t	0.583	0.586	kJ/(s-m-°C)
Tube-side viscosity, μ_t	0.000132	0.000143	kg/(m-s)

Tube-side Convection Resistance	Design	Test	Units
Mass flux $[G/A_t]$	2,140	1,910	kg/(m^2-s)
Reynolds number, $\text{Re}[m_tD_o/A_t\mu_t]$	41,978	38,897	
Tube-side Prandtl number, $\text{Pr}[\mu_t\, c_{p\text{-}ave}/k_t]$	8.78	8.42	
Fanning friction factor, $f\,[1.58\,\ln\text{Re}_t - 3.28)^{-2}]$	0.00546	0.00555	
Nusselt number, Nu_t (from Petukhov and Kirillov)[a]	311	287	
Calculated tube-side film coefficient, h_{tube} $[k_t\text{Nu}_t/D_i]$	7,570	7,006	W/(m^2-°C)
Calculated tube-side film resistance, r_{tube} $[1/h_{tube}]$	0.000130	0.000141	m^2-°C/W

[a] $$\text{Nu} = \frac{\left(\frac{f}{2}\right)\text{Re}_t\,\text{Pr}_t}{1.07 + 12.7\sqrt{\frac{f}{2}}\left(\text{Pr}_t^{2/3} - 1\right)}$$

Tube Wall Resistance	**Design**	**Test**	**Units**
Tube wall resistance, $r_{wall}[(D_o/2k_w)(\ln(D_o/D_i)]$	0.000049	0.000049	m²-°C/W

Shell-side Convection Resistance	**Design**	**Test**	**Units**
Shell-side pressure, P [design/test data]	3.189	3.792	kPa
Shell-side temperature, t_{sat} [steam tables]	27.50	29.92	°C
Tube-side outlet temperature, t_o [design/test data]	17.29	19.32	°C
TTD $[t_{sat} - t_o]$	10.21	10.60	°C
Condensing LMTD $[(t_o - t_i)/\ln\{(t_{sat} - t_i)/(t_{sat} - t_o)\}]$	14.22	15.12	°C

Addative Resistance Method	**Design**	**Test**	**Units**
Shell-side condensate temperature, t_f $[(t_{sat} + t_i)/2]$	17.92	19.53	
Shell-side condensate thermal conductivity, k_f	0.5928	0.5957	W/(m-°C)
Shell-side condensate film density, ρ_f	0.9985	0.9981	kg/m³
Shell-side latent heat of steam, h_{fg} [steam tables]	2,347	2,350	kJ/kg
Shell-side condensate film viscosity, μ_f	0.00107	0.00103	kg/(m-s)
$t_{sat} - t_i$	19.17	20.78	°C
Nusselt parameter, Nu_{shell}' [a]	12.41	12.33	
Nusselt correction, $\mathrm{Nu}_{correct}[\mathrm{Nu}_{shell\text{-}design}'/\mathrm{Nu}_{shell\text{-}test}'$		1.006	
Calculated shell-side film coefficient, h_{shell} [b]	5,844	5,807	W/(m²-°C)
Calculated shell-side film resistance, r_{shell} [c]	0.000171	0.000172	m²-°C/W
Condensing heat trans. rate, clean, U_{clean} [d]	2,779	2,687	W/(m²-°C)
Condensing heat trans. rate, operating, $U_{operating}$ [e]	2,640	2,518	W/(m²-°C)
Condenser performance factor by ARM, PF [f]	95%	94%	
Tube-side fouling resistance [g]	0.000019	0.000025	m²-°C/W

[a] $$\mathrm{Nu}_{shell}' = \left[\frac{k_f^3\ \rho_f^2\ g\ h_{fg}}{\mu_f\left(t_{sat} - t_{CCWin}\right) d_o}\right]^{1/4}$$

[b] $$h_{shell\text{-}design} = \frac{1}{r_{shell\text{-}design}}$$

$$r_{shell\text{-}design} = \frac{1}{U_{clean\text{-}design}} - r_{wall\text{-}design} - \left(\frac{d_o}{d_i}\right) r_{tube\text{-}design}$$

[c] $$h_{shell\text{-}test} = \frac{h_{shell\text{-}design}}{\mathrm{Nu}_{correct}'}$$

$$r_{shell\text{-}test} = \frac{1}{h_{shell\text{-}test}}$$

(Continued)

[d] $U_{clean\text{-}design} = \dfrac{U_{operating}}{PF_{design}}$ $U_{clean\text{-}test} = \dfrac{1}{r_{shell\text{-}test} + r_{wall} + \left(\dfrac{d_o}{d_i}\right) r_{tube\text{-}test}}$

[e] $U = \dfrac{Q}{A\,LMTD}$

[f] $PF = \dfrac{U_{operating}}{U_{clean}}$

[g] $r_{fouling} = \dfrac{1}{U_{operating}} - \dfrac{1}{U_{clean}}$

Heat Exchange Institute Method	**Design**	**Test**	**Units**
HEI constant, C	263	263	
Material correction factor –22 BWG Titanium, F_m	0.870	0.870	
Tube-side inlet temperature	46.99	48.45	°F
Tube-side velocity	7.017	6.265	ft/s
Condensing heat trans. rate, service	465.0	443.4	Btu/(hr-ft²-°F)
HEI temperature correction factor, F_t	0.804	0.819	
Condenser performance factor by HEI, PF[a]	95%	95%	

[a] $PF = \dfrac{U}{C\sqrt{V}\,F_t\,F_m}$

CHAPTER 7

Refer to Figure 7.4.

An FWH is operating at a shell pressure of 285 kPaA. 1207.1 kg/s of condensate passing through the tubes is heated from 99°C to 129°C by extraction steam having an enthalpy of 2736.5 kJ/kg and a drain flow of 50.4 kg/s coming from a higher pressure FWH at a temperature of 128°C. Approximately 0.5% of the extraction steam is vented to the main condenser. The temperature of the drain coming out of the drain cooler is 105°C. Calculate the extraction steam flow rate.

The enthalpy of drains and condensate (i.e., FWH in the cycle below the main feed pump) is taken as saturated liquid because the amount of subcooling is insignificant. Therefore, the enthalpy of the condensate and drains entering and exiting the FWH are as follows from the *ASME Steam Tables Compact Edition.*[2]

$$h_{condensate\text{-}i} = 414.9 \text{ kJ/kg}$$

$$h_{condensate\text{-}o} = 542.0 \text{ kJ/kg}$$

$$h_{drains\text{-}i} = 537.8 \text{ kJ/kg}$$

$$h_{drains\text{-}o} = 440.2 \text{ kJ/kg}$$

The enthalpy of the vent steam corresponds to saturated vapor at the FWH operating pressure. Therefore, from the *ASME Steam Tables Compact Edition*[2]

$$h_{vent} = 2722.3 \text{ kJ/kg}$$

Therefore, from the first law of thermodynamics,

$$m_{drain\text{-}i}h_{drain\text{-}i} + m_{ex}h_{ex} + m_{condensate}h_{condensate\text{-}i} = m_{drain\text{-}o}h_{drain\text{-}o} + m_{vent}h_{vent} + m_{condensate}\,h_{condensate\text{-}o}$$

$$m_{drain\text{-}o} = m_{drain\text{-}i} + m_{ex} - m_{vent} \qquad m_{vent} = 0.005\, m_{ex}$$

$$m_{drain\text{-}o} = m_{drain\text{-}i} + m_{ex}\left(1 - 0.005\right)$$

$$m_{ex}\left(h_{ex} - 0.995\, h_{drain\text{-}o} - 0.005\, h_{vent}\right) = m_{condensate}\left(h_{condensate\text{-}o} - h_{condensate\text{-}i}\right) - m_{drain\text{-}i}\left(h_{drain\text{-}i} - h_{drain\text{-}o}\right)$$

$$m_{ex} = \frac{m_{condensate}\left(h_{condensate\text{-}o} - h_{condensate\text{-}i}\right) - m_{drain\text{-}i}\left(h_{drain\text{-}i} - h_{drain\text{-}o}\right)}{h_{ex} - 0.995\, h_{drain\text{-}o} - 0.005\, h_{vent}}$$

$$m_{ex} = \frac{1207.1\,(542.0 - 414.9) - 50.4\,(537.8 - 440.2)}{2736.5 - 0.995\,(440.2) - 0.005\,(2722.3)} = 65 \text{ kg/s}$$

CHAPTER 8

In 1983, an acceptance test was conducted on the Pickering Nuclear Station Unit 5. In 1999, a surveillance test was conducted to identify any loss in thermal performance in the unit relative to the results of the acceptance test. A HB computer program was developed to model the thermal performance at design conditions. As part of that assessment, a mass balance was calculated using only the FFW flow rate, the extraction enthalpies from the HB program, and the temperatures measured in the condensate and feedwater systems. Table B8.1 shows a comparison among the acceptance test results and the HB program and the acceptance test results and the subsequent surveillance test. The extraction flows shown were calculated as described in Chapter 7. The boundary conditions for the HB program (main condenser back-pressure, etc.) were adjusted to reflect the actual test condition.

TABLE B8.1
Pickering Generating Station Mass Balance Comparison

	1983 Acceptance Test	1999 Surveillance Test	Units
FFW flow rate	762.844	769.243	kg/s
Target value from HB program	765.294	777.467	kg/s
Difference	−0.32%	−1.06%	
Extraction flow to No. 6 FWH	45.308	42.451	kg/s
Target value from HB program	45.478	46.150	kg/s
Difference	−0.37%	−8.02%	
H. P. exhaust flow	717.537	724.777	kg/s
Target value from HB program	719.816	729.303	kg/s
Difference	−0.32%	−0.62%	
Extraction flow to No. 5 FWH	100.156	107.350	kg/s
Target value from HB program	108.490	107.350	kg/s
Difference	−7.68%	0.00%	
L. P. turbine inlet flow	617.381	617.427	kg/s
Target value from HB program	611.326	621.953	kg/s
Difference	0.99%	−0.73%	
Extraction flow to deaerator	21.580	22.491	kg/s
Target value from HB program	22.138	22.916	kg/s
Difference	−2.52%	−1.86%	
Extraction flow to No. 3 FWH	23.233	28.089	kg/s
Target value from HB program	22.306	24.418	kg/s
Difference	4.15%	15.03%	
Extraction flow to No. 2 FWH	24.392	20.727	kg/s
Target value from HB program	23.594	22.868	kg/s
Difference	3.38%	−9.36%	
Extraction flow to No. 1 FWH	36.476	31.603	kg/s

(*Continued*)

TABLE B8.1 (*Continued*)
Pickering Generating Station Mass Balance Comparison

	1983 Acceptance Test	1999 Surveillance Test	Units
Target value from HB program	34.490	40.780	kg/s
Difference	5.76%	–22.50%	
Extraction flow to L. P. FWH	84.101	80.419	kg/s
Target value from HB program	80.390	88.066	kg/s
Difference	4.62%	–8.68%	
L. P. turbine exhaust flow	511.700	514.517	kg/s
Target value from HB program	508.798	510.971	kg/s
Difference	0.57%	0.69%	
Calculated condensate flow[a]	595.801	594.937	kg/s
Target value from HB program	589.188	599.037	kg/s
Difference	1.12%	–0.68%	

[a] Leakage flows were neglected.

CHAPTER 9

A large nuclear station with an NDCT experiences evaporative and drift losses from the CCW of 630 and 25 L/s, respectively. Assuming a desired cycle of concentration of 2.2 to avoid scaling or corrosion, determine the required makeup flow rate. From Section 9.5, where

C = cycles of concentration
M = makeup
B = blowdown
E = evaporation
D = drift

and

$C = 2.2$
$E = 630$ L/s
$D = 25$ L/s.

Therefore,

$$B = \frac{(E + D)}{(C - 1)} = \frac{(630 + 25)}{(2.2 - 1)} = 546 \text{ L/s}$$

and

$$M = C \times B = 2.2(546) = 1201 \text{ L/s}$$

CHAPTER 10

A cooling tower fan is designed to draw 34,830 (m^3/min) of air through fill material with a water loading of 65.1 (m^3/min) while drawing 127 kW. The actual test results indicate a water loading of 65.3 (m^3/min) while drawing 137 kW. Assuming a density of water and air of 1,000 (kg/m^3) and 1.225 (kg/m^3), respectively, determine the design and test *L*/*G* ratio.

$$\left(\frac{L}{G}\right)_{design} = \frac{L_{design}\left(\frac{m^3}{min}\right) \times \rho_{L\text{-}design}\left(\frac{kg}{m^3}\right)}{G_{design}\left(\frac{m^3}{min}\right) \times \rho_{g\text{-}design}\left(\frac{kg}{m^3}\right)}$$

$$= \frac{65.1\left(\frac{m^3}{min}\right) \times 1,000\left(\frac{kg}{m^3}\right)}{34,830\left(\frac{m^3}{min}\right) \times 0.1225\left(\frac{kg}{m^3}\right)} = 1.52$$

$$G_{test} = G_{design}\left(\frac{kW_{test}}{kW_{design}}\right)^{1/3} = 34,830\left(\frac{137}{127}\right)^{1/3} = 35,720\left(\frac{m^3}{min}\right)$$

$$\left(\frac{L}{G}\right)_{test} = \frac{L_{test}\left(\frac{m^3}{min}\right) \times \rho_{L\text{-}test}\left(\frac{kg}{m^3}\right)}{G_{test}\left(\frac{m^3}{min}\right) \times \rho_{g\text{-}test}\left(\frac{kg}{m^3}\right)} = \frac{65.3\left(\frac{m^3}{min}\right) \times 1,000\left(\frac{kg}{m^3}\right)}{35,720\left(\frac{m^3}{min}\right) \times 0.1225\left(\frac{kg}{m^3}\right)} = 1.49$$

CHAPTER 11

One advantage of cooling lakes over cooling towers is that they require lower head CCW pumps. Consider the following comparison between a nuclear power station that requires 2,000 m^3/min of CCW flow. The station situated on a cooling lake may require only 100 kPa of pump head to deliver the CCW through the plant, while the station with cooling towers could require 270 kPa due to the additional CCW piping and the static head required to deliver the CCW to the cooling tower. In both cases, the CCW pump and motor efficiencies are the same at 80% and 95%, respectively. Determine how much additional auxiliary power would be required.

$$Power_{lake} = 0.01667 \frac{2,000\left(m^3/_{min}\right) \times 100\left(kPa\right)}{0.80 \times 0.95} = 4,387\left(kW\right)$$

$$Power_{cooling\ tower} = 0.01667 \frac{2,000\left(m^3/_{min}\right) \times 270\left(kPa\right)}{0.80 \times 0.95} = 11,846\left(kW\right)$$

So, the nuclear power station that is located on a cooling lake has a net electrical output of 7,459 kW greater than the station not located on a cooling lake. If MDCTs are employed, they could easily require an additional 10,000 kW in fan power.

CHAPTER 12

Refer to Figures 12.6 and 12.7.

Calculate the reduction in the CCW temperature entering the main condenser if an FBSP sprays a portion of the water prior to entering the plant and the power required to spray the water.

In 2005, a study was conducted to consider alternatives to reduce the temperature of the CCW entering the main condenser at the Comanche Nuclear Power Plant near Glen Rose, Texas. Among the alternative considered was to spray approximately 25% of the 8327 m^3/min of CCW that was circulated through the plant and the SCR (see Chapter 11).

Assuming that flatbed sprays located 1.524 m above the SCR and similar to those utilized at Rancho Seco that each spray 0.18925 m^3/min at a nozzle pressure of 48.3 kPa was used, the following number of spray nozzles would be required.

$$N_{nozzles} = \frac{0.25 \times 8,327}{0.18925} = 11,000$$

Assuming each nozzle requires a space of 1.524 × 1.524 m or 2.323 m^2, the spray field would require approximately 2.55 hectares. Such a large spray area would require the maximum interference factor of 0.6, as shown in Figure 12.7.

The *LWBT* may be calculated as follows:

$$t_{WB\text{-}local} = t_{WB\text{-}amb} + f\left(t_1 - t_{WB\text{-}amb}\right)$$

where:

$t_{WB\text{-}local}$ = *WBT* at the spray nozzle
$t_{WB\text{-}amb}$ = *WBT* outside the spray region
t_1 = temperature of the CCW entering the spray nozzle
f = interference factor.

Refer to the Chapter 11 example in Appendix A. Assuming an ambient *WBT* of 10°C and a CCW temperature of 27.2°C entering the spray nozzle from SCR, the *LWBT* at the spray nozzle is as follows:

$$t_{WB\text{-}local} = 10 + 0.6\left(27.2 - 10\right) = 20.0°\text{C}$$

From Figure 12.6, assume a spray efficiency of 50% at a wind speed of 6.7 m/s. Calculate the *CWT* from the sprays as follows:

$$\eta = \frac{t_1 - t_2}{t_1 - t_{WB\text{-}local}} \Rightarrow t_2 = t_1 - \eta\left(t_1 - t_{WB\text{-}local}\right) = 23.6°\text{C}$$

where:

η = spray efficiency of one spray nozzle
t_2 = *CWT* of the sprayed water.

Because only 25% of the 8,327 m^3/min of CCW is sprayed, the temperature of the CCW entering the main condenser is calculated as follows:

$$t_{CCW\text{-}in} = \frac{0.75 \times G_{CCW} \times t_1 + 0.25 \times G_{CCW} \times t_2}{G_{CCW}} = 26.4°\text{C}$$

Calculate the power required to spray the water as follows:
The required pump flow is as follows:

$$G_{pump} = 0.25 \times 8{,}327\ \text{m}^3/\text{min} = 2{,}082\ \text{m}^3/\text{min}$$

The required pump head referenced to the lake level and neglecting piping friction losses where the pump and motor efficiencies are 80% and 95%, respectively, is as follows:

$$H_{sprays} = H_{static} + H_{nozzle} = 15\ \text{kPa} + 48.3\ \text{kPa} = 63.2\ \text{kPa}$$

$$Power_{spray} = 0.01667 \frac{G_{sprays} \times H_{sprays}}{\eta_{pump}\ \eta_{motor}} = 2{,}886\ \text{kW}$$

CHAPTER 13

Refer to Figure 13.14.

Calculate the number of spray trees identical to those used at the CGS that would be required to achieve the same reduction in the temperature of the CCW entering the main condenser as in the Chapter 12 example and the power required to spray the water.

Continuing with the Chapter 12 example, in 2005, a study was conducted to consider alternatives to reduce the temperature of the CCW entering the main condenser at the Comanche Nuclear Power Plant near Glen Rose, Texas. Among the alternatives considered was to spray a portion of the 8,327 m³/min of CCW that was circulated through the plant and the SCR (see Chapter 11).

Assume an OSCS with spray trees identical to those described in Section 13.1 with seven 3.81-cm Spraying Systems Whirljet nozzles on each tree delivering a total of 3251.23 m³/min. From Figure 13.15, the efficiency of the OSCS at a wind speed of 6.7 m/s is 62%, and due to the design of the OSCS, there is no interference to the ambient air reaching the nozzles. The operating pressure of the spray trees referenced to the SCR surface is 142.7 kPa. Assuming an ambient WBT of 10°C and a CCW temperature of 27.2°C entering the spray nozzle from SCR, the *CWT* of the CCW leaving the OSCS may be calculated as follows:

$$\eta = \frac{t_1 - t_2}{t_1 - t_{WB\text{-}amb}} \Rightarrow t_2 = t_1 - \eta\left(t_1 - t_{WB\text{-}amb}\right) = 16.5\,°\text{C}$$

where:

η = spray efficiency of the OSCS
t_1 = temperature of the CCW entering the spray nozzle
t_2 = cold-water temperature of the sprayed water
$t_{WB\text{-}amb}$ = *WBT* outside the spray region.

The percentage of the CCW that must be sprayed in order to achieve a reduction of the CCW entering the main condenser from 27.2°C to 26.4°C may be calculated as follows:

$$G_{CCW}\, t_{CCW\text{-}in} = \left(1 - X\right) G_{CCW}\, t_1 + X\, G_{CCW}\, t_2$$

$$t_{CCW\text{-}in} = t_1 + X\left(t_2 - t_1\right)$$

$$X = \frac{t_{CCW\text{-}in} - t_1}{t_2 - t_1} = 7.8\%$$

where:

X = percentage of CCW sprayed.

The following number of spray trees would be required:

$$N_{nozzles} = \frac{0.078 \times 8{,}327}{1.24} = 528$$

Calculate the power required to spray the water as follows:

The required pump flow is as follows:

$$G_{pump} = 0.078 \times 8{,}327 = 650 \text{ m}^3/\text{min}$$

The required pump head referenced to the lake level, neglecting piping friction losses where the pump and motor efficiencies are 80% and 95%, respectively, is as follows:

$$H_{sprays} = 142.7 \text{ kPa}$$

$$Power_{spray} = 0.01667 \frac{G_{sprays} \times H_{sprays}}{\eta_{pump}\, \eta_{motor}} = 2{,}034 \text{ kW}$$

CHAPTER 14

Determine the reduction in *CWT* achievable by converting a counter-flow NDCT to an OSACT.

The main condensers at the Watts Bar Nuclear Plant in Tennessee are of the single pass, multi-pressure, multi-shell main condenser sign with three pressure zones matching the three LP turbine exhausts (see Section 6.2). During periods of high *WBT*, the staff at Unit 1 has experienced shell pressures in the highest-pressure zone exceeding the manufacturer's limit on LP turbine backpressure. In one instance this resulted in a turbine blade failure resulting in an extended outage (see Section 5.3). In 2005, a study was conducted to investigate ways to reduce the CCW temperature entering the main condenser. Alternatives considered included improving the distribution of CCW over the fill material and adding fill material around the perimeter of counter-flow NDCT, adding an MDCT on parallel with the NDCT, and converting the NDCT to an OSACT.

The following table shows a comparison between the critical performance parameters with a *HWT* of 53.3°F and an ambient *WBT* of 26.7°F:

Parameter	**w/o OSA**		**w/OSA**	
Dry air flow	19,651	kg/s	22,070	kg/s
Average water loading	2.536	kg/s-m^2	2.211	kg/s-m^2
Liquid/gas ratio	1.35	dimensionless	1.05	dimensionless
Spray *CWT*			35.56	°C
Basin *CWT*			32.33	°C
Tower *CWT*	33.22	°C	32.72	°C

As one may see, although the *CWT* coming off the sprays is higher than that from the NDCT without the OSA, the increase in the airflow coupled with the reduction in the water loading on the NDCT with the OSA dramatically reduces the *L/G* ratio and thus increases the value of *KaV/L* (see Figure 10.11). The result is a mixed *CWT* that is lower than the NDCT without the OSA.

Figure B14.1 below illustrates the reduction in *CWT* that may be achieved by converting the Watts Bar NDCT to an OSACT.

Applying the equations in the figure, one may determine the annual average reduction in the *CWT* by knowing the monthly average *WBT* as follows:

	Jan	**Feb**	**Mar**	**Apr**	**May**	**Jun**	**Jul**	**Aug**	**Sep**	**Oct**	**Nov**	**Dec**	**Ave**
WBT,°C	1.7	3.3	5.9	11.0	15.7	19.5	20.9	20.6	17.5	11.6	6.1	2.6	
CWT w/o OSA	19.5	20.3	21.6	24.2	26.8	29.0	29.8	29.6	27.8	24.6	21.7	19.9	
CWT w/ OSA	19.1	19.9	21.1	23.7	26.2	28.4	29.2	29.0	27.2	24.0	21.2	19.5	
Difference	0.4	0.4	0.5	0.6	0.6	0.6	0.6	0.6	0.6	0.6	0.5	0.4	0.5

Note that the improvement in *CWT* is the greatest during the periods of the highest *WBT*.

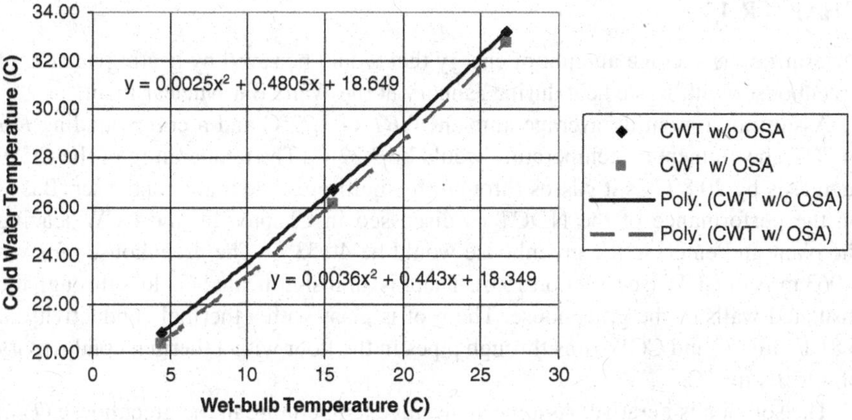

FIGURE B14.1 Reduction in *CWT* achievable by converting the Watts Bar NDCT to an OSACT.

It should be noted that Watts Bar did not modify their NDCT to become an OSACT but instead improved the distribution of CCW over the fill material and added fill material around the perimeter of the NDCT and subsequently procured an LP turbine spindle with ruggedized last stage blades that could withstand higher backpressures.

CHAPTER 15

Determine the average amount of energy that would be saved by heating an acre of greenhouses with waste heat during January at the Watts Bar Nuclear Plant.

Assuming a monthly average ambient *WBT* of 1.72°C and a corresponding *RH* of 76%, the ambient air temperature would be 3.22°C. The temperature of the CCW increases by 20.8°C as it passes through the multi-pressure main condenser. Based on the performance of the NDCT as discussed in Chapter 14, the CCW leaving the plant and entering the greenhouse would be 40.33°C. The greenhouse receives 0.063 m^3/s of CCW (see Section 15.6). There is an insignificant heat loss through the insulated walls of the greenhouse. The roof is glass with a thermal conductivity of 6.81 J/s-m^2-°C and CCW runs through pipes in the floor with a thermal conductivity of 4.15 J/s-m^2-°C.

The solution is iterative. Assume an average temperature in the greenhouse (T_{GH}) and calculate the heat loss through the roof and the heat added by the floor until the two values are equal as follows:

$$\text{Assume } T_{GH} = 17.0°\text{C} \quad \text{(by iteration)}$$

$$Q_{glass} = U_{glass}\,(T_{growing} - t_{amb})\,A = 379{,}900 \text{ J/s-acre}$$

$$\Delta t_{ccw} = \frac{Q_{glass}}{\rho_{ccw} G_{ccw}} = 1.44°\text{C}$$

$$t_{ccw\text{-}out} = t_{CCW\text{-}in} - \Delta t_{ccw} = 38.9°\text{C}$$

$$t_{ccw\text{-}ave} = \frac{t_{ccw\text{-}out} + t_{CCW\text{-}in}}{2} = 39.6°\text{C}$$

$$Q_{glass} = U_{floor}\,(T_{GH} - t_{ccw\text{-}ave})\,A = 379{,}300 \text{ J/s-acre}$$

Therefore, each acre of greenhouse heated by waste heat in January would save 379,600 J/s/acre of energy that would otherwise have been derived from oil or natural gas.

CHAPTER 16

Calculate the heat balance error and the uncertainty of the heat balance error.

In 2005, the staff at the Callaway Nuclear Generating Station in Missouri conducted a heat transfer test on their Train A component cooling system (CCS) HX. The Callaway CCS HX is a shell-and-tube HX similar to a TEMA Type R design with two passes on the tube side furnished by Struthers Wells Corporation. The table below is a listing of the physical data for each bundle.

CCS HX Physical Data

Total number of tubes	N_{tubes}	4,464	–
Percentage of tubes plugged	%	0.0	
Total number of active tubes	$N_{active} = N_{tubes}(1\text{-}\%)$	4,464	–
Number of passes	*Passes*	2	–
Effective tube length	L_{eff}	11.10	m
Outside tube diameter	d_o	0.0191	m
Tube wall thickness	t	0.0012	m
Inside tube diameter	$d_i = d_o - 2t$	0.0166	m
Total effective hot-side area	$A_h = \pi d_o L_{eff} N_t$	2,965	m^2
Tube-side area	$A_c = (d_i/d_o)A_h$	2,578	m^2
Tube wall area	$A_w = (A_h - A_c)/\ln(A_h/A_c)$	2,767	m^2
Tube material		90-10	–
Hot-side fouling resistance	$r_{f,h}$	0.00009	m^2-°C/W
Tube-side fouling resistance	$r_{f,c}$	0.00035	m^2-°C/W
Ratio of hot-side to tube side area	A_h/A_c	1.15	
Tube material thermal conductivity	k_t	147.63	W/m^2-°C
Tube wall resistance	$r_w = (d_o - d_i)/2k_t$	0.00001	m^2-°C)/W

The following table is a listing of the design basis conditions for operating in the normal full power operation with maximum cool-down mode. The properties of the hot-side and cold-side fluids are based on the average temperatures on each side of the HX except for the density that is based on the inlet and outlet temperatures for the cold- and hot-side fluids, respectively.

Design Basis Conditions for Normal Full Power Operation with Maximum Letdown

Heat transfer rate	Q	22,619	kJ/s
Entering hot-side temperature	$T_{h,i}$	48.7	°C
Exiting hot-side temperature	$T_{h,o}$	40.6	°C
Average hot-side temperature	T_h	44.6	°C
Hot-side thermal conductivity	k_h	0.634	kJ/(s-m-°C)
Hot-side absolute viscosity	μ_h	0.00071	kg/(m-s)
Hot-side specific heat	$c_{p,h}$	4.176	kJ/(kg-°C)
Hot-side density	ρ_h	0.991	kg/m^3

(*Continued*)

Design Basis Conditions for Normal Full Power Operation with Maximum Letdown			
Entering water temperature	$t_{c,i}$	35.0	°C
Exiting water temperature	$t_{c,o}$	41.3	°C
Average water temperature	t_c	38.2	°C
Tube-side thermal conductivity	k_c	0.625	kJ/(s-m-°C)
Tube-side absolute viscosity	μ_c	0.00073	kg/(m-s)
Tube-side specific heat	$c_{p,c}$	4.176	kJ/(kg-°C)
Tube-side density	ρ_c	0.993	kg/m^3
Water temperature rise	dT_c	6.3	°C
LMTD correction factor	F	0.9455	–
Tube-side flow rate	m_c	856	kg/s
Hot-side flow rate	m_h	663	kg/s

The overall heat transfer coefficient at design conditions, U_{design}, is calculated for the normal mode of operation in the following table:

Verification of U_{design} for the Normal Mode of Operation			
Greater terminal temperature difference	$\Delta t_1 = T_{h,i} - t_{c,o}$	7.39	°C
Lesser terminal temperature difference	$\Delta t_2 = T_{h,o} - t_{c,i}$	5.56	°C
Log mean temperature difference	$LMTD = (\Delta t_1 - \Delta t_2)/\ln(\Delta t_1/\Delta t_2)$	6.43	°C
Effective mean temperature difference	$EMTD = F(LMTD)$	6.08	°C
Design overall heat transfer coefficient	$U_{design} = Q/(A_h EMTD)$	1,255	kJ/(s-m^2-°C)

The tube-side convection coefficient may be computed from the Petukhov correlation as shown in the following table:

Tube-side Convection Coefficient			
Mass flow rate/tube	$m_t = Passes(m_c/N_t)$	0.38	kg/s
Volumetric flowrate/tube	$V_t = m_t/\rho_\tau$	0.39	M^3/s
Tube area	$a_t = (\pi/4)d_i^2$	0.00022	M^2
Tube velocity	$v_t = V_t/a_t$	1,792	m/s
Prandtl number	$\mathrm{Pr}_c = c_{p,c}\,\mu_c/k_c$	4.88	–
Reynolds number	$\mathrm{Re}_c = \rho v_t d_i/\mu_c$	40,374	–
Fanning friction factor	$f = (1.58 \ln \mathrm{Re}_c - 3.28)^{-2}$	0.00551	–
Nusselt number	$\mathrm{Nu} = ((f/2)\mathrm{Re}_c\,\mathrm{Pr}_c)/(1.07 + 12.7(f/2))^{1/2}(\mathrm{Pr}_c^{2/3}-1))$	233.55	–
Tube-side film coefficient	$h_c = \mathrm{Nu}(k_c/D_i)$	8,814	kJ/(s-m^2-°C)

The hot-side convection coefficient may be computed as follows:

$$h_{h,\,design} = \frac{1}{\dfrac{1}{U_{design}} - r_{f,h} - \left(\dfrac{A_h}{A_w}\right) r_w - \left(\dfrac{A_h}{A_c}\right) r_{f,c} - \left(\dfrac{A_h}{A_c}\right)\dfrac{1}{h_c}} = 6{,}096\ \mathrm{kJ/(s\text{-}m^2\text{-}^\circ C)}$$

The table below shows the results of the CCS test.

CCS Test Results			
Required hot-side flow	g_h	800	L/s
Hot-side inlet temperature	$T_{h,i}$	33.8	°C
Hot-side outlet temperature	$T_{h,o}$	25.6	°C
Average hot-side temperature	T_h	*29.7*	°C
Hot-side thermal conductivity	k_h	0.613	kJ/(s-m-°C)
Hot-side absolute viscosity	μ_h	0.00082	kg/(m-s)
Hot-side specific heat	$c_{p,h}$	4.178	kJ/(kg-°C)
Hot-side density	ρ_h	0.994	kg/m³
Hot-side flow rate	$m_h = g_h\rho_h(60/7.48)$	795	kg/s
Water flow rate	g_t	558	L/s
Tube-side inlet temperature	$t_{c,i}$	19.3	°C
Tube-side outlet temperature	$t_{c,o}$	29.7	°C
Average water temperature	t_c	*24.5*	°C
Tube-side thermal conductivity	k_c	0.604	kJ/(s-m-°C)
Tube-side absolute viscosity	μ_c	0.00092	kg/(m-s)
Tube-side specific heat	$c_{p,c}$	4.181	kJ/(kg-°C)
Tube-side density	ρ_t	0.998	kg/m³
Tube-side flow rate	$m_c = g_t\rho_t(60/7.48)$	557	kg/s
Greater terminal temperature difference	$\Delta t_1 = T_{h,i} - t_{c,i}$	4.07	°C
Lesser terminal temperature difference	$\Delta t_2 = T_{h,o} - t_{c,o}$	6.31	°C
Log mean temperature difference	$LMTD = (\Delta t_1 - \Delta t_2)/\ln(\Delta t_1/\Delta t_2)$	5.11	°C
Ratio of temperature change	$R = (T_{h,i} - T_{h,o})/(t_{c,o} - t_{c,i})$	0.78	
Effectiveness	$P = (t_{c,o} - t_{c,i})/(T_{h,i}\ t_{c,o})$	0.72	
LMTD correction factor	F	0.840	
LMTD (corrected)	$EMTD = F(LMTD)$	4.29	°C
Test heat transfer (hot-side)	$Q_{test,\ h} = m_h c_p(T_{h,i} - T_{h,o})$	27,225	kJ/s
Test heat transfer (tube side)	$Q_{test,\ c} = m_c c_p(t_{c,o} - t_{c,i})$	24,321	kJ/s
Average test heat balance	$Q_{test} = (Q_{test,h} + Q_{test,c})/2$	25,773	kJ/s
Test overall heat transfer coefficient	$U_{test} = dQ_{test}/A_h(EMTD)$	1,912	kJ/(s-m²-°C)
Mass flow rate/tube	$m_t = m_c/PassesN_t$	0.25	kg/s
Volumetric flowrate/tube	$V_t = m_c/\rho_t$	0.25	m³/s
Tube velocity	$v_t = V_t/a_t$	1,161	m/s
Prandtl number	$\mathrm{Pr}_c = c_{p,c}\mu_c/k_c$	6.33	
Reynolds number	$\mathrm{Re}_c = \rho v_t d_i/\mu_c$	20,975	
Fanning friction factor	$f = (1.58 \ln \mathrm{Re}_c - 3.28)^{-2}$	0.00646	
Nusselt number	Nu[a]	152.11	
Tube-side film coefficient	$h_c = \mathrm{Nu}(k_c/d_i)$	5,549	kJ/(s-m²-°C)
Hot-side flow rate	$m_h = m_c(c_{p,c}/c_{p,h})(t_{c,o} - t_{c,i})/(T_{h,i} - T_{h,o})$	710	L/s
Hot-side film coefficient	$h_{h,test}$[b]	5,968	kJ/(s-m²-°C)
Resistances referenced to hot-side area			
Total resistance	$r = 1/U_{test}$	0.000523	(m²-°C)/W
Hot-side convection resistance	$r_h = 1/(\eta_h\ h_h)$	0.000168	(m²-°C)/W

(Continued)

CCS Test Results			
Wall resistance	$r_w = (A_h/A_w)(d_o - d_i)/(2k_t)$	0.000009	(m²-°C)/W
Cold-side convection resistance	$r_c = (A_h/A_c)h_c$	0.000207	(m²-°C)/W
Fouling referenced to hot-side	$r_{f,h,test}{}^{c}$	0.000139	(m²-°C)/W
Tube-side fouling resistance	$r_{f,c} = (A_c/A_h)r_{f,h}$	0.000044	(m²-°C)/W

a $\mathrm{Nu} = \left(\left(f/2\right)\mathrm{Re}_c\mathrm{Pr}_c\right) / \left(1.07 + 12.7\left(f/2\right)^{1/2}\left(\mathrm{Pr}_c{}^{2/3} - 1\right)\right)$

b $$h_{h,test} = \frac{\left[\left(\frac{m_h}{\mu_h}\right)^{0.6}\left(\frac{\mu_h c_{p,h}}{k_h}\right)^{0.333} k_h\right]_{test}}{\left[\left(\frac{m_h}{\mu_h}\right)^{0.6}\left(\frac{\mu_h c_{p,h}}{k_h}\right)^{0.333} k_h\right]_{design}} h_{h,design}$$

c $$r_{f,h,test} = \frac{1}{\frac{1}{U_{test}} - \left(\frac{1}{h_{h,test}}\right) - \left(\frac{A_h}{A_w}\right) r_w - \left(\frac{A_h}{A_c}\right)\frac{1}{h_{c,test}}}$$

Therefore, the total fouling resistance as measured by the test is only approximately 20% of the design value.

Uncertainty Analysis

Refer to Section 16.9.

The following is the frequency and the number of each temperature measuring instrument with the associated Student's "t" for each (see Table 16.1).

Number of temperature measurements	N_t	31
Temperature measurement, Student's "t"	t	2
Number of SW inlet RTDs	J_t	4
No. of sensors, Student's "t"	t	3.182
Number of SW outlet RTDs	J_t	8
No. of sensors, Student's "t"	t	2.306
Number of CCS inlet RTDs	J_t	4
No. of sensors, Student's "t"	t	3.182
Number of CCS outlet RTDs	J_t	8
No. of sensors, Student's "t"	T	2.306

The following tables are the temperature measurements. The table numbers refer to those in Appendix K of Reference 1.

Temperature Measurements			
Calibration bias	b_{cal}	0.111	°C
Installation bias (estimated)	$b_{install}$	0.028	°C

(Continued)

Table 1 – Average, *t(i)*		X^a	S_x^b	$b_{spat\ var}^c$	S_{pv}^e	
Service water inlet temperature-top	$t_{i\text{-}0}$	19.4	0.031	0.0002	0.00002	°C
Service water inlet temperature-right	$t_{i\text{-}90}$	19.2	0.033	0.0003	0.00002	°C
Service water inlet temperature-bottom	$t_{i\text{-}180}$	19.3	0.015	0.0001	0.00000	°C
Service water inlet temperature-left	$t_{i\text{-}270}$	19.3	0.014	0.0000	0.00000	°C
	Ave	19.30			0.00004	°C
	$b_{spat\ var,\ ti}^d$			0.04		
	u_{pv}^f				0.013	

Table 2 – Average, *T(i)*		X^a	S_x^b	$b_{spat\ var}^c$	S_{pv}^e	
Hot-side inlet temperature, No. 1	$T_{i\text{-}0}$	33.7	0.040	0.0004	0.00003	°C
Hot-side inlet temperature, No. 2	$T_{i\text{-}90}$	33.9	0.024	0.0001	0.00001	°C
Hot-side inlet temperature, No. 3	$T_{i\text{-}180}$	33.8	0.011	0.0000	0.00000	°C
Hot-side inlet temperature, No. 4	$T_{i\text{-}270}$	33.8	0.005	0.0000	0.00000	°C
	Ave	33.81			0.00004	°C
	$b_{spat\ var,Ti}^d$			0.04		°C
	u_{pv}^f				0.012	°C

Table 3 – Average, *t(o)*		X^a	S_x^b	$b_{spat\ var}^c$	S_{pv}^e	
Service water outlet temperature, top	$t_{o\text{-}0}$	29.7	0.020	0.0001	0.00000	°C
Service water outlet temperature, top-right	$t_{o\text{-}45}$	29.8	0.022	0.0001	0.00000	°C
Service water outlet temperature, right	$t_{o\text{-}90}$	29.7	0.003	0.0000	0.00000	°C
Service water outlet temperature, bottom-right	$t_{o\text{-}135}$	29.7	0.005	0.0000	0.00000	°C
Service water outlet temperature, bottom	$t_{o\text{-}180}$	29.7	0.001	0.0000	0.00000	°C
Service water outlet temperature, bottom-left	$t_{o\text{-}225}$	29.6	0.045	0.0006	0.00002	°C
Service water outlet temperature, left	$t_{o\text{-}275}$	29.8	0.043	0.0005	0.00002	°C
Service water outlet temperature, top-left	$t_{o\text{-}315}$	29.7	0.003	0.0000	0.00000	°C
	Ave	29.73			0.00004	°C
	$b_{spat\ var,to}^d$			0.03		°C
	u_{pv}^f				0.013	°C

Table 4 – T_o		X^a	S_x^b	$b_{spat\ var}^c$	S_{pv}^e	
Hot-side outlet temperature, top	$T_{o\text{-}0}$	25.6	0.019	0.0001	0.00000	°C
Hot-side outlet temperature, top-right	$T_{o\text{-}45}$	25.7	0.018	0.0001	0.00000	°C
Hot-side outlet temperature, right	$T_{o\text{-}90}$	25.6	0.009	0.0000	0.00000	°C
Hot-side outlet temperature, bottom-right	$T_{o\text{-}135}$	25.6	0.010	0.0000	0.00000	°C
Hot-side outlet temperature, bottom	$T_{o\text{-}180}$	25.7	0.023	0.0001	0.00000	°C
Hot-side outlet temperature, bottom-left	$T_{o\text{-}225}$	25.6	0.022	0.0001	0.00000	°C
Hot-side outlet temperature, left	$T_{o\text{-}275}$	25.6	0.002	0.0000	0.00000	°C
Hot-side outlet temperature, top-left	$T_{o\text{-}315}$	25.6	0.003	0.0000	0.00000	°C
	Ave	25.61			0.00002	°C
Bias uncertainty, *b*(*spat* var)	$b_{spat\ var,To}^d$			0.02		
	u_{pv}^f				0.008	°C

Table 6 – Average Temperature Measurements		Ave. Temp.	$b_{spat\ var}$			
Service water inlet temperature	t_i	19.3	0.040			°C
Service water outlet temperature	t_o	33.8	0.031			°C
Hot-side inlet temperature	T_i	29.7	0.038			°C
Hot-side outlet temperature	T_o	25.6	0.019			°C
Table 7 – Temperature Uncertainty due to Process Variations						°C
Service water inlet temperature	$u_{pv,ti}$	0.013				°C
Service water outlet temperature	$u_{pv,to}$	0.013				°C
Hot-side inlet temperature	$u_{pv,Ti}$	0.012				°C
Hot-side outlet temperature	$u_{pv,To}$	0.008				°C
Table 8 – Overall Measurement Uncertainty for Temperature	b_{cal}	$b_{install}$	$b_{spat\ var}$	u_{pv}	$u_{overall}$[g]	°C
Service water inlet temperature	0.111	0.028	0.040	0.013	0.122	°C
Service water outlet temperature	0.111	0.028	0.031	0.013	0.119	°C
Hot-side inlet temperature	0.111	0.028	0.038	0.012	0.121	°C
Hot-side outlet temperature	0.111	0.028	0.019	0.008	0.116	°C

[a] $X = \frac{1}{N_t}\sum_{i=1}^{N} X_i$

[b] $S_X = \sqrt{\sum_{i=1}^{N_t} \frac{(X_k - X)^2}{N_t - 1}}$

[c] $\left(\frac{1}{J_t}\right)\frac{(T - T_{ave})^2}{J_t - 1}$

[d] $b_{spat\,var} = t\sqrt{\left(\frac{1}{J_t}\right)\sum_{i=1}^{J}\frac{(t - t_{ave})^2}{J_t - 1}}$

[e] $\frac{1}{J_t}\frac{S_X{}^2}{N_{pv}}$

[f] $u_{pv} = 2\sqrt{\sum_{i=1}^{J}\frac{1}{J_t}\frac{S_X{}^2}{N_{pv}}}$

[g] $u_{overall} = \sqrt{b_{cal}{}^2 + b_{spat\,var}{}^2 + u_{pv}{}^2}$

The following tables are the average standard deviations and uncertainties for flow measurements.

Number of flow measurements	J_f	**1**		
Hot-side flow measurement calibration uncertainty	$u_{f\text{-}cal\text{-}H}$	2.0		%
Cold-side flow measurement calibration uncertainty	$u_{f\text{-}cal\text{-}C}$	2.0		%
Table 5 – Average Flow Measurement			S_X	
Cold stream flow rate	G_c	558	18.9	L/s
Hot stream flow rate	G_H	800	25.2	L/s

(*Continued*)

Number of flow measurements	J_f	1		
Mass flow rate, m_c per coil	m_c	557[a]	18.8	kg/s
Mass flow rate	m_H	795[a]	25.0	kg/s
Table 9 – Flow Measurement Uncertainty, Calibration, etc.				
Service water flow measurement	$u_{m,c\text{-cal}}$	11.1[b]		kg/s
Hot-side flow measurement	$u_{m,h\text{-cal}}$	15.9[b]		kg/s
Table 10 – Flow Measurement Uncertainty due to Process Variances				
Service water flow measurement	$u_{m,c\text{-}pv}$	9.7[c]		kg/s
Hot-side flow measurement	$u_{m,h\text{-}pv}$	12.9[c]		kg/s
Table 11 – Overall Flow Measurement Uncertainty				
Service water flow measurement	$u_{m,c\text{-overall}}$	14.8[d]		kg/s
Hot-side flow measurement	$u_{m,h\text{-overall}}$	20.5[d]		kg/s

[a] $m = G\rho$

[b] $u_{cal} = u_{f\text{-}cal} \times m$

[c] $s_{pv} = 2\sqrt{\sum_{i=1}^{J} \frac{1}{J_t} \frac{s_X^{\,2}}{N_{pv}}}$

[d] $u_{m\text{-}overall} = \sqrt{u_{m\text{-}cal}^{\;2} + u_{m\text{-}pv}^{\;2}}$

The following is an assessment of the uncertainty in the rate of heat transfer:

Table 12 – Uncertainty in Q_H	$u_{Q\text{-}H}$	**Units**	$\theta_{Q\text{-}H}$	**Units**	$(u_{Q\text{-}H} \times \theta_{Q\text{-}H})^2$	**Units**
Hot stream inlet temperature, T_i	0.121	°C	−3.312[a]	kJ/(s-°C)	162,667	kJ/s²
Hot stream outlet temperature, T_o	0.116	°C	3.312[b]	kJ/(s-°C)	149,354	kJ/s²
Hot stream mass flow rate, m_H	20.48	kg/s	17,222[c]	kJ/s	124,367	kJ/s²
Hot stream specific heat, $C_{p\text{-}H}$	0.0418	kJ/(kg-°C)	3,279[d]	kJ/(s-°C)/s	18,784	kJ/s²
Total uncertainty in the hot stream heat transfer rate, $u_{Q\text{-}H}$					675[e]	kJ/s
Table 13 – Uncertainty in Q_C	$u_{Q\text{-}C}$	Units	$\theta_{Q\text{-}C}$	Units	$(u_{Q\text{-}C} \times \theta_{Q\text{-}C})^2$	Units
Cold stream inlet temperature, t_i	0.122	°C	−2327[a]	kJ/(s-°C)	80,421	kJ/s²
Cold stream outlet temperature, t_o	0.119	°C	2,327[b]	kJ/(s-°C)	76,974	kJ/s²
Cold stream mass flow rate, m_C	14.78	kg/s	60,631[c]	kJ/s	802,860	kJ/s²
Cold stream specific heat, $C_{p\text{-}C}$	0.0418	kJ/(kg-°C)	8,090[d]	(kg-°C)/s	114,352	kJ/s²
Total uncertainty in the cold stream heat transfer rate, $u_{Q\text{-}C}$					1,037[e]	kJ/s

[a] $\theta_{Q,T_{in}} = -m\,c_p$

[b] $\theta_{Q,T_{out}} = m\,c_p$

[c] $\theta_{Q,M} = c_p\left(T_{in} - T_{out}\right)$

[d] $\theta_{Q,c_p} = m\left(T_{in} - T_{out}\right)$

[e] $u_Q = \sqrt{\sum (u_Q \times \theta)^2}$

The following is an assessment of the heat balance between the hot and cold streams:

Hot Stream Range			
Upper limit of the heat transfer rate	$Q_h + u_{Q,h}$	27,900	kJ/s
Heat transfer rate	Q_h	27,225	kJ/s
Lower limit of the heat transfer rate	$Q_h - u_{Q,h}$	26,550	kJ/s
Cold Stream Range			
Upper limit of the heat transfer rate	$Q_c + u_{Q,c}$	25,357	kJ/s
Heat transfer rate	Q_c	24,321	kJ/s
Lower limit of the heat transfer rate	$Q_c + u_{Q,c}$	23,284	kJ/s
Weighted Average			
Weighted average heat transfer rate	Q_{ave}	26,361	kJ/s
Uncertainty of the average heat transfer rate	u_{Qave}	565	kJ/s
Upper limit of the heat transfer rate	$Q_{ave} + u_{Qave}$	26,926	kJ/s
Lower limit of the heat transfer rate	$Q_{ave} - u_{Qave}$	25,795	kJ/s

One may note that there is no overlap in the ranges of uncertainty as $Q_c + u_{Q,c} < Q_h - u_{Q,h}$.

The following is an assessment of the uncertainty of the rate of heat transfer:

Hot-side heat transfer rate	Q_H	27,225	kJ/s
Cold-side heat transfer rate	Q_c	24,321	kJ/s
Total uncertainty in Q_H	$u_{Q,h}$	675	kJ/s
Total uncertainty in Q_C	$u_{Q,c}$	1,037	kJ/s
Heat balance error (referenced to hot-side)	HBE	10.7%	
Uncertainty of heat balance error	u_{HBE}	4.4%	

Although the uncertainty of the test is very good, the test fails to meet the criteria of $u_{HBE} \geq HBE$, that is an indication of a good test. However, the imbalance in the computed heat transfer rate only slightly exceeds the anticipated range of 3%–10% as indicated in Reference1.

CHAPTER 17

In December 2009, the staff at the Virgil C. Summer Nuclear Generating Station conducted two heat transfer tests on one of their four RBCU. The tests were conducted one day apart with the SW flow rate during the second test being half that of the first test. The following is the physical data for the RBCU:

RBCU Physical Data			Units
Number of coil sections in parallel	C	8	
Number of rows	N_r	8	
Number of serpentines	S	2	
Effective tube length	L_{eff}	3.35	m
Number of tubes per row	N_t	16	
Outside tube diameter	d_o	0.0163	m
Tube wall thickness	t	0.0012	m
Fin thickness	δ	0.0021	m
Number of fins per unit of length	N_f	74	fins/m
Bundle depth	D	0.30	m
Face height	B	0.61	m

The following physical parameters may be calculated.

Additional RBCU Physical Data			Units
Face area	$FA = L_{eff}B$	2.04	m^2
Inside tube diameter	$d_i = d_o - 2t$	0.01377	m
Tube-side area	$A_t = \pi d_i L_{eff} N_r N_t$	18.6	m^2
Total cross-sectional area of plate fins	$A_{c,f} = BD$	0.19	m^2
Total cross-sectional area of tubes	$A_{c,t} = (\pi/4)d_o^2 N_t N_{rows}$	0.03	m^2
Net area of each plate	$A_{net,f} = 2(A_{c,f} - A_{c,t})$	0.32	m^2
Number of plates	$NP = L_{eff}N_r$	814	
Total area of plate fins	$A_f = A_{net,f}\,NP$	259	m^2
Net outside area of tubes (prime area)	$A_p = \pi d_o N_t N_r L_{eff}(1 - N_f\delta)$	11	m^2
Total effective area of the finned surface	$A_h = A_f + A_p$	270	m^2
Ratio of the finned side to inside area	A_h/A_t	14.54	dimensionless
Area of the finned surface per fin	$a_o = A_f/N_r N_t NP$	0.00249	m^2/fin
Effective fin diameter	$d_{fin} = ((2/\pi)a_o + d_o^2)^{0.5}$	0.04299	m
Area of the finned side per fin	$a_h = A_h/N_r N_t N_f L_{eff}$	0.00259	m^2/fin
Area of the prime per fin	$a_p = \pi d_o(1/N_f - \delta)$	0.00010	m^2/fin
Area of the fin per fin	$a_f = a_h - a_p$	0.00249	m^2/fin
Outside tube area	$A_{t,o} = (d_o/d_i)A_t$	21.9	m^2
Tube wall area	$A_w = (A_{t,o} - A_t)/\ln(A_{t,o}/A_t)$	20.2	m^2
Tube and fin thermal conductivity (Cu 122)	k_t	0.000064	kJ/(s-m-°C)
Tube wall resistance	$r_w = [d_o \ln(A_h/A_t)]/2k_t$	2.04	(s-m-°C)/kJ

The fin efficiency may be calculated as follows:

Fin Efficiency			**Units**
Characteristic fin length	$l = \delta/2$	0.00107	m
Assumed air-side film coefficient	h_h	75.5	kJ/(s-m-°C)
Biot number	$Bi = h_h l/k_{fin}$	0.0002	
Fin radius	$r_2 = d_{fin}/2$	0.0215	m
Characteristic fin radius	$r_{2c} = r_2 + \delta/2$	0.0226	m
Fin height	$L = r_2 - d_o/2$	0.0134	m
Equivalent fin height	$L_c = L + \delta/2$	0.0144	m
Radius ratio (consult Figure 17.2 below[2])	$r_{2c} = (d_o/2r_1)$	2.78	
x-axis	$L_c^{1.5}[h_h/k_{fin}L_c\delta]^{0.5}$	0.14745	
Fin efficiency	$\eta_{fin,h}$	0.950	
Surface efficiency	$\eta_h = (a_p + a_f\eta_{fin,h})/a_h$	0.952	

The following parameters were either taken from the HX datasheet or calculated as shown:

Design HX Data			**Units**
Total service water flow rate	G	41	L/s
Service water flow rate per coil	$g_c = G/C$	5.13	L/s
Entering water temp.	$t_{c,i}$	29.44	°C
Exiting water temp.	$t_{c,o}$	32.00	°C
Average water temp.	$t_c = (t_{c,i} + t_{c,o})/2$	30.72	°C
Tube-side thermal conductivity	k_c	0.614	kJ/(s-m-°C)
Tube-side absolute viscosity	μ_c	0.00081	kg/(m-s)
Tube-side specific heat	$c_{p,c}$	4.178	kJ/(kg-°C)
Tube-side density	ρ_c	0.994	kg/m³
Tube-side temperature rise	$dt_c = t_{c,o} - t_{c,i}$	2.56	°C
Tube-side heat transfer rate	$Q_c = g_c c_{p,c}\rho_c dt_c$	54	kJ/s
Entering air dry-bulb temp	$T_{h,i}$	37.78	°C
Exiting air dry-bulb temp	$T_{h,o}$	31.53	°C
Average air dry-bulb temp	T_h	34.65	°C
Hot-side thermal conductivity	k_h	0.620	kJ/(s-m-°C)
Hot-side absolute viscosity	μ_h	0.00076	kg/(m-s)
Hot-side specific heat	$c_{p,h}$	4.177	kJ/(kg-°C)
LMTD correction factor	F	0.989	
Tube-side flow rate per coil	$m_{c\text{-}coil} = g_c\rho_c$	5.10	kg/s
Face velocity out of air side	FV	3.76	m/s
Relative humidity	RH	40%	%
Total atmospheric pressure	P	100.66	kPa

The overall heat transfer coefficient at design conditions, U_{design}, is calculated as follows:

Design Heat Transfer Coefficient			**Units**
Greater terminal temperature difference	$\Delta T_1 = T_{h,i} - t_{c,o}$	5.78	°C
Lesser terminal temperature difference	$\Delta T_2 = T_{h,o} - t_{c,i}$	2.08	°C
Log mean temperature difference	$LMTD = (\Delta T_1 - \Delta T_2)/\ln(\Delta T_1/\Delta T_2)$	3.62	°C
Effective mean temperature difference	$EMTD = F\ (LMTD)$	3.58	°C
Design overall heat transfer coefficient	$U_{design} = Q/(A_h EMTD)$	56.31	kJ/(s-m-°C)

The tube-side convection coefficient may be computed from the Petukhov correlation in Chapter 16 as follows:

Mass flow rate/tube	$m_t = m_{c\text{-}coil}/N_t$	0.159	kg/s
Volumetric flow rate/tube	$V_t = m_t/\rho_t$	0.00016	m³/s
Tube area	$a_t = \pi d_i^2$	0.00015	m²
Tube velocity	$v_t = V_t/a_t$	1.08	m/s
Prandtl number	$\mathrm{Pr}_c = c_{p,c}\mu_c/k_c$	5.50	
Reynolds number	$\mathrm{Re}_c = \rho v_t d_i/\mu_c$	18,235	
Fanning friction factor	$f = (1.58\ \ln \mathrm{Re}_c - 3.28)^{-2}$	0.00669	
Nusselt number	$\mathrm{Nu} = ((f/2)\mathrm{Re}_c\mathrm{Pr}_c)/(1.07 + 12.7(f/2)^{1/2}\ (\mathrm{Pr}_c^{2/3} - 1))$	127.81	
Tube-side film coefficient	$h_c = \mathrm{Nu}(k_c/d_i)$	5,701	kJ/(s-m²-°C)

The hot-side convection coefficient may be computed as follows:

$$h_{h,\,design} = \frac{1/\eta_h}{\dfrac{1}{U_{design}} - r_{f,h} - r_w - \left(\dfrac{A_h}{A_c}\right) r_{f,c} - \left(\dfrac{A_h}{A_c}\right)\dfrac{1}{h_c}} = 75.76\ \text{kJ/s-m}^2\text{-°C}$$

The design air-side mass flow rate may be calculated as follows:

Volumetric flow rate per coil	V	62	m³/s
Saturation pressure at $T_{h,i}$	ASME Steam Tables	6.59	kPa
Partial pressure of the water vapor in the air	$P_{wv} = RH\ P_{sat}$	2.63	kPa
Partial pressure of the air	$P_a = P - P_{wv}$	98.03	kPa
Humidity ratio	$w = (M_{wv}/M_a)\ (P_{wv}/P_a)$	0.017	
Mass flow rate of air	$m_a = [P_a\ 60V/R(T + 460)]$	8.427	kg/s
Mass of the water vapor	$m_{wv} = wm_a$	0.141	kg/s
Total mass flow rate on the air side	$m_h = m_a + m_{wv}$	8.567	kg/s
Specific heat of dry air	$c_{p\text{-}DA}$	1.005	kJ/(kg-°C)
Specific heat of water vapor	$c_{p\text{-}WV}$	4.178	kJ/(kg-°C)
Specific heat of moist air	$c_{p\text{-}MA} = (c_{p\text{-}DA}m_a + c_{p\text{-}WV}\ m_{wv})/m_h$	1.057	kJ/(kg-°C)
Density of air	$\rho_a = m_h/V$	1.157	kg/m³
Hot side flow measurement per coil	$m_h = V\rho_a$	8.567	kg/s
Air-side heat transfer rate	$Q_h = m_h c_p(T_{h,i} - T_{h,o})$	56.60	kJ/s

The HX test data is as follows:

HX Test Data		Test 1	Test 2	Units
Total service water flow rate	G	82.95	39.80	L/s
Service water flow rate per coil	$g_c = G/C$	10.37	4.98	L/s
Entering water temp	$t_{c,i}$	15.42	27.99	°C
Exiting water temp	$t_{c,o}$	17.94	31.27	°C
Average water temp	$t_c = (t_{c,i} + t_{c,o})/2$	16.68	29.63	°C
Tube-side thermal conductivity	k_c	0.591	0.612	kJ/(s-m-°C)
Tube-side absolute viscosity	μ_c	0.00111	0.00082	kg/(m-s)
Tube-side specific heat	$c_{p,c}$	4.187	4.178	kJ/(kg-°C)
Tube-side density	ρ_c	0.999	0.995	kg/m³
Tube-side flow rate per coil	$m_{c\text{-}coil} = g_c\rho_c$	10.36	4.96	kg/s
Tube-side heat transfer rate	$Q_c = g_c c_{p,c}\rho_c dt_c$	109	68	kJ/s
Inlet relative humidity	RH	22.19%	11.12%	%
Total volumetric flow rate	V_{total}	54.23	54.52	m³/s
Volumetric flow rate per coil	$V_{coil} = V_{total}/C$	6.78	6.81	m³/s
Entering air dry-bulb temperature	$T_{h,i}$	33.59	40.13	°C
Leaving air dry-bulb temperature	$T_{h,o}$	18.32	30.45	°C
Average air dry-bulb temperature	$T_h = (T_{h,i} + T_{h,o})/2$	25.95	35.29	°C
Hot-side thermal conductivity	k_h	0.607	0.621	kJ/(s-m-°C)
Hot-side absolute viscosity	μ_h	0.00089	0.00075	kg/(m-s)
Hot-side specific heat	$c_{p,h}$	4.180	4.177	kJ/(kg-°C)
Hot-side density of dry air	ρ_h	1.245	1.230	kg/m³
Total pressure	P	102.39	105.01	kPa
Saturation pressure at inlet temperature	P_{sat}	5.23	7.48	kPa
Partial pressure of water	$P_{wv} = RH \times P_{sat}$	1.16	0.83	kPa
Molecular weight of water	M_w	18	18	
Molecular weight of air	M_a	29	29	
Specific heat of air	$c_{p\text{-}a}$	1.057	1.057	kJ/(kg-°C)
Specific heat of water	$c_{p\text{-}w}$	4.180	4.177	kJ/(kg-°C)
Partial pressure of air	$P_a = P - P_{wv}$	101.23	104.18	kPa
Humidity ratio	$w = (M_{wv}/M_a)\,(P_{wv}/P_a)$	0.0071	0.0050	
Mass flow rate of air per coil	$m_a = [P_a\,60V_{coil}/R(T + 460)]$	7.790	7.891	kg/s
Mass flow rate of water vapor per coil	$m_{w\text{-}coil} = w\,m_a$	0.055	0.039	kg/s
Total mass flow rate per coil	$m_h = m_a + m_{w\text{-}coil}$	7.845	7.931	kg/s
Specific heat of moist air	$c_{p\text{-}ma} = (c_{p\text{-}DA}\,m_a + c_{p\text{-}WV}\,m_{wv})/m_h$	1.079	1.072	kJ/(kg-°C)
Test heat transfer (hot side)	$Q_h = m_h c_p (T_{h,i} - T_{h,o})$	129	82.3	kg/s
Average heat transfer rate	$Q_{test,ave} = (Q_c + Q_h)/2$	116	73.1	kJ/s
Greater terminal temperature difference	$\Delta T_1 = T_{h,i} - t_{c,o}$	15.64	8.86	°C
Lesser terminal temperature difference	$\Delta T_2 = T_{h,o} - t_{c,i}$	2.89	2.46	°C

(Continued)

HX Test Data		Test 1	Test 2	Units
Log mean temperature difference	$LMTD = (\Delta T_1 - \Delta T_2)/\ln(\Delta T_1/\Delta T_2)$	7.56	4.99	°C
Ratio of temperature change	$R = (T_{h,i} - T_{h,o})/(t_{c,o} - t_{c,i})$	6.06	2.96	
Effectiveness	$P = (t_{c,o} - t_{c,i})/(T_{h,i} - t_{c,i})$	0.14	0.27	
LMTD correction factor	F	0.975	0.954	
EMTD	$EMTD = F(LMTD)$	7.37	4.76	°C
Test overall heat transfer coefficient	$U_{test} = Q_{ave}/(A_h EMTD)$	57.08	54.29	kJ/(s-m²-°C)
Average tube temperature	$t_{tube} = (T_h + t_c)/2$	21.32	32.46	°C
Wall resistance	$r_w = [d_o \ln(A_h/A_t)]/2k_t$	0.00002	0.00002	m²-°C/W
Mass flow rate/tube	$m_t = m_{c\text{-}coil}/S\,N_t$	0.3237	0.1549	kg/s
Volumetric flow rate/tube	$V_t = m_t/\rho_t$	0.324	0.156	m³/s
Tube velocity	$v_t = V_t/a_t$	2,177	1,046	m/s
Prandtl number	$Pr_c = c_{p,c}\mu_c/k_c$	7.84	5.62	
Reynolds number	$Re_c = \rho v_t D_i/\mu_c$	27,088	17,385	
Fanning friction factor	$f = (1.58 \ln Re_c - 3.28)^{-2}$	0.00606	0.00678	
Nusselt number	Nu[a]	205.4	124.1	
Tube-side film coefficient	$h_c = Nu(k_c/D_i)$	8,812	5,519	kJ/(s-m²-°C)
Hot-side flow rate	$m_{h\text{-}coil} = g_h \rho_h$	7.845	7.931	kg/s
Hot-side film coefficient	h_h[b]	66.7	72.2	kJ/(s-m²-°C)
Total resistance	$r = 1/U_{test}$	0.01752	0.01842	(m²-°C)/W
Hot-side convection resistance	$r_h = 1/(\eta_h h_h)$	0.01576	0.01456	(m²-°C)/W
Wall resistance	$r_w = [d_o \ln(A_h/A_t)]/2k_t$	0.00002	0.00002	(m²-°C)/W
Cold-side convection resistance	$R_c = (A_h/A_c)h_c$	0.00165	0.00263	(m²-°C)/W
Fouling referenced to hot side	$r_{f,h}$[c]	0.00009	0.00121	(m²-°C)/W
Fouling referenced to tube side	$r_{f,c} = (A_c/A_h)r_{f,h}$	0.00001	0.00008	(m²-°C)/W

The test fouling resistance may be calculated as follows:

a $$Nu = \left(\left(f/2\right)Re_c Pr_c\right)/\left(1.07 + 12.7\left(f/2\right)^{1/2}\left(Pr_c^{2/3} - 1\right)\right)$$

b $$h_{h,test} = \frac{\left[\left(\frac{m_h}{\mu_h}\right)^{0.681}\left(\frac{\mu_h c_{p,h}}{k_h}\right)^{0.333} k_h\right]_{test}}{\left[\left(\frac{m_h}{\mu_h}\right)^{0.681}\left(\frac{\mu_h c_{p,h}}{k_h}\right)^{0.333} k_h\right]_{design}} h_{h,design}$$

c $$r_{f,test} = \frac{1}{\frac{1}{U_{test}} - \left(\frac{1}{\eta_h h_{h,test}}\right) - r_w - \left(\frac{A_h}{A_c}\right)\frac{1}{h_{c,test}}}$$

CHAPTER 18

Refer to Section 18.4. Calculate the constant C in the Colburn analogy equation.

A PHX has the following dimensions:

$$L_H = 1.7145\,\text{m}$$

$$L_w = 0.8446\,\text{m}$$

$$L_{cp} = 0.7969\,\text{m}$$

$$N_p = 191$$

$$\Delta X = 0.0005\,\text{m}$$

$$\phi = 1.3$$

$$N_{cp} = (N_p - 1)/2 = (191-1)/2 = 95$$

$$A_{eff} = \phi\, L_H\, L_W\, N_p = 1.3 \times 1.7145 \times 0.8446 \times 191 = 360\,\text{m}^2$$

$$b = \left({L_{cp}}\big/{N_p} \right) - \Delta X = \left({0.7969}\big/{191} \right) - 0.0005 = 0.003664\,\text{m}$$

$$D_e = \frac{4 L_w b}{(2L_w + 2\varphi b)} = \frac{4 \times 0.8446 \times 0.003664}{(2 \times 0.8446 + 2 \times 1.3 \times 0.003664)} = 0.00729\,\text{m}$$

The design operating conditions are as follows:

CWT entering, $t_i = 27.8°C$
CWT exiting, $t_o = 32.8°C$
HWT entering, $T_i = 37.8°C$
HWT exiting, $T_o = 28.6°C$
Coldwater flow rate, $m_c = 267$ kg/s
Hot water flow rate, $m_h = 132$ kg/s
Plate wall resistance, $r_w = 0.000024$ m^2-°C/W.

The properties of the fluids are as follows:

Cold-side density, $\rho_c = 995$ kg/m^3
Cold -side Specific Heat, $c_{pc\text{-}ave} = 4.178$ kJ/kg-°C
Cold-side viscosity, $\mu_c = 0.81$ g/m-s
Cold-side thermal conductivity, $k_c = 0.613$ kJ/s-m-°C
Hot-side density, $\rho_h = 994$ kg/m^3
Hot-side specific heat, $c_{ph\text{-}ave} = 4.177$ J/kg-°C

Hot-side viscosity, $\mu_h = 0.78$ kg/m-s
Hot-side thermal conductivity, $k_h = 0.618$ kJ/s-m-°C.

Therefore

$$\Delta t_1 = T_i - t_o = 5.00°\text{C}$$

$$\Delta t_2 = T_o - t_i = 0.84°\text{C}$$

$$LMTD = \frac{\Delta t_1 - \Delta t_2}{\ln\left(\frac{\Delta t_1}{\Delta t_2}\right)} = 2.33\,°\text{C}$$

$$Q_h = m_h c_p \left(T_i - T_o\right) = 5{,}578 \text{ kJ / s}$$

$$Q_c = m_c c_p \left(t_o - t_i\right) = 5{,}044 \text{ kJ / s}$$

$$Q_{ave} = \left(Q_{c\text{-}h} + Q_c\right)/2 = 5{,}311 \text{ kJ/s}$$

$$U_{req} = \frac{Q_{ave}}{A_{eff}\ LMTD} = 6{,}337 \text{ W/m}^2\text{-°C}$$

Calculate the mass flux, G, as follows:

$$G_c = \frac{m_c}{N_{cp}\ b\ L_W \phi} = 699 \text{ kg/m}^2\text{-s}$$

$$G_h = \frac{m_h}{N_{cp}\ b\ L_W\ \phi} = 345 \text{ kg/n}^2\text{-s}$$

The Reynolds numbers are

$$\text{Re}_c = D_e G_c / \mu_c = 6{,}253$$

$$\text{Re}_h = D_e G_h / \mu_h = 3{,}240$$

The Prandtl numbers are

$$\text{Pr}_c = \frac{\mu_c c_{pc}}{k_c} = 5.55$$

$$\text{Pr}_h = \frac{\mu_h c_{ph}}{k_h} = 5.24$$

Therefore,

$$C = \frac{\dfrac{D_e}{\text{Re}_c^{3/4}\ \text{Pr}_c^{1/3}\ k_c} + \dfrac{D_e}{\text{Re}_h^{3/4}\ \text{Pr}_h^{1/3}\ k_h}}{\dfrac{A\,(LMTD)}{Q} - r_w} = 0.189$$

CHAPTER 19

From June 2002 to May 2014, the staff at the V. C. Summer Nuclear Generating Station conducted heat transfer tests on all four of their RBCU. The physical data for the RBCU may be found in the example problem for Chapter 17.

The following tables provide the results of the RBCU heat transfer tests on the indicated days.

RBCU 1A Heat Transfer Test Results

		June 3, 2002	May 28, 2005	June 6, 2008	May 22, 2014	Units
Air volumetric flow rate	V	54	58	57	53	m³/s
Service water flow measurement	G_c	42	80	80	80	L/s
Service water inlet temperature	t_i	23.10	21.86	23.75	24.11	°C
Service water outlet temperature	t_o	27.37	24.19	25.69	27.23	°C
Hot-side inlet temperature	T_i	38.94	36.22	35.84	44.69	°C
Hot-side outlet temperature	T_o	26.52	24.42	25.95	27.83	°C

RBCU 2A Heat Transfer Test Results

		June 3, 2002	May 28, 2005	December 7, 2009	May 22, 2014	Units
Air volumetric flow rate	V	55	57	51	53	m³/s
Service water flow measurement	G_c	43	78	80	79	L/s
Service water inlet temperature	t_i	22.98	21.91	23.76	23.82	°C
Service water outlet temperature	t_o	27.27	24.18	25.68	27.17	°C
Hot-side inlet temperature	T_i	39.42	36.45	36.58	45.84	°C
Hot-side outlet temperature	T_o	26.51	24.34	25.77	27.61	°C

RBCU 1B Heat Transfer Test Results

		June 3, 2002	May 28, 2005	December 7, 2009	December 8, 2009	May 22, 2014	Units
Air volumetric flow rate	V	55	55	54	55	49	m³/s
Service water flow measurement	G_c	38	80	83	40	80	L/s
Service water inlet temperature	t_i	23.01	21.81	15.42	27.99	24.64	°C
Service water outlet temperature	t_o	27.86	23.88	17.94	31.27	27.92	°C
Hot-side inlet temperature	T_i	38.87	34.93	33.59	40.13	46.64	°C
Hot-side outlet temperature	T_o	26.87	24.14	18.32	30.45	28.68	°C

RBCU 2B Heat Transfer Test Results

		June 3, 2002	May 28, 2005	June 6, 2008	May 22, 2014	Units
Air volumetric flow rate	V	56	56	53	52	m³/s
Service water flow measurement	G_c	37	79	81	83	L/s
Service water inlet temperature	t_i	22.98	21.80	23.58	24.67	°C
Service water outlet temperature	t_o	28.03	23.97	26.01	28.35	°C
Hot-side inlet temperature	T_i	39.14	36.02	39.56	47.53	°C
Hot-side outlet temperature	T_o	27.02	24.21	26.04	29.01	°C

The following tables show the calculated results using the procedure described in Chapter 17.

RBCU 1A Calculated Results

		June 3, 2002	May 28, 2005	June 6, 2008	May 22, 2014	Units
Mass flow rate, m_c per coil	m_c	5.26	10.03	10.02	9.99	kg/s
Mass flow rate, $m(h)$	m_h	7.67	8.37	8.12	7.33	kg/s
Heat transfer rate	Q_c	94.02	98.07	81.65	130.82	kJ/s
Weighted average heat transfer rate	Q_{ave}	94.59	98.58	81.58	126.49	kJ/s
Heat balance error (referenced to hot side)	*HBE*	1.74%	1.20%	−1.22%	−5.31%	%
Effective mean temperature difference	*EMTD*	6.61	6.06	5.14	8.79	°C
Overall test heat transfer coefficient	U_{test}	52.8	59.6	58.5	52.8	kJ/(s-m-°C)

RBCU 2A Calculated Results

		June 3, 2002	May 28, 2005	December 7, 2009	May 22, 2014	Units
Mass flow rate, m_c per coil	m_c	5.43	9.72	10.00	9.86	kg/s
Mass flow rate, $m(h)$	m_h	7.76	8.20	7.32	7.30	kg/s
Heat transfer rate	Q_c	97.61	92.28	80.48	138.31	kJ/s
Weighted average heat transfer rate	Q_{ave}	98.88	91.22	80.40	134.59	kJ/s
Heat balance error (referenced to hot side)	*HBE*	3.01%	–2.20%	–1.19%	–3.40%	%
Effective mean temperature difference	*EMTD*	6.90	5.66	5.20	9.23	°C
Overall test heat transfer coefficient	U_{test}	52.8	59.6	56.8	53.9	kJ/(s-m-°C)

RBCU 1B Calculated Results

		June 3, 2002	May 28, 2005	December 7, 2009	December 8, 2009	May 22, 2014	Units
Mass flow rate, m_c per coil	m_c	4.80	9.98	10.39	4.96	9.99	kg/s
Mass flow rate, $m(h)$	m_h	8.08	8.01	7.85	7.93	6.91	kg/s
Heat transfer rate	Q_c	97.44	86.62	109.45	67.93	136.90	kJ/s
Weighted average heat transfer rate	Q_{ave}	97.44	86.72	115.50	73.10	127.69	kJ/s
Heat balance error (referenced to hot side)	*HBE*	0.01%	0.25%	8.69%	11.90%	–9.83%	%
Effective mean temperature difference	*EMTD*	6.75	5.54	7.47	4.94	9.47	°C
Overall test heat transfer coefficient	U_{test}	53.4	57.4	56.8	53.9	49.4	kJ/(s-m-°C)

RBCU 2B Calculated Results

		June 3, 2002	May 28, 2005	June 6, 2008	May 22, 2014	Units
Mass flow rate, m_c per coil	m_c	4.62	9.81	10.07	10.34	kg/s
Mass flow rate, $m(h)$	m_h	8.31	8.06	7.42	7.24	kg/s
Heat transfer rate	Q_c	97.68	89.30	102.39	159.25	kJ/s
Weighted average heat transfer rate	Q_{ave}	98.88	92.73	101.47	144.39	kJ/s
Heat balance error (referenced to hot side)	*HBE*	3.52%	6.58%	–1.64%	–18.16%	%

(Continued)

RBCU 2B Calculated Results

		June 3, 2002	May 28, 2005	June 6, 2008	May 22, 2014	Units
Effective mean temperature difference	*EMTD*	6.91	5.92	6.42	9.87	°C
Overall test heat transfer coefficient	U_{test}	52.8	57.9	57.9	53.9	kJ/(s-m-°C)

As stated in Section 19.8 and shown in Figure 19.5, previous heat transfer tests have demonstrated that the *EMTD* is a linear function of the cold-stream heat transfer rate.

The last piece of required information to determine the *NTU* is the effective heat transfer surface area, $A_{h\text{-}eff}$, which is obtained by conducting a flow and *dP* test along with the final heat transfer test.

Figure B19.1 below is from Figure 19.6 expressed in SI units.

$$\%_{unplugged} = 100 \times (0.000059 \times dP^2 - 0.0048356 \times dP + 1.693)$$

$$= 100 \times (0.000059 \times 232.15^2 - 0.0046356 \times 232.15 + 1.693) = 88.8\%$$

Then,

$$A_{h\text{-}eff} = \%_{unplugged} \times A_h = 0.888 \times 271.7 = 241.3 \text{ m}^2$$

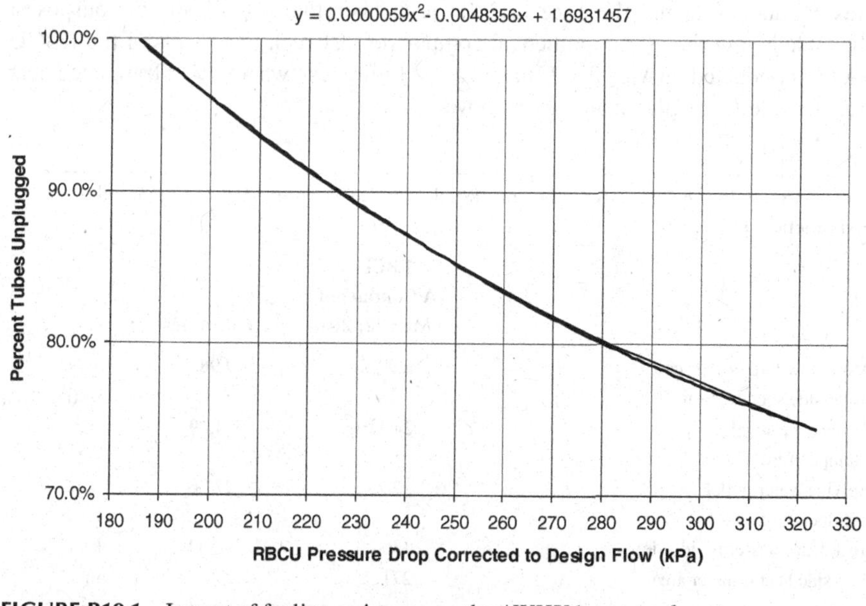

FIGURE B19.1 Impact of fouling resistance on the AWHX heat transfer rate.

It is noted that the *EMTD* thus calculated is slightly different from that calculated from the May 22, 2014 RBCU 1A heat transfer test data. This difference should be immaterial as long as the same equation is used going forward for future tests. Therefore,

$$U = \frac{q_c}{A_h\ EMTD} = \frac{130{,}820\ \text{J/s}}{271.7\left(\text{m}^2\right)\times 8.68\ (°\text{C})} = 55.47\left(\text{J/s}\right)$$

and

$$NTU = \frac{UA_{h\text{-}eff}}{m_c\ c_{p_c}} = \frac{0.05547\left(\text{kJ/s}\right)\ \times 241.3\ \text{ft}^2}{9.99\left(\text{kg/s}\right)4.180\left(\text{kJ}\big/_{\text{kg-°C}}\right)} = 0.32$$

Therefore, the heat transfer rate for some future test may be estimated as follows:

$$(Q^* - u^*)_{future} = \left(\frac{NTU_{future}}{NTU_{test}}\right)(Q^* - u^*)_{test}$$

For the RBCU 1A test conducted on May 22, 2014, the heat transfer rate was calculated for design basis LOCA conditions less the uncertainty of the test to be 39,028 kJ/s, whereas the required value was 26,560 kJ/s, a margin of 47%. The corresponding calculated heat transfer rate for design basis LOCA conditions less the uncertainty of the test conducted on June 6, 2006 was only 29,827 kJ/s, so obviously timely cleaning of the HX was performed. It would be useful to compare calculated heat transfer rate for the test conducted on June 6, 2006 with that calculated on May 22, 2014 by using the procedure discussed herein. Unfortunately, no *dP* test was conducted during the June 6, 2006 test. Therefore, for illustrative purposes, the table below shows a hypothetical comparison between the results of the RBCU 1A test conducted on May 22, 2014 and some future test, where the extrapolated heat transfer rate is calculated as shown above.

Comparison between Calculated Efficiency Results for RBCU 1A on May 22, 2014 and a Hypothetical Future Test

		RBCU 1A Conducted on May 22, 2014	Future Test	
Mass flow rate, m_c per coil	m_c	9.99	9.98	kg/s
Tube-side specific heat	$c_{p,c}$	4.180	4.180	kJ/(kg-°C)
Service water inlet temperature	t_i	24.11	24.89	°C
Service water outlet temperature	t_o	27.23	27.06	°C
Test heat transfer (cold side)	$q_{test,c}$	130.28	90.52	kJ/s
Hot-side heat transfer area	A_h	271.7	271.7	m^2

(Continued)

Comparison between Calculated Efficiency Results for RBCU 1A on May 22, 2014 and a Hypothetical Future Test

		RBCU 1A Conducted on May 22, 2014	Future Test	
Effective mean temperature difference	*EMTD*	8.68	6.01	°C
Overall test heat transfer coefficient	U_{test}	55.5	55.5	kJ/(s-m-°C)
dP corrected to 2,000 gal/min	*dP*	232.15	242.76	kPa
Percentage of tubes unplugged		88.80%	86.64%	%
Effective surface area	$A_{h\text{-}eff}$	241.3	235.4	m²
Number of transfer units	*NTU*	0.32	0.312	–
Extrapolated LOCA heat transfer rate-uncertainty	$Q^* - u$	39,028	38,090	kJ/s

CHAPTER 20

Refer to Section 20.5. Calculate the required MFP turbine steam flow and the MFP turbine condenser condensate outlet temperature.

Enthalpies in Figure B20.1 are taken from PEPSE:

$$\text{Work (J/s)} = 9.806\ \frac{G\ (\text{L/s})\ H\ (\text{m})\ s.g.}{\eta_{pump}}$$

$$= 9.806\ \frac{(2{,}468.5)\ (678.8)\ (0.853)}{(0.9)} = 15{,}573{,}000\ (\text{J/s})$$

$$m_{HRH} = \frac{\text{Work}\left(\text{kJ}/\text{s}\right)}{\left(h_{HRH} - h_{MFPT\text{-}exh}\right)\left(\text{kJ}/\text{kg}\right)} = \frac{15{,}573}{2985.5 - 2550.5} = 35.8\left(\text{kg}/\text{s}\right)$$

$$Duty\left(\text{kJ}/\text{s}\right) = m_{steam}\left(\text{kg}/\text{s}\right)\left(h_{exhaust} - h_{drain}\right)\left(\text{kJ}/\text{kg}\right)$$

$$= 35.8(2550.5 - 321.3) = 7.981 \times 10^4\ \left(\text{kJ}/\text{s}\right)$$

$$Duty\left(\text{kJ}/\text{s}\right) = m_{cond}\left(\text{kg}/\text{s}\right)\left(h_{cond\text{-}out} - h_{cond\text{-}in}\right)\left(\text{kJ}/\text{kg}\right)$$

$$h_{cond\text{-}out} = \frac{m_{cond} h_{cond\text{-}in} + Duty}{m_{cond}} = \frac{(1{,}185)(235.9) + 79{,}810}{1{,}185} = 303.2\left(\text{kJ}/\text{kg}\right)$$

$$t_{cond\text{-}out} = 72.4\,°\text{C}$$

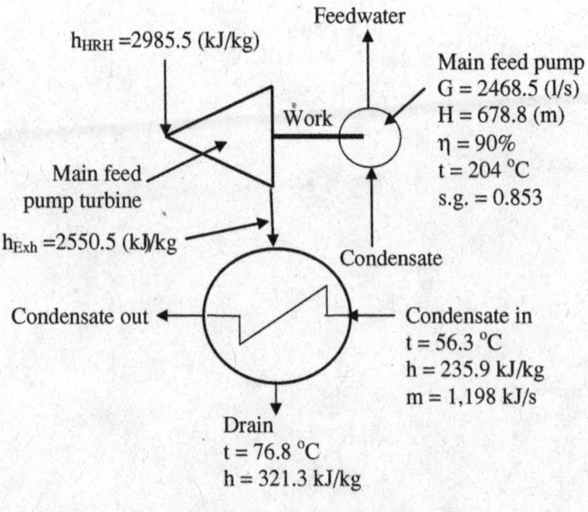

FIGURE B20.1 Main feed pump turbine and condenser.

REFERENCES

1. *ASME Steam Tables Compact Edition*, American Society of Mechanical Engineers, New York, 1993, 2006.
2. Meyer, C. A., et al., *ASME Steam Tables*, 6th ed., American Society of Mechanical Engineers, New York, 1993.

Index

Note: Page numbers in italic and bold refer to figures and tables, respectively.

S

T

Printed in the United States
by Baker & Taylor Publisher Services